Valorization of Residues from Energy Conversion of Biomass for Advanced and Sustainable Material Applications

Valorization of Residues from Energy Conversion of Biomass for Advanced and Sustainable Material Applications

Editors

Dirk Enke
Hossein Beidaghy Dizaji
Volker Lenz
Thomas Zeng

MDPI • Basel • Beijing • Wuhan • Barcelona • Belgrade • Manchester • Tokyo • Cluj • Tianjin

Editors

Dirk Enke
Institute of Chemical Technology
Universität Leipzig
Leipzig
Germany

Hossein Beidaghy Dizaji
Thermochemical Conversion Department
DBFZ Deutsches Biomasseforschungszentrum gemeinnützige GmbH
Leipzig
Germany

Volker Lenz
Thermochemical Conversion Department
DBFZ Deutsches Biomasseforschungszentrum gemeinnützige GmbH
Leipzig
Germany

Thomas Zeng
Thermochemical Conversion Department
DBFZ Deutsches Biomasseforschungszentrum gemeinnützige GmbH
Leipzig
Germany

Editorial Office
MDPI
St. Alban-Anlage 66
4052 Basel, Switzerland

This is a reprint of articles from the Special Issue published online in the open access journal *Sustainability* (ISSN 2071-1050) (available at: www.mdpi.com/journal/sustainability/special_issues/Material_Applications).

For citation purposes, cite each article independently as indicated on the article page online and as indicated below:

LastName, A.A.; LastName, B.B.; LastName, C.C. Article Title. *Journal Name* **Year**, *Volume Number*, Page Range.

ISBN 978-3-0365-4216-4 (Hbk)
ISBN 978-3-0365-4215-7 (PDF)

© 2022 by the authors. Articles in this book are Open Access and distributed under the Creative Commons Attribution (CC BY) license, which allows users to download, copy and build upon published articles, as long as the author and publisher are properly credited, which ensures maximum dissemination and a wider impact of our publications.

The book as a whole is distributed by MDPI under the terms and conditions of the Creative Commons license CC BY-NC-ND.

Contents

Hossein Beidaghy Dizaji, Thomas Zeng, Volker Lenz and Dirk Enke
Valorization of Residues from Energy Conversion of Biomass for Advanced and Sustainable Material Applications
Reprinted from: 2022, 14, 4939, doi:10.3390/su14094939 1

Marzieh Bagheri, Marcus Öhman and Elisabeth Wetterlund
Techno-Economic Analysis of Scenarios on Energy and Phosphorus Recovery from Mono- and Co-Combustion of Municipal Sewage Sludge
Reprinted from: 2022, 14, 2603, doi:10.3390/su14052603 7

Garima Singh, Hossein Beidaghy Dizaji, Hariprasad Puttuswamy and Satyawati Sharma
Biogenic Nanosilica Synthesis Employing Agro-Waste Rice Straw and Its Application Study in Photocatalytic Degradation of Cationic Dye
Reprinted from: 2022, 14, 539, doi:10.3390/su14010539 33

Shengtai Yan, Dezheng Yin, Fang He, Junmeng Cai, Thomas Schliermann and Frank Behrendt
Characteristics of Smoldering on Moist Rice Husk for Silica Production
Reprinted from: 2021, 14, 317, doi:10.3390/su14010317 49

Katja Oehmichen, Stefan Majer and Daniela Thrän
Biomethane from Manure, Agricultural Residues and Biowaste—GHG Mitigation Potential from Residue-Based Biomethane in the European Transport Sector
Reprinted from: 2021, 13, 14007, doi:10.3390/su132414007 63

Kawthar Frikha, Lionel Limousy, Muhammad Bilal Arif, Nicolas Thevenin, Lionel Ruidavets and Mohamed Zbair et al.
Exhausted Grape Marc Derived Biochars: Effect of Pyrolysis Temperature on the Yield and Quality of Biochar for Soil Amendment
Reprinted from: 2021, 13, 11187, doi:10.3390/su132011187 77

Ji Yeon Park, Yang Mo Gu, Seon Young Park, Ee Taek Hwang, Byoung-In Sang and Jinyoung Chun et al.
Two-Stage Continuous Process for the Extraction of Silica from Rice Husk Using Attrition Ball Milling and Alkaline Leaching Methods
Reprinted from: 2021, 13, 7350, doi:10.3390/su13137350 95

Hans Bachmaier, Daniel Kuptz and Hans Hartmann
Wood Ashes from Grate-Fired Heat and Power Plants: Evaluation of Nutrient and Heavy Metal Contents
Reprinted from: 2021, 13, 5482, doi:10.3390/su13105482 107

Huan Li, Huawei Mou, Nan Zhao, Yaohong Yu, Quan Hong and Mperejekumana Philbert et al.
Nitrogen Migration during Pyrolysis of Raw and Acid Leached Maize Straw
Reprinted from: 2021, 13, 3786, doi:10.3390/su13073786 125

Kudzai Mugadza, Annegret Stark, Patrick G. Ndungu and Vincent O. Nyamori
Effects of Ionic Liquid and Biomass Sources on Carbon Nanotube Physical and Electrochemical Properties
Reprinted from: 2021, 13, 2977, doi:10.3390/su13052977 141

Hyun Jin Jung, Hyun Kwak, Jinyoung Chun and Kyeong Keun Oh
Alkaline Fractionation and Subsequent Production of Nano-Structured Silica and Cellulose Nano-Fibrils for the Comprehensive Utilization of Rice Husk
Reprinted from: **2021**, *13*, 1951, doi:10.3390/su13041951 . . . . **153**

Ncamisile Nondumiso Maseko, Denise Schneider, Susan Wassersleben, Dirk Enke, Samuel Ayodele Iwarere and Jonathan Pocock et al.
The Production of Biogenic Silica from Different South African Agricultural Residues through a Thermo-Chemical Treatment Method
Reprinted from: **2021**, *13*, 577, doi:10.3390/su13020577 . . . . **171**

Jinyoung Chun and Jin Hyung Lee
Recent Progress on the Development of Engineered Silica Particles Derived from Rice Husk
Reprinted from: **2020**, *12*, 10683, doi:10.3390/su122410683 . . . . **185**

Editorial

Valorization of Residues from Energy Conversion of Biomass for Advanced and Sustainable Material Applications

Hossein Beidaghy Dizaji [1,*], Thomas Zeng [1], Volker Lenz [1] and Dirk Enke [2]

1. Department of Thermo-Chemical Conversion, DBFZ Deutsches Biomasseforschungszentrum Gemeinnützige GmbH, 04347 Leipzig, Germany; thomas.zeng@dbfz.de (T.Z.); volker.lenz@dbfz.de (V.L.)
2. Institute of Chemical Technology, Universität Leipzig, 04103 Leipzig, Germany; dirk.enke@uni-leipzig.de
* Correspondence: hossein.beidaghy@dbfz.de

Citation: Beidaghy Dizaji, H.; Zeng, T.; Lenz, V.; Enke, D. Valorization of Residues from Energy Conversion of Biomass for Advanced and Sustainable Material Applications. *Sustainability* 2022, *14*, 4939. https://doi.org/10.3390/su14094939

Received: 16 March 2022
Accepted: 14 April 2022
Published: 20 April 2022

Publisher's Note: MDPI stays neutral with regard to jurisdictional claims in published maps and institutional affiliations.

Copyright: © 2022 by the authors. Licensee MDPI, Basel, Switzerland. This article is an open access article distributed under the terms and conditions of the Creative Commons Attribution (CC BY) license (https://creativecommons.org/licenses/by/4.0/).

1. Introduction

The reduction in greenhouse gas (GHG) emissions by shifting towards renewable energy sources to control global warming is one of the main challenges of the 21st century [1]. Recent studies have shown that bioenergy and bio-economy can positively contribute to this emissions reduction [2–4]. Bioenergy can also address the volatility of wind and solar energy and play a role in circular economy as one of the important goals of the European Green Deal [5]. In general, biomass can be classified into six different material classes, which are wood and woody biomass, herbaceous agricultural biomass, aquatic biomass, animal and human biomass wastes, contaminated biomass and industrial biomass wastes (semi-biomass), and biomass mixtures [6].

In recent years, agricultural residues as sustainable energy sources have attracted a lot of research attention. This is due to their low material cost, lack of conflict as a food source, as well as their annual production capacity and distribution in both developed and developing countries [7,8].

These assorted residues are characterized by a relatively high ash content, which can reach up to 20 times the ash content in woody biomass, implying an overall increased slag formation tendency [9]. Ash-forming elements are generally Si, Ca, Mg, K, Na, P, S, Cl, Al, Fe, and Mn in varying amounts [9,10]. Therefore, the ash obtained from thermochemical conversion of biogenic residues and wastes can be employed in different material applications. In particular, Si-rich biomass assortments such as rice husk and rice straw have been increasingly investigated in the last few years [11,12]. Schneider considered the potential of regional feedstocks in Germany in the production of high-quality biogenic silica at the laboratory scale [13]. Schliermann et al. [14] investigated the quality of biogenic silica produced from combustion of rice husk in bench-scale biomass boilers. Their study overcame several challenges in operation of boilers and handling of the generated ashes. Further considerations on Si-rich biomass fuels are also reflected in several publications of the current Special Issue [15–21].

However, the composition of biomass ashes is not homogeneous due to their different origin and depending on fuel processing as well as thermochemical conversion conditions [22], and as a result, defining a specific application for all the biomass ashes is challenging [23]. The potential utilization of biomass ashes is mainly influenced by contaminations such as heavy metals and their slag formation tendencies [24]. Furthermore, the thermochemical conversion process should also be controlled in order to modify the ash composition for specific applications [25]. Considering diverse biomass ash compositions, there are different applications for these materials including cement and concrete production, soil stabilization, filler in asphalt, synthetic aggregate, catalysis, semiconducting materials, energy storage, drug delivery, electrochemical applications and batteries, carbon capture, etc. [23,26,27]. In this respect, any modification of fuel composition using a pre-treatment processes as well as different thermochemical conversion technologies can influence the ash characteristics [23,28].

2. The Special Issue

The present Special Issue entitled "Valorization of Residues from Energy Conversion of Biomass for Advanced and Sustainable Material Applications" includes 12 publications from several authors from Germany, Korea, South Africa, France, China, India, and Sweden.

The publications cover a wide range of topics related to biomass conversion and ash-related aspects. In this respect, Chun and Lee [17] provided an overview on recent developments in the purification of silica components and the production of engineered biogenic silica in a bottom-up process employing liquid silicate extracted from rice husk. The products can be used in various applications including heterogeneous catalysts, CO_2 capture, adsorbents for aqueous pollutants, biomolecule delivery, and cosmetic ingredients. Maseko et al. [16] investigated the quality of biogenic silica produced from the thermochemical conversion of different South African agricultural residues such as sugarcane leaves, pith and fiber, and maize leaves. Their investigation proved that silica with high purity and specific surface area can be produced from chemically pre-treated biomass fuels, especially from sugarcane-based input materials. Park et al. [21] extracted biogenic silica from rice husk using a novel two-stage attrition ball milling and alkaline leaching processes. The purity of the produced silica was comparable with the products generated from common thermochemical conversion processes. However, the advantages of the introduced synthesized method were a low production cost and higher productivity. Yan et al. [20] investigated a smoldering process, which is a slow, low-temperature, and flameless burning process to synthesize high-quality biogenic silica from rice husk. Their results showed a maximum conversion temperature of only 560 °C in the naturally piled fuel bed, which is lower than regularly reported combustion temperatures in different biomass combustion technologies. Produced biogenic silica had porous and amorphous nature due to the low conversion temperature. This publication reported a higher ash porosity generated from the smoldering of untreated rice husk as compared to the ashes produced from combustion processes in the literature. Therefore, smoldering shows great potential for the industrial production of high-quality biogenic silica from untreated silica-rich biomass fuels, and further investigations are required in this field. Singh et al. studied photocatalytic degradation of cationic dye using silica nanoparticles (SiNPs) synthesized from the combustion of rice straw followed by a sol-gel route. In this investigation, the effects of combustion temperature and ash crystallinity on the quality of SiNPs were studied, and it was shown that a higher combustion temperature leads to SiNPs with larger particle size with a crystalline nature. However, under controlled combustion temperatures, amorphous SiNPs with an average particle size lower than 30 nm can be synthesized from rice straw ash, which showed promising photocatalytic properties. Jung et al. [15] produced high-quality nano-structured silica as well as cellulose nano-fibrils through an alkaline fractionation process of rice husk. Mugadza et al. [29] synthesized nitrogen-doped carbon nanotubes (N-CNTs) from biomass as carbon precursors for supercapacitors and electrochemical applications. They showed that the initial biomass fuel can influence physicochemical and electrochemical properties of N-CNTs. Nitrogen migration behaviors during the pyrolysis of untreated and chemically pre-treated maize straw were investigated by Li et al. [18]. They proved that an appropriate chemical pre-treatment and pyrolysis temperature would guarantee the fixation of N in the generated biochar, which prevents NOx emission during the pyrolysis process. The produced N-enhanced biochar can be employed as N fertilizer in order to improve soil quality and production yield in agriculture [30]. Furthermore, soil amendment by biochar is one of the potential applications of biomass fuels to reduce the global carbon emission, improve soil quality, and increase soil carbon sequestration [31–34]. Frikha et al. [35] studied the effect of pyrolysis temperature on the yield and quality of biochar produced from grape marc. Grape marc, as a winery waste, is produced throughout the winemaking process, which has 10–30 wt.% of the grape fresh weight in wet basis.

Their investigation revealed that pyrolysis temperature influences pyrolysis yields, as well as thermal stability, specific surface area, mineral composition, and ash content of the biochar. Bachmaier et al. [36] conducted comprehensive research work on analyzing different nutrients and pollutant contents of different ash fractions obtained from various biomass heating plants in Bavaria, Germany in order to use the ashes as fertilizers. As prescribed by the German Fertilizer Ordinance, the study showed that the ashes from waste wood can contain an elevated content of heavy metals. In addition, fuel quality and combustion conditions modified the concentrations of heavy metals in the ashes to ensure an appropriate ash quality.

The recovery of some of the ash-forming elements such as P is important, as it is one of the critical and irreplaceable elements in human nutrition with a limited resource. However, the annual consumption of P in the agriculture sector is 20 million tons [37]. Therefore, recovery of this element besides the energetic utilization of biomass fuels is very beneficial. In this respect, an economic feasibility investigation of energy and P recovery from municipal sewage sludge was conducted by Bagheri et al. [38], considering 16 different technology scenarios of investments in new combustion plants. This study provides insights into economic performance and required financial support for energetic utilization and P recovery from municipal sewage sludge in Europe.

In order to decrease GHG emissions in a sustainable way in the transport sector, production of biomethane from manure, agricultural residues, and biowaste can have a profound impact [39]. This aspect of biomass application was considered in the publication by Oehmichen et al. [40] by assessing the market. The study showed that different advanced biofuel pathways have significantly different GHG mitigation costs. Furthermore, the magnitude of this mitigation is influenced by the type of substrate used in biomethane production processes, GHG emissions from the fossil energy carrier substituted by biomethane, and the calculation method. These aspects should be considered in future developments in this field.

3. Summary and Future Prospects

The focus of this Special Issue was on biomass ash valorization with respect to their potential for various material applications. Most of the publications in this Special Issue focused on the production of biogenic silica with different properties. Additionally, some of the publications considered application of biomass ashes and biochar as a fertilizer, for soil amendment and recovery of ash forming elements such as N and P as well as the application of biomass feedstocks in biofuel production.

Accordingly, ashes produced from the thermochemical conversion of agricultural residues have high potential to be utilized for different material applications. However, local availability as well as scaling up the process and life-cycle assessment should be considered prior to the utilization of these materials. Furthermore, densification as a mechanical pre-treatment can be crucial to improve the fuel properties, while purification of some of the ash forming elements such as calcium, potassium and prosperous should also not be disregarded in future investigations.

Author Contributions: H.B.D. prepared draft of this Special Issue editorial. All the editors (D.E., H.B.D., T.Z. and V.L.) read, revised, and approved the content for publication. All authors have read and agreed to the published version of the manuscript.

Funding: This research received no external funding.

Institutional Review Board Statement: Not applicable.

Informed Consent Statement: Not applicable.

Data Availability Statement: Not applicable.

Acknowledgments: The editors found the task of editing the submitted manuscripts for this Special Issue very stimulating and rewarding. The editors would like to express their sincere gratitude to the staff and reviewers of the *Sustainability* journal for their effort and input. The financial support of DBFZ Deutsches Biomasseforschungszentrum gemeinnützige GmbH is highly appreciated. The editors also thank all the international researchers who accepted to support the project by submitting their manuscripts to this Special Issue.

Conflicts of Interest: The authors declare no conflict of interest.

References

1. Dotzauer, M.; Oehmichen, K.; Thrän, D.; Weber, C. Empirical greenhouse gas assessment for flexible bioenergy in interaction with the German power sector. *Renew. Energy* **2022**, *181*, 1100–1109. [CrossRef]
2. Sulaiman, C.; Abdul-Rahim, A.S.; Ofozor, C.A. Does wood biomass energy use reduce CO_2 emissions in European Union member countries? Evidence from 27 members. *J. Clean. Prod.* **2020**, *253*, 119996. [CrossRef]
3. Jonsson, R.; Rinaldi, F.; Pilli, R.; Fiorese, G.; Hurmekoski, E.; Cazzaniga, N.; Robert, N.; Camia, A. Boosting the EU forest-based bioeconomy: Market, climate, and employment impacts. *Technol. Forecast. Soc. Chang.* **2021**, *163*, 120478. [CrossRef]
4. Saidur, R.; Abdelaziz, E.A.; Demirbas, A.; Hossain, M.S.; Mekhilef, S. A review on biomass as a fuel for boilers. *Renew. Sustain. Energy Rev.* **2011**, *15*, 2262–2289. [CrossRef]
5. ForschungsVerbund Erneuerbare Energien. Bioenergie für Eine Konsistente Klimaschutz- und Energiepolitik—Empfehlungen des FVEE. Available online: https://www.dbfz.de/fileadmin//user_upload/Referenzen/Statements/2020_08_25_FVEE-Bioenergie.pdf (accessed on 19 April 2022).
6. Vassilev, S.V.; Baxter, D.; Andersen, L.K.; Vassileva, C.G. An overview of the chemical composition of biomass. *Fuel* **2010**, *89*, 913–933. [CrossRef]
7. Kaltschmitt, M. Renewable Energy Renewable Energy from Biomass renewable energy from Biomass, Introduction. In *Renewable Energy Systems*; Kaltschmitt, M., Themelis, N.J., Bronicki, L.Y., Söder, L., Vega, L.A., Eds.; Springer: New York, NY, USA, 2013; pp. 1393–1396, ISBN 978-1-4614-5819-7.
8. IEA. *World Energy Outlook*; IEA: Paris, France, 2021. Available online: https://www.iea.org/reports/world-energy-outlook-2021 (accessed on 19 April 2022).
9. Boström, D.; Skoglund, N.; Grimm, A.; Boman, C.; Öhman, M.; Broström, M.; Backman, R. Ash transformation chemistry during combustion of biomass. *Energy Fuels* **2011**, *26*, 85–93. [CrossRef]
10. Vassilev, S.V.; Baxter, D.; Andersen, L.K.; Vassileva, C.G. An overview of the composition and application of biomass ash. Part 1. Phase–mineral and chemical composition and classification. *Fuel* **2013**, *105*, 40–76. [CrossRef]
11. Beidaghy Dizaji, H.; Zeng, T.; Hölzig, H.; Bauer, J.; Klöß, G.; Enke, D. Ash transformation mechanism during combustion of rice husk and rice straw. *Fuel* **2022**, *307*, 121768. [CrossRef]
12. Beidaghy Dizaji, H.; Zeng, T.; Enke, D. New fuel indexes to predict ash behavior for biogenic silica production. *Fuel* **2022**, *310*, 122345. [CrossRef]
13. Schneider, D. Biogenic Silica from Regional Feedstocks—Sustainable Synthesis and Characterization. Ph.D. Thesis, Universität Leipzig, Leipzig, Germany, 2019.
14. Schliermann, T.; Hartmann, I.; Beidaghy Dizaji, H.; Zeng, T.; Schneider, D.; Wassersleben, S.; Enke, D.; Jobst, T.; Lange, A.; Roelofs, F.; et al. High quality biogenic silica from combined energetic and material utilization of agricultural residues. In Proceedings of the 7th International Symposium of Energy from Biomass and Waste, Venice, Italy, 15–18 October 2018.
15. Jung, H.; Kwak, H.; Chun, J.; Oh, K. Alkaline fractionation and subsequent production of nano-structured silica and cellulose nano-fibrils for the comprehensive utilization of rice husk. *Sustainability* **2021**, *13*, 1951. [CrossRef]
16. Maseko, N.N.; Schneider, D.; Wassersleben, S.; Enke, D.; Iwarere, S.A.; Pocock, J.; Stark, A. The production of biogenic silica from different south african agricultural residues through a thermo-chemical treatment method. *Sustainability* **2021**, *13*, 577. [CrossRef]
17. Chun, J.; Lee, J.H. Recent progress on the development of engineered silica particles derived from rice husk. *Sustainability* **2020**, *12*, 10683. [CrossRef]
18. Li, H.; Mou, H.; Zhao, N.; Yu, Y.; Hong, Q.; Philbert, M.; Zhou, Y.; Dizaji, H.B.; Dong, R. Nitrogen migration during pyrolysis of raw and acid leached maize straw. *Sustainability* **2021**, *13*, 3786. [CrossRef]
19. Singh, G.; Beidaghy Dizaji, H.; Puttuswamy, H.; Sharma, S. Biogenic nanosilica synthesis employing agro-waste rice straw and its application study in photocatalytic degradation of cationic dye. *Sustainability* **2022**, *14*, 539. [CrossRef]
20. Yan, S.; Yin, D.; He, F.; Cai, J.; Schliermann, T.; Behrendt, F. Characteristics of smoldering on moist rice husk for silica production. *Sustainability* **2022**, *14*, 317. [CrossRef]
21. Park, J.Y.; Gu, Y.M.; Park, S.Y.; Hwang, E.T.; Sang, B.-I.; Chun, J.; Lee, J.H. Two-stage continuous process for the extraction of silica from rice husk using attrition ball milling and alkaline leaching methods. *Sustainability* **2021**, *13*, 7350. [CrossRef]
22. Mlonka-Mędrala, A.; Magdziarz, A.; Gajek, M.; Nowińska, K.; Nowak, W. Alkali metals association in biomass and their impact on ash melting behaviour. *Fuel* **2020**, *261*, 116421. [CrossRef]
23. James, A.; Thring, R.; Helle, S.; Ghuman, H. Ash Management Review—Applications of Biomass Bottom Ash. *Energies* **2012**, *5*, 3856–3873. [CrossRef]

24. Khan, A.A.; de Jong, W.; Jansens, P.J.; Spliethoff, H. Biomass combustion in fluidized bed boilers: Potential problems and remedies. *Fuel Process. Technol.* **2009**, *90*, 21–50. [CrossRef]
25. Beidaghy Dizaji, H.; Zeng, T.; Hartmann, I.; Enke, D.; Schliermann, T.; Lenz, V.; Bidabadi, M. Generation of high quality biogenic silica by combustion of rice husk and rice straw combined with pre- and post-treatment strategies—A review. *Appl. Sci.* **2019**, *9*, 1083. [CrossRef]
26. Shen, Y. Rice husk silica derived nanomaterials for sustainable applications. *Renew. Sustain. Energy Rev.* **2017**, *80*, 453–466. [CrossRef]
27. Pode, R. Potential applications of rice husk ash waste from rice husk biomass power plant. *Renew. Sustain. Energy Rev.* **2016**, *53*, 1468–1485. [CrossRef]
28. Zareihassangheshlaghi, A.; Beidaghy Dizaji, H.; Zeng, T.; Huth, P.; Ruf, T.; Denecke, R.; Enke, D. Behavior of Metal Impurities on Surface and Bulk of Biogenic Silica from Rice Husk Combustion and the Impact on Ash-Melting Tendency. *ACS Sustain. Chem. Eng.* **2020**, *8*, 10369–10379. [CrossRef]
29. Mugadza, K.; Stark, A.; Ndungu, P.G.; Nyamori, V.O. Effects of ionic liquid and biomass sources on carbon nanotube physical and electrochemical properties. *Sustainability* **2021**, *13*, 2977. [CrossRef]
30. Chan, K.Y.; van Zwieten, L.; Meszaros, I.; Downie, A.; Joseph, S. Using poultry litter biochars as soil amendments. *Soil Res.* **2008**, *46*, 437. [CrossRef]
31. Schmidt, H.-P.; Kammann, C.; Niggli, C.; Evangelou, M.W.H.; Mackie, K.A.; Abiven, S. Biochar and biochar-compost as soil amendments to a vineyard soil: Influences on plant growth, nutrient uptake, plant health and grape quality. *Agric. Ecosyst. Environ.* **2014**, *191*, 117–123. [CrossRef]
32. Bachmann, H.J.; Bucheli, T.D.; Dieguez-Alonso, A.; Fabbri, D.; Knicker, H.; Schmidt, H.-P.; Ulbricht, A.; Becker, R.; Buscaroli, A.; Buerge, D.; et al. Toward the standardization of biochar analysis: The COST Action TD1107 Interlaboratory Comparison. *J. Agric. Food Chem.* **2016**, *64*, 513–527. [CrossRef]
33. Hagemann, N.; Joseph, S.; Schmidt, H.-P.; Kammann, C.I.; Harter, J.; Borch, T.; Young, R.B.; Varga, K.; Taherymoosavi, S.; Elliott, K.W.; et al. Organic coating on biochar explains its nutrient retention and stimulation of soil fertility. *Nat. Commun.* **2017**, *8*, 1089. [CrossRef]
34. Joseph, S.; Cowie, A.L.; van Zwieten, L.; Bolan, N.; Budai, A.; Buss, W.; Cayuela, M.L.; Graber, E.R.; Ippolito, J.A.; Kuzyakov, Y.; et al. How biochar works, and when it doesn't: A review of mechanisms controlling soil and plant responses to biochar. *GCB Bioenergy* **2021**, *13*, 1731–1764. [CrossRef]
35. Frikha, K.; Limousy, L.; Arif, M.B.; Thevenin, N.; Ruidavets, L.; Zbair, M.; Bennici, S. Exhausted grape marc derived biochars: Effect of pyrolysis temperature on the yield and quality of biochar for soil amendment. *Sustainability* **2021**, *13*, 11187. [CrossRef]
36. Bachmaier, H.; Kuptz, D.; Hartmann, H. Wood ashes from grate-fired heat and power plants: Evaluation of nutrient and heavy metal contents. *Sustainability* **2021**, *13*, 5482. [CrossRef]
37. Mayer, B.K.; Baker, L.A.; Boyer, T.H.; Drechsel, P.; Gifford, M.; Hanjra, M.A.; Parameswaran, P.; Stoltzfus, J.; Westerhoff, P.; Rittmann, B.E. Total value of phosphorus recovery. *Environ. Sci. Technol.* **2016**, *50*, 6606–6620. [CrossRef] [PubMed]
38. Bagheri, M.; Öhman, M.; Wetterlund, E. Techno-economic analysis of scenarios on energy and phosphorus recovery from mono- and co-combustion of municipal sewage sludge. *Sustainability* **2022**, *14*, 2603. [CrossRef]
39. Oehmichen, K.; Thrän, D. Fostering renewable energy provision from manure in Germany—Where to implement GHG emission reduction incentives. *Energy Policy* **2017**, *110*, 471–477. [CrossRef]
40. Oehmichen, K.; Majer, S.; Thrän, D. Biomethane from manure, agricultural residues and biowaste—GHG mitigation potential from residue-based biomethane in the European transport sector. *Sustainability* **2021**, *13*, 14007. [CrossRef]

Article

Techno-Economic Analysis of Scenarios on Energy and Phosphorus Recovery from Mono- and Co-Combustion of Municipal Sewage Sludge

Marzieh Bagheri *, Marcus Öhman and Elisabeth Wetterlund

Division of Energy Science, Luleå University of Technology, 97187 Luleå, Sweden; marcus.ohman@ltu.se (M.Ö.); elisabeth.wetterlund@ltu.se (E.W.)
* Correspondence: marzieh.bagheri@ltu.se

Abstract: This study evaluates the techno-economic feasibility of energy and phosphorus (P) fertilizer (PF) recovery from municipal sewage sludge (MSS) through incineration in new combustion plants. We evaluated the economic impact of five critical process design choices: (1) boiler type, (2) fuel (MSS mono-combustion/co-combustion with wheat straw), (3) production scale (10/100 MW), (4) products (heat, electricity, PF), and (5) ash destination. Aspen Plus modeling provided mass and energy balances of each technology scenario. The economic feasibility was evaluated by calculating the minimum selling price of the products, as well as the MSS gate fees required to reach profitability. The dependency on key boundary conditions (operating time, market prices, policy support) was also evaluated. The results showed a significant dependency on both energy and fertilizer market prices and on financial support in the form of an MSS gate fee. Heat was preferred over combined heat and power (CHP), which was feasible only on the largest scale (100 MW) at maximum annual operating time (8000 h/y). Co-combustion showed lower heat recovery cost (19–30 €/MWh) than mono-combustion (29–66 €/MWh) due to 25–35% lower energy demand and 17–25% higher fuel heating value. Co-combustion also showed promising performance for P recovery, as PF could be recovered without ash post-treatment and sold at a competitive price, and co-combustion could be applicable also in smaller cities. When implementing ash post-treatment, the final cost of ash-based PF was more than four times the price of commercial PF. In conclusion, investment in a new combustion plant for MSS treatment appears conditional to gate fees unless the boundary conditions would change significantly.

Keywords: municipal sewage sludge; energy recovery; phosphorus recovery; techno-economic analysis; mono-combustion; co-combustion

Citation: Bagheri, M.; Öhman, M.; Wetterlund, E. Techno-Economic Analysis of Scenarios on Energy and Phosphorus Recovery from Mono- and Co-Combustion of Municipal Sewage Sludge. *Sustainability* 2022, 14, 2603. https://doi.org/10.3390/su14052603

Academic Editors: Dirk Enke, Hossein Beidaghy Dizaji, Volker Lenz and Thomas Zeng

Received: 2 February 2022
Accepted: 18 February 2022
Published: 23 February 2022

Publisher's Note: MDPI stays neutral with regard to jurisdictional claims in published maps and institutional affiliations.

Copyright: © 2022 by the authors. Licensee MDPI, Basel, Switzerland. This article is an open access article distributed under the terms and conditions of the Creative Commons Attribution (CC BY) license (https://creativecommons.org/licenses/by/4.0/).

1. Introduction

Phosphorus (P) is a critical and irreplaceable element in human nutrition. However, the primary P resource, phosphate rock, is limited and geographically concentrated to a few regions, and the dependency of agriculture on mineral P causes increases in fertilizer price and uncertainty in the P market [1–3]. The world's economic reserves of high quality, low extraction cost P have been estimated at about 17 billion tons [4], and the agriculture sector alone consumes 20 million tons on P each year [5]. The P market experienced an 800% price increase in 2008, and even after the peak dropped, the new P price was twice that of before 2008 [6]. Therefore, efforts to recover P from P-rich wastes have been intensified [7]. MSS, the solid waste residue of wastewater treatment plants, is considered one of the most promising P-rich sources due to both high P concentration and large volumes [8,9]. It is expected that 17–31% of the currently used mineral fertilizer could be substituted by P from biogenic materials, mainly MSS, via advanced technologies by 2030 [10,11]. Land application and composting of MSS are simple P recovery methods and the dominant practice in Europe for disposal or recovery [12]. However, using those methods also leads to the inclusion of heavy metals (HMs), pathogens, and pharmaceuticals with the nutrients [13,14].

Conversely, the combustion of MSS encompasses advantages such as significant volume reduction, energy recovery, and the destruction of organic contaminants and pathogens without necessarily impairing the P recovery opportunity [15,16]. Energy and P recovery from MSS is directly related to the United Nations Sustainable Development Goals 7 and 12, where replacement of non-renewable energy with green energy resources and reuse of valuable material is crucial for a sustainable society. Sewage sludge ash (SSA) typically contains 10–25 wt.% P_2O_5, which is comparable with phosphate rock (5–40 wt.% P_2O_5) [17], and allows for 5–10 times higher P recovery compared to recovery from MSS and leachates due to higher concentration of P [8,18,19].

Primary drawbacks of MSS mono-combustion are high moisture content [20], simultaneous potential accumulation of P and HMs in the ash [17], and low plant availability of the dominant P species in the SSA apatite [21]. The first issue makes MSS mono-combustion inefficient from an energy perspective, while the latter two make direct land application of SSA less suitable and limit the possible substitution of phosphate rock in commercial fertilizer production to 10–20% [22]. Thermochemical treatment and wet extraction methods have been used to transfer ash-based P to a more water-soluble species with lower HMs content [23–25]. Extracted P by the wet process still needs post-treatment to achieve an adequate quality regarding HMs content and plant availability [26]. However, Herzel et al. [27] showed that thermochemical treatment of the SSA under reducing conditions with Na- or K-based alkaline additives, such as NaOH, Na_2CO_3, K_2CO_3, and KOH, results in a marketable fertilizer, as indicated by the ash-based PF solubility in neutral ammonium citrate. Similarly, in the thermochemical conversion process Ash Dec, the SSA is treated with alkaline additives in a rotary kiln at 1000 °C [27]. This process leads to plant-available P species and more than 90% decontamination of HMs. Ash Dec has been applied commercially and has shown high P recovery, with a final product that does not need further post-processing [19,28,29].

Alternatively, the plant availability of SSA-based P can be improved by altering the ash formation during the combustion process, which can also decrease the accumulation of P and HMs in the same ash fraction. According to previous studies, co-combustion of MSS and K-rich agricultural fuels with low moisture (e.g., wheat straw) may directly both provide more plant-available phosphates in the combustion process [24,30,31] and eliminate the need for energy-intensive drying. Furthermore, the type of combustion technology can affect how P and HMs accumulate in different ash fractions, thereby stimulating HMs separation [32].

Energy recovery from MSS via combustion with subsequent P recovery from SSA requires expensive technology [33,34], and the possibility of moving from disposal to recovery for MSS is substantially tied to economic feasibility [35]. Sustainable MSS management thus needs to be built around technology that enables a high recovery rate, cost-effectiveness, and a marketable output that can compete with conventional products and preferably be used directly. Various techno-economic analysis studies have been conducted on either P or energy recovery from MSS rather than SSA. The benefits of P recovery by considering the environmental externalities have been shown by [36]. The economic feasibility of P recovery from wastewater and sludge for a pilot-test condition was evaluated by comparing the experiment's operation costs (mainly energy and chemical costs) with the market price of fertilizer [37]. Horttanainen et al. reported a 2.5–10 years payback period of heat and power generation through MSS combustion in different technical conditions [15]. In general, the focus of previous studies has been either on various technical options for P and energy recovery or on comparison between different available options for either P or energy recovery [5,35–38]. However, techno-economic investigations of combined energy and P recovery through mature technologies, requirements of final products' marketability by calculation of the minimum selling prices and gate fees in various market conditions, and comparisons with different ash handling practices have, to the best of our knowledge, not yet been addressed together.

This study aims to explore the economic feasibility of energy and P recovery from MSS through combustion under different technology, operation, and market conditions. We consider investment in a new combustion plant for different technology design options regarding combustion technology, plant scale, fuel composition, energy and material outputs, and the final destinations of the ash residue. Specific objectives are to investigate (i) to what extent the economic feasibility can be affected by the technical design, (ii) the requirement of financial support, and (iii) necessary energy and fertilizer market conditions. The objectives are addressed by developing a techno-economic analysis that evaluates the minimum selling price of sewage sludge-based energy and PF. The assessment is performed for a number of technology scenarios designed to investigate the inherent relations between heat, power, and PF production, as well as for different economic variations designed to investigate the influence of non-technical and operational parameters. The results provide insights into economic performance and required financial support to produce energy carriers from MSS and replace mineral P products with sludge-based ones in Europe.

2. Technology Scenarios

The following sections describe the main perspectives behind the scenario selections. Sixteen technology scenarios were developed based on variations of (1) boiler type and size, (2) fuel composition, (3) final destination of the ash residue, and (4) outputs from the plant. Figure 1 gives a schematic overview of the plants, with descriptions in the following sections.

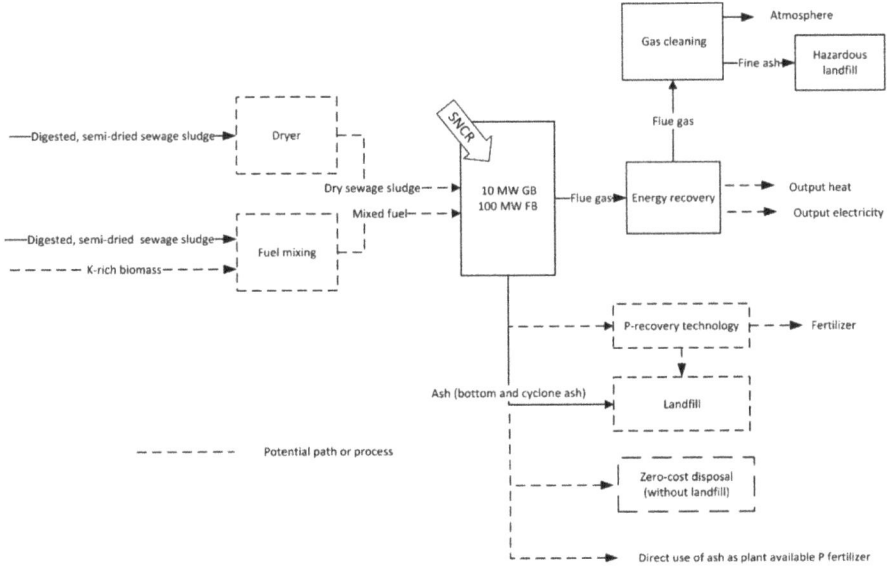

Figure 1. Schematic overview of variations in the modeled technology scenarios.

2.1. Boiler Type, Size, and Ash Distribution

Fluidized bed (FB) boilers and grate boilers (GB) respectively constitute the most common type of MSS and municipal solid waste combustion incinerators [39–41]. The selection of boiler technology for waste and biomass is affected by a number of parameters, such as the need for storage, fuel characteristics and preparation, combustion efficiency, emission, and other region-related conditions such as availability of specific biomass types and potential policy instruments related to, e.g., renewable energy [42]. The optimal boiler type selection is beyond the scope of this paper, and both FB and GB were selected as commercially available options.

Because of the low energy density on a mass basis of biomass compared to fossil fuels and the potentially high transportation costs associated with the large spatial distribution of biomass resources, biomass combustion plants in Europe usually have a small scale (10–15 MW), compared to, e.g., coal power plants [40]. Since the GB is the most common combustion technology for biomass and solid waste in Europe [41], GB scenarios were designed for 10 MW boiler capacity. FB is commonly applied for MSS combustion at a larger scale than GB [31,34,39]. In order to also consider the effects of economy-of-scale and to represent larger CHP plants, FB technology was thus selected for scenarios with 100 MW boiler capacity.

To comply with emission limits, SO_2, HCl, NO_x, HMs, and particle emissions must be carefully controlled; therefore, the same gas cleaning system that is particularly suggested for MSS-burned plants is considered in all scenarios [34,43,44]. In this system, a cyclone extracts fly ash, and an electrostatic precipitator separates the fly ash left after the cyclone [45]. Subsequently, a wet scrubber using ($Ca(OH)_2$) is applied for desulfurization. Lime or urea is directly injected into the boiler to reduce nitrogen oxides to eliminate nitrogen through selective non-catalytic reduction (SNCR) [46]. Finally, coke is used for Hg, Cd, dioxins, and furans separation, and the flue gas is polished from fine ash particles through a fabric filter capture [34,47]. As shown in Figure 1, the fly ash collected in the cyclone and bottom ash are discharged together, whereas fine ash is discharged separately.

2.2. Fuel Composition

The high moisture content of MSS causes a low heating value, lower furnace temperature, and/or high energy demand for drying to the moisture content required by the thermochemical process [38,48]. The solid content of digested MSS after mechanical dewatering can typically be found in the ranges of 20–28% (belt presses), 20–35% (decanting centrifuges), and 28–45% (filter presses), respectively [49,50]. At least 28–33 wt.% of solid is typically needed in theory to initiate auto-thermal combustion of sewage sludge without auxiliary fuel [17]; however, 40–50 wt.% of solid material is the minimum practical requirement for MSS incineration [51]. Total energy gain from sludge combustion must be considered against the drying demand for higher solid content. Therefore, waste incineration plants typically accept sludge with 60 wt.% of solid material and higher for co-incineration [16,34] to not disturb their positive energy balance. For sludge, mono-combustion positive energy balance occurs when solid content is more than 70 wt.% [20].

Besides the mentioned issues regarding moisture content, MSS undergoes a sticky phase at a solid content of 55–70 wt.% [52], which makes it difficult to handle and feed to the incinerator. Since combustion plants are usually centralized and sludge feed is completely or partially transferred to the plant from different wastewater treatment plants, sludge with higher solid content is favorable unless it crosses the sticky phase zone. To meet these practical requirements, this study assumed that, in the case of mono-combustion, digested semi-dried MSS with 50 wt.% solid content enters the combustion plant, dried to 80 wt.% solid by a hot air dryer.

There is an opportunity to decrease the energy demand for drying and enhance P recovery by blending MSS with agricultural residues with low moisture content. Through thermochemical post-treatment with K and Na additives apatite ($Ca_5(PO_4)_3OH$) in SSA, which is poorly plant-available, alters to a more plant-available form of P such as $CaNaPO_4$ [22]. Agriculture residues can be K-rich and applicable as K additives [53]. It has been shown that the co-combustion of MSS with K-rich agricultural residue, such as wheat straw (WS), analogously transfers the ash formation pathway toward K-bearing phosphates instead of the Ca/Fe/Al phosphates otherwise dominate the SSA [30]. Typical characteristics of MSS and WS are shown in Table 1 [54,55]. Häggström et al. examined the co-combustion of MSS with various agriculture residues in different mass ratios. They showed desirable P species ($Ca_9MgK(PO_4)_7$ and $CaKPO_4$) only forms when MSS is in a low share of the fuel mixture

(90% WS and 10% MSS) [30]. They also showed that in this fuel mixture, at least 42% of P in the ash was found in a plant-available form (28% $Ca_9MgK(PO_4)_7$ and 14% $CaKPO_4$) [30].

Table 1. Typical MSS and WS characteristics and ash composition (ar: as received, db: dry base) [54,55].

Fuel Characteristics	WS		MSS	
	min	max	min	max
Moisture content (wt% ar)	8.3	17.4	53	77.6
Ash content (wt%,db)	3.06	9.55	29	54.5
Volatile material (VM) (wt%,db)	71.1	81.2	39.7	60.4
Fixed carbon (FC) (wt%,db)	14.9	19.3	2.1	11.5
C (wt%,db)	42.9	48.3	23.1	36.5
H (wt%,db)	3.1	5.96	2.6	5.3
O (wt%,db)	38.3	45.6	10.3	32.5
N (wt%,db)	0.28	1.54	1.4	5.6
S (wt%,db)	0.03	0.29	0.5	1.88
Cl (mg/kg)	200	500	300	800
High heating value (MJ/kg,db)	16.4	20.7	7.2	16.7
Ash Analyses (mg/kg ash)	WS		MSS	
	min	max	min	max
Si	217,371	339,380	72,778	200,000
Al	1535	3017	30,278	116,000
Fe	1259	4057	100,000	233,645
Pb	n.a	n.a	22	424
Mn	387	697	421	1938
Ca	67,896	109,348	37,400	94,884
Mg	12,062	20,505	3673	13,256
Na	5564	7419	1430	8500
K	9132	152,747	1102	18,600
P	7419	16,583	48,900	105,030

In summary, by co-combusting MSS with carefully selected agricultural residues, a plant-available ash-based PF can potentially be obtained without subsequent post-treatment while simultaneously avoiding energy-intensive and expensive drying. Therefore, both co-combustion of an optimal fuel mixture (90% WS, 10% MSS) and mono-combustion (100% MSS) were considered in the technology scenarios in this study.

2.3. The Final Destinations of the Ash Residue

Although significant volume reduction is one of the main advantages of MSS combustion [22,56], it is not a zero-waste operation. Thus, combustion plants always have an output in the form of ash, which is normally not associated with any economic gains, but typically instead incurs a cost for the plant owner. In the mono-combustion scenarios in this study, the effect of three different ash destinations on the economic performance of the combustion plant was investigated: landfilling, transfer of the ash to another industrial process (zero-cost disposal), and implementation of a post-treatment process.

On a dry base, MSS usually contains around 30–50% ash [34] which mainly contains SiO_2, CaO, Al_2O_3, Fe_2O_3, MgO, and P_2O_5 (see Table 1) [57]. The high amount of CaO and SiO_2 enables the ash to be a partial alternative ingredient of building materials [58,59]. This option removes the ash disposal cost, but it may eliminate the opportunity to recover the P instead. The Ash Dec process was selected for the post-treatment as it is a commercially available and approved process for producing fertilizer from SSA without further treatment [19,29,30].

As mentioned in Section 2.2, a well-selected fuel mixture was hypothesized to yield ash with plant-available P without any requirement for post-treatment in the co-combustion case. Research is still ongoing on different fuel mixtures, and how much of the P in the ash could be sellable as a competitor to commercial fertilizers [33]. Here, it was assumed that 50% of P [30] in the ash of the given fuel mixture has fertilizer value and that it would be directly sellable as plant-available PF in the applicable cases (co-combustion direct PF, see below).

2.4. Products

Three different potential products were considered from the combustion plant: heat for district heating, electricity, and PF, in different combinations. Of those, electricity and heat are already well-established market products, while full-scale production PF from SSA has yet to reach its full potential [13,57].

2.5. Scenario Summary

Table 2 summarizes the modeled scenarios, as described in the previous sections, regarding the four selected technology variations.

Table 2. Overview of modeled technology scenarios. For details, see the text.

Scenario	Boiler Capacity (MW)	Boiler Type	Fuel (MSS/WS wt%,db)	Ash Destination				Energy Recovery	
				Ash Landfill	Zero-cost Ash Handling [a]	Ash Post-Treatment [b]	Direct Use	Heat	Electricity
(1)-10MW-mono-heat only-with landfill	10	GB	100/0	yes	no	no	no	yes	no
(2)-10MW-mono-heat only-without landfill	10	GB	100/0	no	yes	no	no	yes	no
(3)-10MW- mono-heat and PF	10	GB	100/0	no	no	yes	no	yes	no
(4)-10MW-mono-CHP with landfill	10	GB	100/0	yes	no	no	no	yes	yes
(5)-10MW-mono-CHP-without landfill	10	GB	100/0	no	yes	no	no	yes	yes
(6)-10MW-mono-CHP and PF	10	GB	100/0	no	no	yes	no	yes	yes
(7)-10MW-co-heat-direct PF	10	GB	10/90	no	no	no	yes [c]	yes	no
(8)-10MW-co-CHP-direct PF	10	GB	10/90	no	no	no	yes [c]	yes	yes
(9)-100MW-mono-heat only-with land fill	100	FB	100/0	yes	no	no	no	yes	no
(10)-100MW-mono-heat only-without landfill	100	FB	100/0	no	yes	no	no	yes	no
(11)-100MW-mono-heat and PF	100	FB	100/0	no	no	yes	no	yes	no
(12)-100MW-mono-CHP and landfill	100	FB	100/0	yes	no	no	no	yes	yes
(13)-100MW-mono-CHP without landfill	100	FB	100/0	no	yes	no	no	yes	yes
(14)-100MW-mono-CHP and PF	100	FB	100/0	no	no	yes	no	yes	yes
(15)-100MW-co-heat-direct PF	100	FB	10/90	no	no	no	yes [c]	yes	no
(16)-100MW-co-CHP-direct PF	100	FB	10/90	no	no	no	yes [c]	yes	yes

[a] Labeled "without landfill" in the scenarios, as there is no landfill cost in this case except the landfill of the residue of flue gas cleaning system. The ash was assumed to be transferred to another industrial process at zero cost. [b] Ash post-treatment using Ash Dec. The waste of ash post-treatment process is landfilled. [c] Co-combustion scenarios include the landfill cost of the fine ash.

3. Methodology and Input Data

The different combustion plant configurations in the technology scenarios were modeled using Aspen Plus® in order to obtain mass and energy balances. The obtained balances were used as the basis for the economic evaluations to calculate the required minimum selling prices (MSP) of the considered products (energy carriers and PF) for different operation and market conditions.

3.1. Process Modeling

A combustion plant flowsheet model, partly based on Salman et al. [59], was developed using the commercial software Aspen Plus® to obtain mass and energy balances for the technology scenarios (Figure 2). The characteristics of digested MSS and WS taken in this study are shown in Table 3, which falls in the typical range of characteristics of MSS and WS by considering the data shown in Table 1. Digested MSS is dried by a hot air dryer before entering the boiler for plant configurations involving a dryer. MSS was modeled as a non-conventional solid through ultimate and proximate analyses according to data in Table 3. For this reason, the boiler section was modeled as two reactors: an RYield reactor that decomposes the MSS to its constituent elements, in tandem with an RGibbs reactor that simulates the combustion reactions considering thermodynamic equilibrium by minimizing the Gibbs free energy. These two reactors are connected with a heat stream to include the decomposition energy in the combustion. The steam generation section contains three heat exchangers to represent the economizer, evaporator, and superheater, where the combustion products from the boiler section pass on one side of the heat exchangers, and boiler feed water flows on the other side. Flowrates of hot air for the dryer, excess air for combustion, and boiler feedwater were calculated based on the mass and energy balance using the calculator box and design spec options in Aspen Plus®. Calculations for unknown variables were calculated using in-line Fortran.

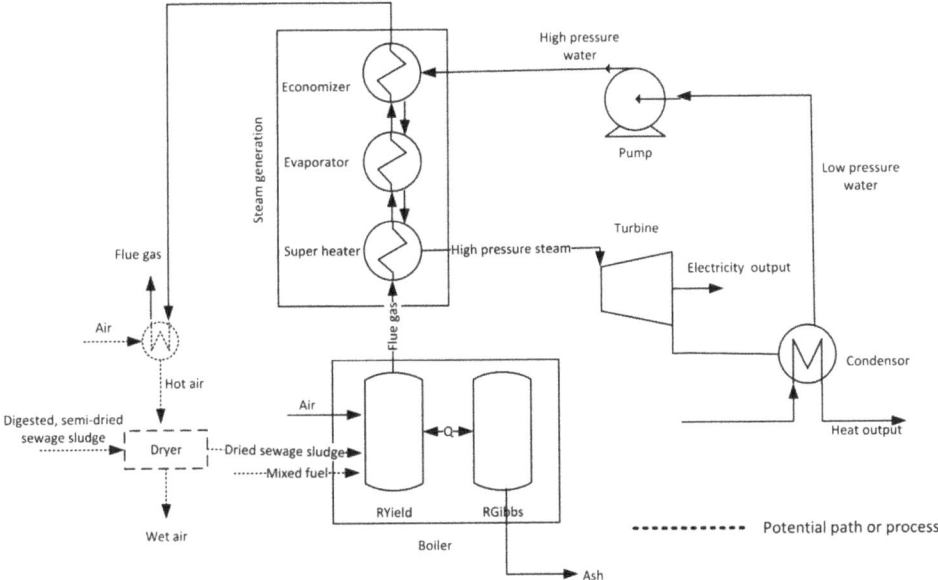

Figure 2. Simplified process flow diagram of the modeled combustion plant. Boxes represent modules containing two or more pieces of equipment.

Table 3. Data applied in the simulations.

Fuel Characteristics		
	MSS [a]	WS [b]
Proximate	%	%
Moisture	50	10.25
FC	8.6	18.68
VM	53	77.2
Ash	38.4	4.12
Ultimate	wt%,db	wt%,db
Ash	38.4	4.1
C	31.1	46.56
H	4.2	5.68
N	3.3	0.43
Cl	0.9	0.2
S	1.1	0.064
O	21	42.9
P_2O_5 in the ash	21	2.7
High heating value (MJ/kg)	13.9	18.8
Process Data		
Dryer output moisture content (%)		20
Combustion temperature (°C)		900
Steam pressure (bar)		45
Steam temperature (°C)		450
Stack temperature (°C)		120
Excess air (%)	GB	80
	FB	30
District heating temperature (°C)	inlet	40
	outlet	95

[a] Digested MSS based on the work of [44]. [b] Wheat straw based on the work of [30].

All the P in the MSS transfers to the SSA in combustion [32], although it can be collected from different ash fractions; therefore, in this study, the entire fuel P content was considered in the calculation. A black box method based on data from the literature [29] was used for the ash post-treatment for P recovery through the Ash Dec process. Detailed data can be found in Appendix A (Table A1).

3.2. Economic Analysis

The economic viability of energy recovery and PF production from MSS was evaluated by calculating the final products' MSP (Equation (1)), which indicates the break-even point for the investment. In addition, a feedstock gate fee (receiving fee paid by local authorities to the waste processing facility, per unit of inlet waste) was applied to indicate the financial dependency of MSS-based products on external support. The gate fee per ton of received MSS was thus considered as potential revenue, when applicable. The fuel price for the MSS price was correspondingly set to zero.

$$MSP_i = \frac{Annualized\ CAPEX + OPEX - revenue_j}{annual\ production\ of\ i} \quad (1)$$

where MSP_i is the minimum selling price of the target product i, where product j represents other sold products. The total capital investment cost (CAPEX) of each scenario was estimated from the literature. Details of included equipment can be found in Appendix A (Table A2). The CAPEX was inflation-adjusted to €$_{2020}$ using the chemical engineering plant cost index and an exchange rate of 0.85 €/US$. The reference costs were also adjusted to the desired scale through Equation (2).

$$Cost\ in\ desired\ scale = Cost\ in\ reference\ scale \times \left(\frac{desired\ scale}{reference\ scale}\right)^{0.8} \quad (2)$$

The $CAPEX$ was annualized applying an annuity factor of 10%, corresponding to an internal rate of return of 8% and 20 years of lifetime. The costs associated with land, design, engineering, and construction were excluded from the investment costs. Regarding operation cost ($OPEX$) of each scenario, labor, scheduled maintenance, routine component/equipment replacement, and insurance costs were considered in the fixed operation costs, which were set as 4% of the CAPEX [60,61]. The variable operation costs include chemicals required for the flue gas cleaning and Ash Dec process, energy (electricity and natural gas), disposal costs (ash and hazardous waste), and fuel price (WS) (see Table A1 in Appendix A). The 50–100% increase in fuel price based on WS price was also considered to capture the effect of fuel preparation costs. Hazardous waste refers to the flue gas cleaning waste containing fine ash with high HMs concentrations. Based on Hermann et al. [29], it was assumed that the fine ash constitutes 3% of the total ash in all scenarios. Produced electricity, in CHP scenarios, first covers internal demands, and then the rest can be sold to the market.

The gate fee for the target product i manifests the gap between MSP_i and the estimated market price i, which must be compensated by external revenue based on inlet feedstock Equation (3). Table 4 contains the details of MSP and gate fee calculations for each scenario.

$$Gate\ fee_i = \frac{MSP_i - Market\ price_i}{Requried\ sewage\ sludge\ per\ unit\ of\ i} \quad (3)$$

Operation and Market Variations

Energy costs and revenues for heat, electricity, and PF can be subject to significant volatility, following various market conditions, which was reflected here by applying different prices. The impact of electricity and heat market prices on the plant's economic viability was further evaluated for three different annual utilization hours; 3500, 5000, and 8000 h/year, where 5000 h/year was used as the base case. Table 5 summarises the applied energy and chemical prices, as well as variations in prices and other varied operational parameters.

Table 4. MSP and gate fee calculation for the scenarios.

Scenarios	Scenario Number	MSP	Gate Fee
Heat-only [a]	1,2,9,10	$MSP_{heat} = \frac{Annualized\ CAPEX + OPEX}{Annual\ production\ of\ heat}$	$Gate\ fee_{heat} = \frac{MSP_{heat} - Heat\ market\ price}{t_{sewage\ sludge}\ per\ MWh_{heat}}$
CHP [b]	4,5,12,13	$MSP_{heat} = \frac{Annualized\ CAPEX + OPEX - electricity\ revenue}{Annual\ production\ of\ heat}$	
Heat+ direct PF [c]	7,15	$MSP_{heat} = \frac{Annualized\ CAPEX + OPEX - direct\ PF\ revenue}{Annual\ production\ of\ heat}$	
CHP+ direct PF [c]	8,16	$MSP_{heat} = \frac{Annualized\ CAPEX + OPEX - electricity\ revenue - direct\ PF\ revenue}{Annual\ production\ of\ heat}$	
Heat+PF [d]	3,7,11,15	$MSP_{PF} = \frac{Annualized\ CAPEX + OPEX - heat\ revenue}{Annual\ production\ of\ P\ fertilizer}$ $MSP_{heat} = \frac{Annualized\ CAPEX + OPEX - Recoverd\ PF\ revenue}{Annual\ production\ of\ heat}$	$Gate\ fee_{PF} = \frac{MSP_{PF} - PF\ market\ price}{t_{sewage\ sludge}\ per\ t_{PF}}$
CHP+PF [e]	6,8,14,16	$MSP_{PF} = \frac{Annualized\ CAPEX + OPEX - heat\ revenue - electricity\ revenue}{Annual\ production\ of\ P\ fertilizer}$ $MSP_{heat} = \frac{Annualized\ CAPEX + OPEX - electricity\ revenue - Recoverd\ PF\ revenue}{Annual\ production\ of\ heat}$	

[a] No products besides heat. [b] The MSP was calculated for the output heat as the electricity market can be regarded as a more distinct market than heat. [c] Co-combustion of MSS with WS. A total of 50% of the P in the produced ash was assumed to be sellable at the market price of a commercial fertilizer (triple superphosphate), based on [30]. The MSP was calculated for the output heat for all scenarios as the electricity market can be regarded as a more distinct market than heat. [d] The MSP was calculated for both PF and heat. For MSP_{PF}, the estimated market heat price is taken since the SSA-based PF market is unclear. However, for comparison with other scenarios, MSP_{heat} is calculated when produced P was assumed sellable at the market price of commercial fertilizer (triple superphosphate) [62]. [e] The MSP was calculated for both PF and heat. For MSP_{PF}, electricity and estimated heat market prices are taken since the SSA-based PF market is unclear. However, for comparison with other scenarios, MSP heat is calculated when produced P was assumed sellable at the market price of commercial fertilizer (triple superphosphate) [62].

Table 5. Energy and chemical prices and operational parameters used in the economic evaluations.

Parameter	Unit	Base Value	Variation	Reference
Varied Parameters				
Electricity selling price [a]	€/MWh	40	10/40/60	[63,64]
Electricity buying price [b]	€/MWh	81	40/81/140	
Heat selling price [c]	€/MWh	-	5/30	[15]
WS [d]	€/t	14	21–28	[65]
Annual plant operation time	h/y	5000	3500/5000/8000	
Fixed parameters				
Natural gas price [e]	€/MWh	31	-	[66]
Na_2SO_4	€/t	80	-	
$Ca(OH)_2$	€/t	90	-	
Coke	€/t	400	-	[28,29]
NaOH	€/t	90	-	
NH_3 25%	€/t	150	-	
Water	€/m^3	0.5	-	[67]
Ash landfilling	€/t	50	-	[28]
Hazardous waste landfilling [f]	€/t	120	-	
Sewage sludge	€/t	0	-	-
Commercial fertilizer (triple superphosphate) [g]	€/t	267	-	[62]

[a] Electricity wholesale market price in Europe, variation covers minimum, mean, and maximum electricity wholesale market prices during 2020. [b] Average national price in the EU without taxes applicable for the first semester of each year for medium-sized industrial consumers [63]. [c] A ratio of 0.5 between heat and electricity market prices was assumed. [d] Fuel preparation cost (e.g., mixing and pelletizing) in the co-combustion plant was considered from 50% to 100% increase compared to WS price reported by [65]. [e] Natural gas is included in the energy balance of the Ash Dec process, which was incorporated as a black box. Natural gas price is fixed based on non-household consumers in the first half of 2020 in the EU [66]. [f] Flue gas cleaning waste that contains HMs in high concentration. In all scenarios, the hazardous waste was assumed at 3% of the total ash [29]. [g] Used to estimate the value of the ash P in the direct PF scenarios. The sellable P in the ash was assumed to amount to 50% of the total P of the ash.

4. Results

4.1. Mass and Energy Balances

The results from the Aspen Plus® simulations are summarized in Table 6. The heat demand for drying in the mono-combustion scenarios (1–6, 9–14) accounts for 27–36% of the total output heat. Contrarily, the co-combustion scenarios (7, 8, 15, 16) eliminate the drying demand and result in a fuel mix heating value 32% higher than for the pure MSS fuel. The co-combustion thus shows a dual-energy advantage compared to the mono-combustion: no dryer needed and an overall higher heat production potential. Regarding ash residues and P production, the results show that mono-combustion yields a higher output of P per ton of fuel (scenarios 1–6, 9–14) than co-combustion (scenarios 7, 8, 15, 16). This demonstrates the presence of technical and financial trade-offs regarding fuel selection for a combustion plant. It is worth noting that local limiting factors, such as city size or availability of suitable biomass fuels for co-combustion, would obviously also affect the fuel selection.

Table 6. Summary of resulting mass and energy flows for all simulated scenarios.

Scenario	Heat Output (MWh)	Electricity Output (MWh)	Drying Energy (MWh)	Electricity Demand (MWh)	Fuel [a] (t/h)	Ash (t/h)	P Production (kg/h)
(1)-10MW-mono-heat only-with landfill	6.6	0	1.8	0.61	5.29	1.01	0
(2)-10MW-mono-heat only-without landfill	6.6	0	1.8	0.61	5.29	1.01	0
(3)-10MW- mono-heat and PF	6.6	0	1.8	0.65	5.29	1.01	90
(4)-10MW-mono-CHP- with landfill	4.9	1.35	1.8	0.61	5.29	1.01	0
(5)-10MW-mono-CHP- without landfill	4.9	1.35	1.8	0.61	5.29	1.01	0
(6)-10MW-mono-CHP and PF	4.9	1.35	1.8	0.65	5.29	1.01	90
(7)-10MW-co-heat+direct PF	8.7	0	0	0.46	2.25	0.14	1.39 [b]
(8)-10MW-co-CHP+direct PF	7.0	1.43	0	0.46	2.25	0.14	1.39 [b]
(9)-100MW-mono-heat only-with landfill	67	0	19	6.17	53.7	10.3	0
(10)-100MW-mono-heat only-without landfill	67	0	19	6.17	53.7	10.3	0
(11)-100MW-mono-heat and PF	67	0	19	6.58	53.7	10.3	919
(12)-100MW-mono-CHP and landfill	53	14.0	19	6.17	53.7	10.3	0
(13)-100MW-mono-CHP- without landfill	53	14.0	19	6.17	53.7	10.3	0
(14)-100MW-mono-CHP and PF	53	14.0	19	6.58	53.7	10.3	919
(15)-100MW-co-heat+direct PF	89	0	0	4.67	22.6	1.40	13.8 [b]
(16)-100MW-co -CHP+ direct PF	71	18.2	0	4.67	22.6	1.40	13.8 [b]

[a] As received. [b] P balance calculation is performed based on the assumption that 50% of the P content of the ash is as useful as commercial PF.

4.2. Economic Results

Table 7 summarises the resulting investment costs, operation costs, heat and PF MSPs, and gate fees for all analyzed technology scenarios, for the base parameter values used in the evaluations, and for the minimum and maximum applied heat and electricity market prices (according to Table 5). The following sections explore the results per group of technology scenarios and for all economic parameter variations. 'Direct PF' scenarios are explored in relation to landfilling and zero-cost ash handling ('without landfill') for heat-only and CHP scenarios (Section 4.2.1 Heat—Only and Heat+Direct PF Product Scenarios and Section 4.2.2), while the 'PF' scenarios (with ash post-treatment) are explored separately (Section 4.2.4 PF Production Scenarios).

4.2.1. Heat—Only and Heat+Direct PF Product Scenarios

This group of scenarios (1, 2, 7, 9, 10, 15) found the lowest heat MSPs in the 100 and 10 MW co-combustion scenarios (15 and 7), at 19 and 27 €/MWh, respectively. For these scenarios, the lower investment, energy, and chemicals costs of the co-combustion thus had a bigger impact than economy-of-scale effects (compare to scenario 10, with the lowest MSP of the mono-combustion scenarios, at 29 €/MWh). Planning for ash destination is critical for the plant's economic performance since ash landfilling, as a common approach, accounts for 17% to 25% of the operation cost in the studied mono-combustion plants. Even with zero-cost disposal ('without landfill' scenarios), the heat MSP for mono-combustion was still 52% to 63% higher than for co-combustion, for 10 and 100 MW, respectively, because of the drying demand.

Conversely, the contribution of direct PF recovery in the co-combustion scenarios ('direct PF') was negligible, with revenue covering less than 0.1% of the operation costs. This is due to the low initial P content of the ash. Even if all the ash P would be plant-available, the revenue would still cover less than 1% of the operation costs. A change of the fuel cost (co-combustion) of 50–100% caused an increase in the heat MSP of 4–11% (10 MW) and 20–21% (100 MW), respectively.

Table 7. Summarizing economic results for all scenarios, for the base parameter values (average electricity prices, operation time of 5000 h/y). MSPs and gate fees are given for the minimum (5 €/MWh) and maximum (30 €/MWh) heat market prices.

Scenarios	Specific Investment Cost (€/kW [a])	Specific Operation Cost (€/MWh [b])	MSP$_{heat}$ (€/MWh)	MSP$_{PF}$ (€/kg PF)	Gate Fee$_{heat}$ [c] (€/t)	Gate Fee$_{PF}$ [c] (€/t)
(1)-10MW-mono-heat only-with landfill	1187	28	53	-	60–29	-
(2)-10MW-mono-heat only-without landfill	1187	19	44	-	49–17	-
(3)-10MW- mono-heat and PF	2175	82	77 [d]	5.6–3.8	90–59	92–61
(4)-10MW-mono-CHP- with landfill	2004	71	66	-	57–34	-
(5)-10MW-mono-CHP- without landfill	2004	60	55	-	47–23	-
(6)-10MW-mono-CHP and PF	3328	111	98 [d]	5.5–4.1	87–64	90–66
(7)-10MW-co-heat+direct PF	619	14	27	59–0	86–0	731–0
(8)-10MW-co-CHP+direct PF	1050	35	30	55–0.3	78–0	682–0.9
(9)-100MW-mono-heat only-with landfill	646	23	37	-	40–9	-
(10)-100MW-mono-heat only-without landfill	646	15	29	-	30–0	-
(11)-100MW-mono-heat and PF	1271	56	51 [d]	3.7–1.9	57–26	60–29
(12)-100MW-mono-CHP and landfill	1352	52	47	-	41–16	-
(13)-100MW-mono-CHP- without landfill	1352	42	36	-	31–6	-
(14)-100MW-mono-CHP and PF	2143	77	64 [d]	3.8–2.4	58–33	61–36
(15)-100MW-co-heat+direct PF	371	11	19	40–0	59–0	487–0
(16)-100MW-co-CHP+direct PF	902	31	23	41–0	57–0	506–0

[a] Boiler capacity. [b] Output heat. [c] Given as €/t received MSS. [d] It is assumed that the produced PF is sold at market price to calculate heat MSP in a uniform method.

Figure 3 indicates the sensitivity of the economic performance to the annual utilization hours and the electricity buying price. The difference between the heat MSPs and the heat market prices depicts how far the plant is from economic viability. Longer annual utilization time, larger plant size, and lower electricity buying price are all factors contributing to a decreasing heat MSP.

However, city size and transportation distance are local limiting elements for scaling MSS combustion plants. Assuming MSS production per capita is 19 kg/inhabitant, y dry base [68] with 8000 h/year operation, a city with 1.1 million inhabitants would be needed to provide sludge for one 10 MW mono-combustion plant and 11 million inhabitants for 100 MW. Therefore, the applicability of 10 MW mono-combustion plants would narrow down to relatively few cities, from a European perspective, and 100 MW mono-combustion plants would inevitably be centralized plants, the economic feasibility of which would be bound to transportation costs affected by distance and moisture content. For co-combustion, the scale of the plant would be less sensible to MSS supply and population, but the availability of suitable biomass and transportation costs would be instead ruling factors for decision makers.

Although co-combustion resulted in lower MSP, none of the scenarios were cost-efficient without a high heat market price or the existence of a gate fee. Depending on the fuel, the heat market price, plant size, and ash handling options, a gate fee between 0 and 86 € per ton of received MSS was needed to achieve the economic feasibility of heat production from sewage sludge. Co-combustion, with high solid content, low total amount of fuel, and low share of sewage, resulted in higher required gate fee per ton of MSS to fill the gap between production cost and market price. Therefore, the external financial support to prompt investments in co-combustion for P recovery should focus on output energy rather than the feedstock. Otherwise, for decision makers, mono-combustion with a lower gate fee gets priority, even though it excludes P recovery.

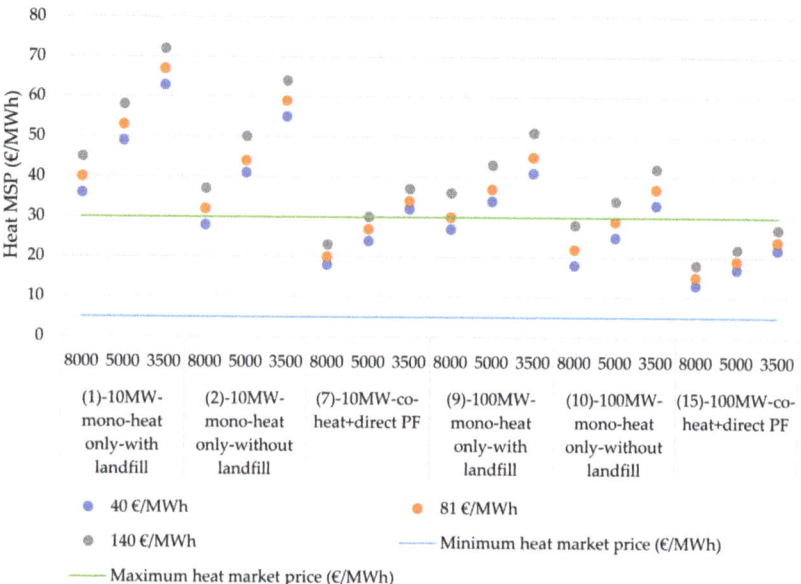

Figure 3. Heat MSPs (€/MWh) for varying annual utilization hours (8000, 5000, 3500 h/year) and buying electricity prices (40, 81, 140 €/MWh).

4.2.2. CHP and CHP+direct PF Product Scenarios

In this group of scenarios (4, 5, 8, 12, 13, 16), electricity production accounted for a 20% increase in investment cost for 10 MW and a 40% increase for 100 MW plants compared to the corresponding heat-only plants, with a subsequent higher heat MSP. However, at low heat market price (5 €/MWh), a slightly lower gate fee was in fact needed compared to in the corresponding heat-only and heat+direct PF scenarios due to the electricity production revenues. This advantage of CHP declined with the higher heat price (30 €/MWh).

Similar to the heat-only scenarios, the MSP for mono-combustion was higher than for co-combustion, while co-combustion required a higher gate fee per ton of received MSS at the low heat price. When changing the fuel cost by 50–100% for co-combustion, the heat MSPs increased by 7–13% (10 MW) and 13–21% (100 MW), respectively. Electricity market price is the main trigger of investing in electricity production in MSS combustion plants. Figure 4 shows the annual CAPEX related to electricity production versus the economic benefits of either supplying internal demand (saving revenue) or selling to the market (selling revenue) in 10 MW and 100 MW mono-/co-combustion plants respectively. Generally, a plant with high annual utilization hours is economically preferable, yet the level of economic benefits as the main motivation of investment builds upon the market prices. In relative terms, saving revenues contribute more to the mono-combustion plants' economic performance due to the higher energy demand compared to co-combustion. Conversely, the higher electricity production for co-combustion results in higher selling revenues.

Although end-use electricity prices are linked to various factors such as wholesale market price (taken as selling price in this study), it is reported that the coupling of wholesale and end-use electricity prices is not close in many countries [69]. Therefore, the analysis consists of the comparison of extreme prices. According to the result, even if the buying and selling electricity prices are the same, plants still need to have high utilization hours to be economically feasible. Otherwise, electricity production is economically inefficient, or the benefits of electricity production only compensate for the costs. The economic feasibility of the 10 MW plant shows more flexibility toward lower market prices caused by the scaling method. This method is common in pre-feasibility that is in favor of small-scale equipment. In the extreme prices, a high electricity buying price stimulates investment in CHP in mono-combustion plants, and a high electricity selling price promotes investment in CHP in co-combustion plants, especially in 100 MW plants.

Figure 5 summarizes the heat MSPs of CHP scenarios (4, 5, 8, 12, 13, 16) for different annual utilization hours. When the utilization time is 8000 h/ year, 100 MW CHP plants show better economic performance than heat-only production (See Figure 3), while the reverse is true for lower annual operation time. The reason is that the internal electricity demand is covered by produced electricity, and the cost difference between buying and selling electricity prices makes a significant saving revenue. For a given boiler capacity, co-combustion has lower heat MSP and less sensitivity to the utilization hours due to several reasons such as higher heat revenue, the reduction in ash landfilling cost from operation costs, and lower energy cost in comparison to mono-combustion.

4.2.3. Comparing Heat-Only to CHP Scenarios

Figure 6 explores in more detail the share of different costs and revenues in all CHP and heat-only scenarios for the base parameter values (Table 5) except 'PF' scenarios (3,6,11,14). When comparing the corresponding co- and mono-combustion scenarios, in particular, the CAPEX and the avoided cost for landfilling contribute to the lower heat MSP. The general economic benefits of co-combustion regarding the deduction of investment on the dryer and drying energy demand may provide a better economic opportunity for energy and P recovery from sewage sludge. However, the mixing of MSS with WS dilutes the MSS P content and PF revenue. The increase in CAPEX for electricity production outweighs the economic benefits of electricity production unless the plant has a 100 MW boiler with 8000 h per years of operation (see Figure 4).

4.2.4. PF Production Scenarios

Part A of Figure 7 shows the PF MSPs in relation to the market heat price for the ash post-treatment ('PF') scenarios (3, 6, 11, 14) and part B for co-combustion ('direct PF') scenarios (7, 8, 15, 16). Here, electricity production revenues were found to only marginally affect the financial driving force for PF production from SSA, and only at very low heat market prices in 10 MW plants. For all the instances with ash post-treatment, the PF MSP was, in fact, found to be up to a magnitude higher than the considered commercial fertilizer price. The 100 MW heat-only scenario (11) resulted in the lowest PF MSP with 1.95 €/kg at a market heat price of 30 €/MWh, which can be compared to the commercial fertilizer price of 0.44 €/kg. Economically speaking, P recovery from SSA in the mono-combustion plant through post-treatment in all cases resulted in uncompetitive final prices and inefficient investment with heavy reliance on the gate fee (Table 7).

The direct PF production scenarios (7, 8, 15, 16) entail an opportunity for being a cost-effective P recovery strategy at a high market heat price. In these cases, the economic feasibility of the plant was independent of P production revenue since the heat revenue covered the annual costs, and the final PF could be sold at the market price. However, P recovery through post-treatment required external financial support in all cases. Despite the high initial P content of ash in mono-combustion, which is important to obtain feasibility of post-treatment implementation, the final product was several times more expensive than the commercial fertilizer product.

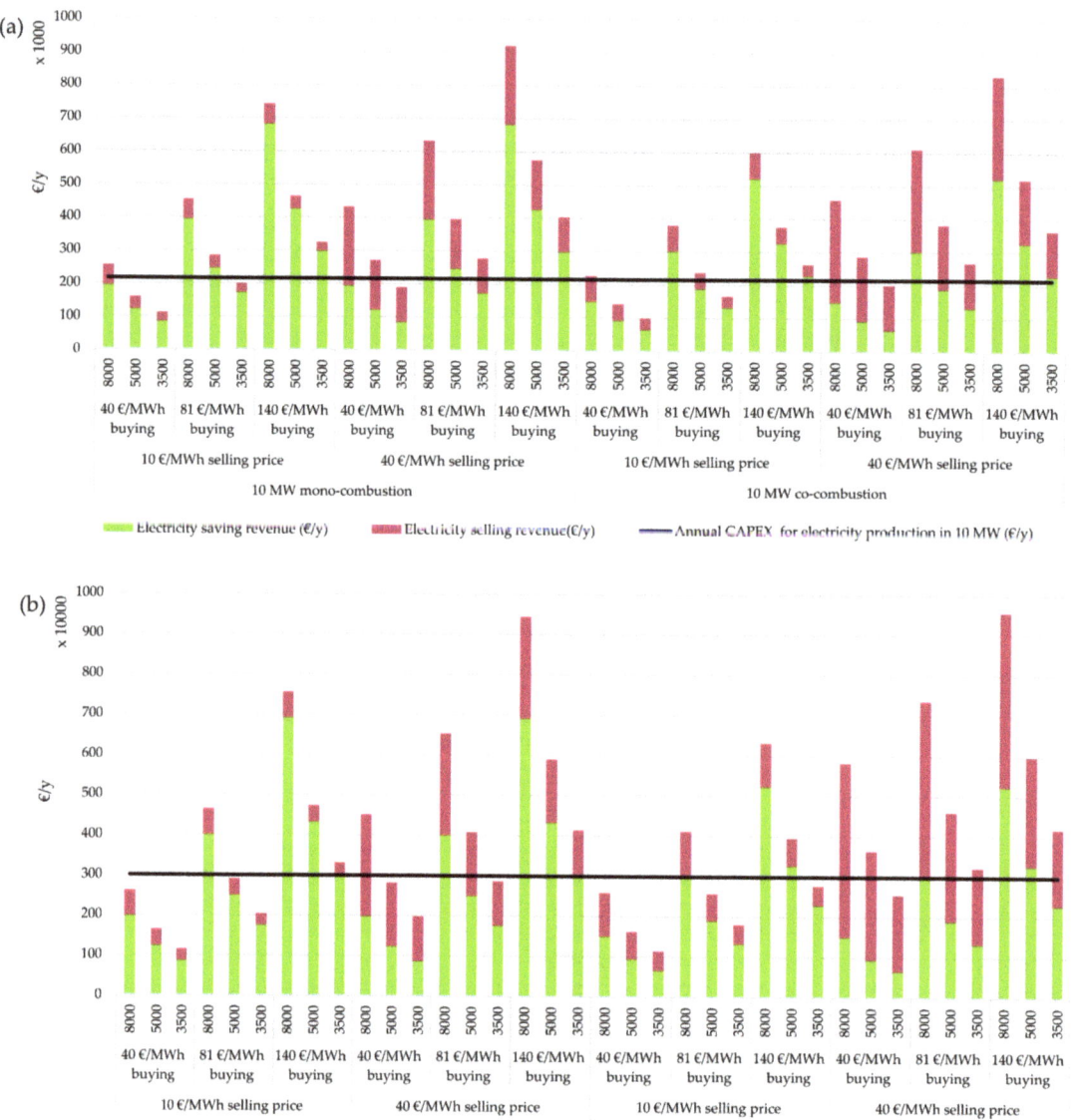

Figure 4. Total annual costs versus economic benefits of electricity production in the (**a**) 10 MW and (**b**) 100 MW mono-/co-combustion scenarios, respectively and for different electricity buying prices (40, 81, 140 €/MWh) and selling prices (10, 40, 60 €/MWh).

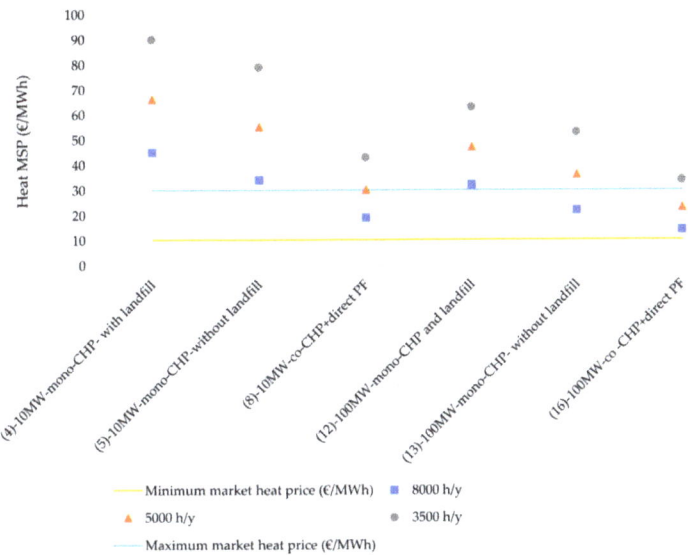

Figure 5. Heat MSP (€/MWh) for varying annual utilization hours for base value for electricity price.

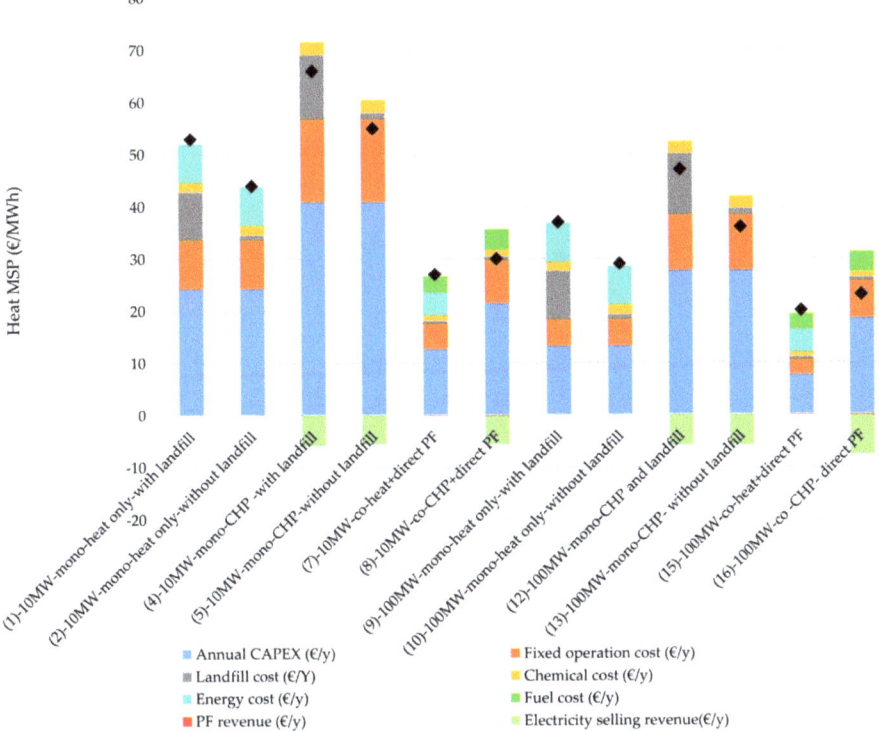

Figure 6. Cost and revenue break-down of the heat MSP in heat-only and CHP plants, at an annual utilization time of 5000 h/year.

Figure 8 shows the range of heat MSPs in the PF production scenarios that would be required to reach the point where ash-based PF can be sold at the market price. With the same utilization hours, electricity production impaired the economic feasibility of P recovery from MSS in both 10 and 100 MW plants. By selling ash-based PF at the market price, none of the cases were economically feasible regarding energy market prices. For example, the resulting heat-based gate fee for 100 MW heat-only and PF (scenario 11) was 26 €/t $_{received\ MSS}$, while the corresponding 100 MW heat-only landfill scenario (9) resulted in a gate fee of 9 €/t $_{received\ MSS}$.

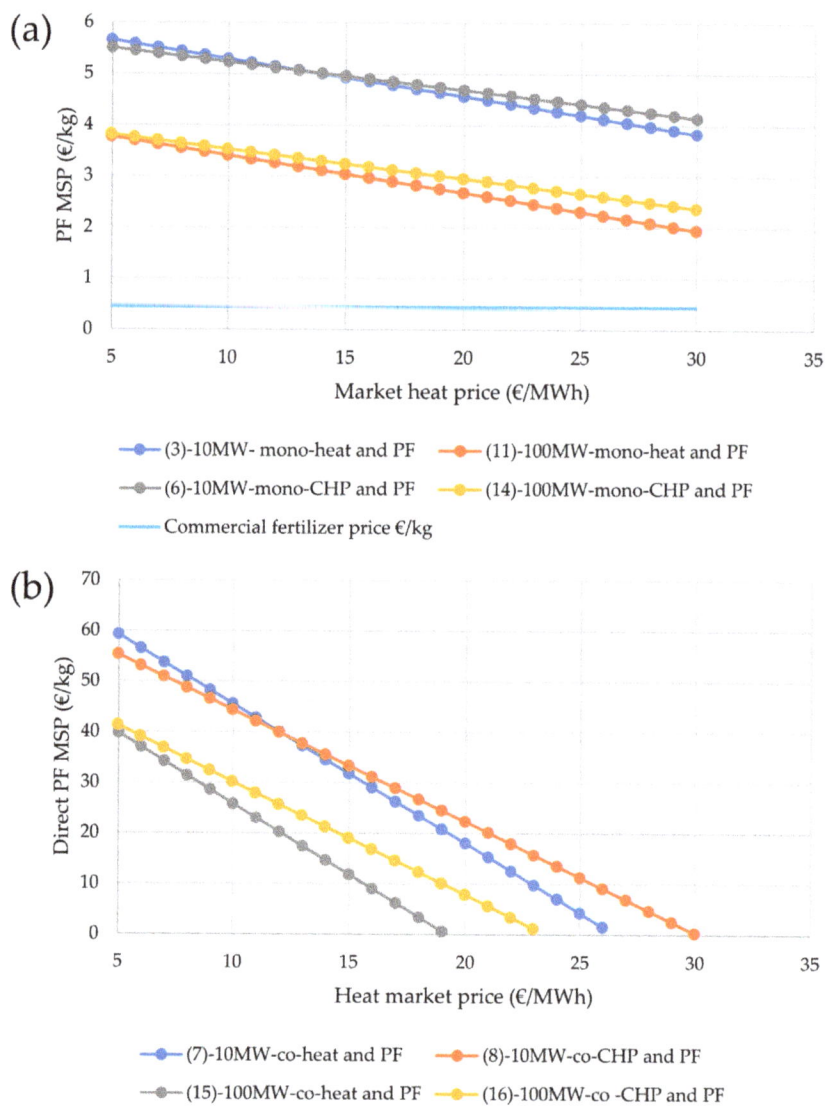

Figure 7. Sewage sludge-based fertilizer (PF) MSP (€/kg), for the base parameter values (average electricity prices, operation time of 5000 h/y), for (**a**) PF produced via ash post-treatment ('PF' scenarios), and (**b**) PF produced via co-combustion ('direct PF' scenarios).

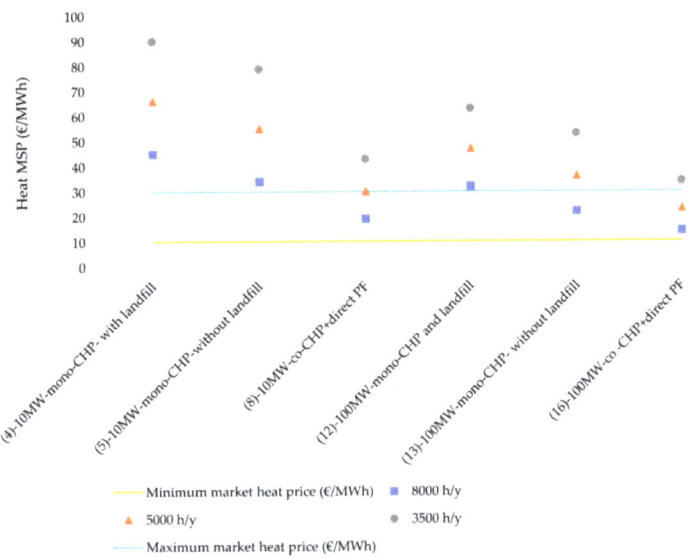

Figure 5. Heat MSP (€/MWh) for varying annual utilization hours for base value for electricity price.

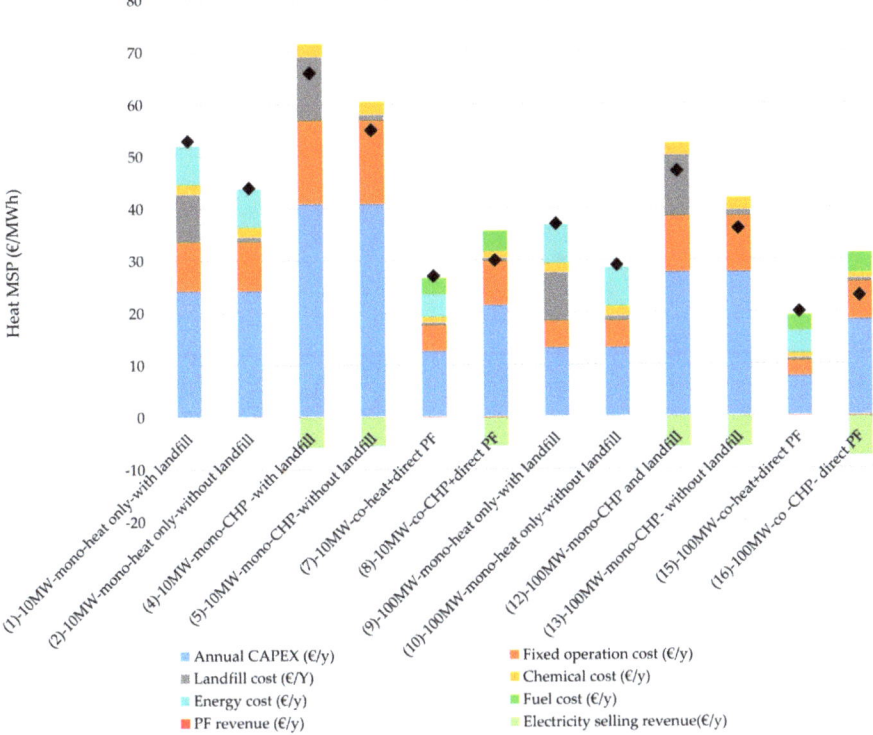

Figure 6. Cost and revenue break-down of the heat MSP in heat-only and CHP plants, at an annual utilization time of 5000 h/year.

Figure 8 shows the range of heat MSPs in the PF production scenarios that would be required to reach the point where ash-based PF can be sold at the market price. With the same utilization hours, electricity production impaired the economic feasibility of P recovery from MSS in both 10 and 100 MW plants. By selling ash-based PF at the market price, none of the cases were economically feasible regarding energy market prices. For example, the resulting heat-based gate fee for 100 MW heat-only and PF (scenario 11) was 26 €/t $_{received\ MSS}$, while the corresponding 100 MW heat-only landfill scenario (9) resulted in a gate fee of 9 €/t $_{received\ MSS}$.

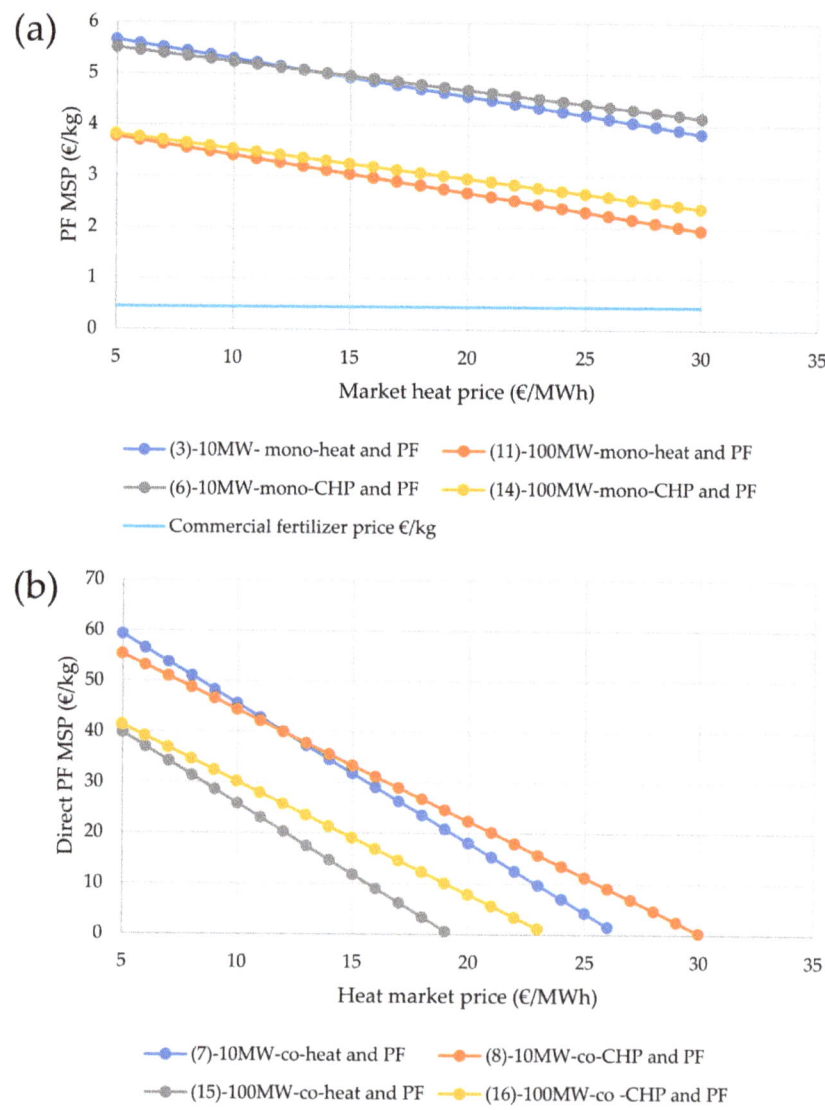

Figure 7. Sewage sludge-based fertilizer (PF) MSP (€/kg), for the base parameter values (average electricity prices, operation time of 5000 h/y), for (**a**) PF produced via ash post-treatment ('PF' scenarios), and (**b**) PF produced via co-combustion ('direct PF' scenarios).

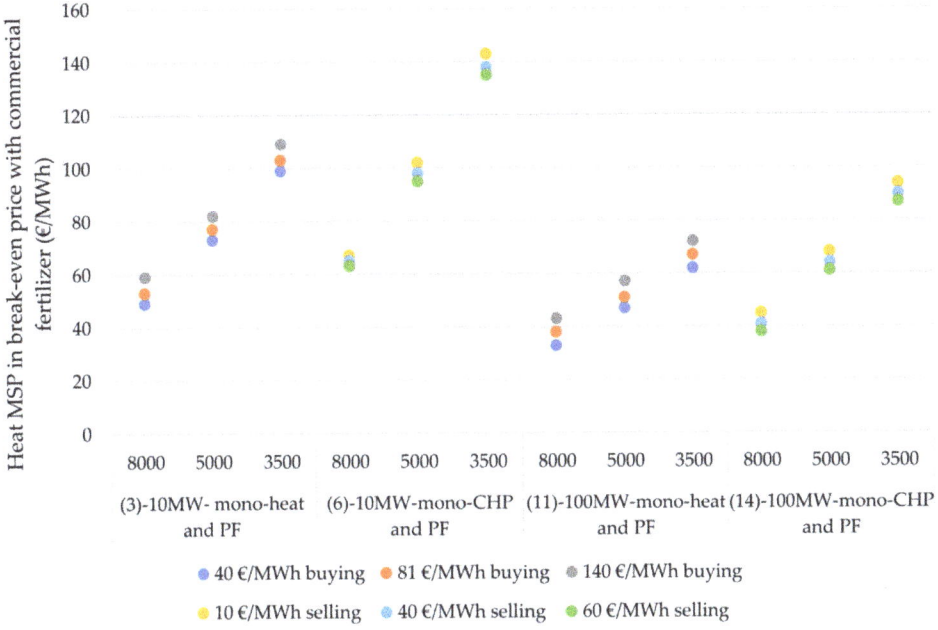

Figure 8. Required heat MSP (€/MWh) to reach break-even for ash-based versus commercial P fertilizer, for varying annual utilization hours (8000, 5000, 3500 h/year) and electricity prices.

5. Discussion

The development of MSS mono-combustion plants to recover both P and energy imposes higher financial responsibility on policymakers compared to only energy recovery or direct PF production through co-combustion development. P recovery from MSS ash through post-treatment appears to be an inefficient investment, and external financial support would be crucial to circulate P from MSS to the food production process.

The studied fluctuation of energy market prices along with different utilization hours can change production costs by 50%. Green energy subsidies or other financial support can be used as tools to aid investments in energy recovery from MSS and buffer against market uncertainty. On the other hand, the feedstock of the combustion plant must be semi-dried; otherwise, the process itself consumes most of the combustion heat to evaporate the moisture content. Therefore, the energy revenue of the plant drops dramatically, and the gate fee turns into the main income of the plant rather than the return on the recovered energy. MSS management strategy is conditional on the decision maker's perspective. When decision makers consider MSS as a waste to dispose of, the mono-combustion plant would be the most efficient option because of the lower required gate fee. However, when policymakers aim to recirculate P from MSS into the food production system, the advantages of co-combustion may outweigh the implementation of ash post-treatment in the mono-combustion plants.

Moreover, heat is an economically more favorable form of energy than electricity due to the MSS quality and local scalability barriers of MSS combustion plants. However, market heat prices depend on several aspects, the most prominent being the local heat demand, the availability of district heating distribution system, and alternative heat production costs (e.g., the presence of low-cost waste heat), and this adds more economic uncertainty to energy recovery from MSS mono-combustion.

Although 100 MW plants show better economic performance, the scalability of studied plants depends on the size of the collection area (availability of sludge), availability of desired biomass (low moisture and high K content) for fulfilling co-combustions goals (eliminate drying demand and direct PF production), and transportation demand. Costs related to centralization and transportation are unavoidable for 100 MW mono-combustion due to cities' populations. Therefore, trade-offs between the benefits of large-scale plants and transport costs require further investigation.

The direct PF revenue was insignificant in the studied scenarios due to the low total P in the co-combustion ash, but it reduces the complexity of the MSS treatment system while containing both energy and P recovery. Direct PF also contains K, S, Ca, and Mg, which are valuable for agriculture. The effect of these added values must be considered in the economic evaluation of direct PF. For this, better knowledge of plant availability of P and HMs content of direct-produced PF is needed.

Another opportunity that can improve the economic feasibility of energy and P recovery from MSS is the adaptation of existing incineration infrastructure to either mono- or co-combustion. However, switching the fuel of an existing plant may be accompanied by both investment and operational costs, related to, e.g., fuel handling, flue gas cleaning, and auxiliary boiler capacity to cover the difference in output energy when MSS is substituted with high heat value fuel. For mono-combustion, existing waste incineration plants are favorable since they are already equipped with advanced flue gas cleaning. For co-combustion, the partial introduction of MSS to boilers combusting K-rich biomass could both reduce bed agglomeration problems and the risk of alkali-related fouling and corrosion [70], i.e., significant challenges of K-rich biomass combustion. Therefore, further investigation is needed on the costs of using existing plants versus investing in dedicated new mono- or co-combustion plants for MSS.

6. Conclusions

Techno-economic analysis was performed to evaluate the feasibility of energy and P recovery from municipal sewage sludge (MSS) considering 16 different technology scenarios of investments in new combustion plants. The scenarios covered variations in five technical affecting aspects: (a) type of boiler (FB; GB), (b) fuel mixture (100% MSS; 10% MSS mixed with 90% wheat straw (WS)), (c) co-products (heat; electricity; P fertilizer (PF)), (d) economy-of-scale (10 MW; 100 MW boiler), and (e) final ash destinations (landfilling; zero-cost disposal; PF production by either ash post-treatment using Ash Dec or direct ash utilization).

Co-combustion improved the economic viability of energy and P recovery in the studied plants due to (i) elimination of drying demand, which consumed 25–35% of the output heat in the mono-combustion plants, (ii) removal of the ash landfill cost that accounted for 17–25% of the annual cost of the mono-combustion plants, and (iii) increased fuel mix heating value (32% higher). However, the availability of the WS could be a limiting factor.

None of the studied cases were economically feasible without either the revenue of a gate fee paid by the local authority for received MSS or simultaneous high plant capacity and high revenues from sold energy carriers when the market prices are high or green energy subsidies are available. Therefore, the economic feasibility of the given scenarios is interconnected with volatile revenues from sold energy carriers (heat and/or electricity) and external financial support for waste disposal. The over-reliance of the economies of scale to improve the economic performance of the given plants is conditional to MSS and WS availability (population or transportation).

The heat was, in general, the economically favorable energy carrier recoverable from MSS, the exception being the 100 MW plant with 8000 h/year of operation. In this case, electricity production benefits (saving and selling) improved the economic feasibility of the combustion plant. P recovery through Ash Dec post-treatment in mono-combustion plants had a higher PF yield than direct PF production from co-combustion due to the higher P concentration in the ash. However, the production cost was still four times higher than the commercial fertilizer price in the best-performing case (100 MW mono-combustion with heat production). In addition to being less dependent on the size of cities for the supply of sufficient quantities of MSS, 10 and 100 MW co-combustion were the only cases where the economic feasibility was independent of the PF price at a high heat market price. Consequently, ash-based PF could potentially be sold at a competitive market price, thus stimulating the marketability of P recovered from MSS. Of particular importance are conditions related to energy markets, policies for energy and P recovery from MSS, or drastically increased prices of mineral fertilizer due to, e.g., fertilizer shortages on the market. The findings shed light on the importance of less energy-intense drying technologies and further study on the co-combustion of MSS and agriculture residue.

Author Contributions: Conceptualization, M.B., M.Ö. and E.W.; formal analysis, M.B.; funding acquisition, M.Ö. and E.W.; investigation, M.B.; methodology, M.B., M.Ö. and E.W.; project administration, M.B., M.Ö. and E.W.; software, M.B.; supervision, M.Ö. and E.W.; validation, M.B.; visualization, M.B.; writing—original draft, M.B.; writing—review and editing, M.B., M.Ö. and E.W.; All authors have read and agreed to the published version of the manuscript.

Funding: The work has been carried out under the auspices of Graduate School in Energy Systems, financed by the Swedish Energy Agency. Economic support from the Swedish Research Council Formas, within the national research program Sustainable Spatial Planning, is also gratefully acknowledged (dnr. 2018-00194), as is support from Bio4Energy, a strategic research environment appointed by the Swedish government.

Institutional Review Board Statement: Not applicable.

Informed Consent Statement: Not applicable.

Data Availability Statement: Data are available upon request.

Conflicts of Interest: The authors declare no conflict of interest.

Abbreviations

ar	As received
CAPEX	Total capital costs
db	Dry base
FB	Fluidized bed
FC	Fixed carbon
GB	Grate boiler
HMs	Heavy metals
h	Hour
MSP	Minimum selling price
MSS	Municipal sewage sludge
OPEX	Operational costs
P	Phosphorus
PF	Phosphorus fertilizer
SNCR	Selective non-catalytic reduction
SSA	Sewage sludge ash
VM	Volatile material
WS	Wheat straw

Appendix A

Table A1 provides the input and output material flows for Ash Dec technology based on one ton of phosphate (P_2O_5) and the corresponding cost for each flow, investment cost [29], and chemicals needed for the flue gas cleaning system. Table A2 presents the investment cost of equipment in a combustion plant.

Table A1. Ash Dec inlet and outlet material flow and the chemicals cost.

Input Elements	Unit	Amount	Reference Per Feed	Price Unit	Price	Ref
Hot ash	kg/h	1725	0	-	0	
Na_2SO_4	kg/h	640	0.37	€/t	80	
$Ca(OH)_2$	kg/h	26	0.02	€/t	90	
Electricity	kWh/kg ash	150	0.04	€/MWh	81 [a]	[28,29]
Natural gas	kWh/kg ash	670	0.39	€/MWh	31 [b]	
Calcined fertilizer	kg/h	2273	1.32			
Waste (filter residue, concentrated metals)	kg/h	43	0.03			
Water	Liter/t waste		0.3	€/m³	0.5	[67]
Flue Gas Cleaning System	Unit	Amount	Price unit	Price		Ref
Coke	g/kg TS	0.3	€/t	400		
Lime	g/kg TS	5	€/t	90		
NaOH	g/kg TS	16.5	€/t	90		
Electricity for flue gas cleaning	kWh/kg TS	0.23	€/MWh	81 [a]		[28]
$Ca(OH)_2$	kg/kg off gas	0.005	€/t	90		
NH_3 25%	g/kg TS	16.5	€/t	150		
Ash landfill			€/t	50		
Hazardous waste landfill			€/t	120		
Wheat straw			€/t	13.65		[65]

[a] Average national price in Euro without taxes applicable for the first semester of each year for medium-sized industrial consumers [71]. [b] Natural gas prices for non-household consumers [66].

Table A2. Equipment cost of MSS combustion plant (inflation-adjusted to €$_{2020}$ using the Chemical Engineering Plant cost index and an exchange rate of 0.85 €/US$).

Element	Base Capacity	Capacity Unit	Value (1000 € 2020)	Description	Ref
Dryer	100	MW	1467	includes conveyor to and from the dryer	[72]
GB	150	MW	24,641	Steam generation cost and cyclone are included	[73]
FB	355	MW	49,566	Steam generation cost and cyclone are included	[72]
Turbine	275	MW	63,578	Generator cost in included	
Fuel conveyor	17	MW	96	It is included in co-combustion scenarios	[73]
Ash container and conveyor	17	MW	145		[73]
Electrostatic precipitator	18	Tons of waste/h	1909	Particle's remover in gas cleaning system	[67,74]
Wet scrubber	18	Tons of waste/h	5967	included of the water treatment system	[67,74]
Bag filter	18	Tons of waste/h	2625		[67,74]
SNCR	18	Tons of waste/h	1193	To remove NOx from flue gas	[67,74]
Ash Dec	30,000	Tons of ash/y	18,600		Contact with company

References

1. Cordell, D.; White, S. Peak phosphorus: Clarifying the key issues of a vigorous debate about long-term phosphorus security. *Sustainability* **2011**, *3*, 2027–2049. [CrossRef]
2. Bia, A.; Koziel, J.A.; Jama-rodze, A. Waste to phosphorus: A transdisciplinary solution to P recovery from wastewater based on the TRIZ approach. *J. Environ. Manag.* **2021**, *287*. [CrossRef]
3. Smol, M. The importance of sustainable phosphorus management in the circular economy (CE) model: The Polish case study. *J. Mater. Cycles Waste Manag.* **2019**, *21*, 227–238. [CrossRef]

4. Meng, X.; Huang, Q.; Xu, J.; Gao, H.; Yan, J. A review of phosphorus recovery from different thermal treatment products of sewage sludge. *Waste Dispos. Sustain. Energy* **2019**, *1*, 99–115. [CrossRef]
5. Mayer, B.K.; Baker, L.A.; Boyer, T.H.; Drechsel, P.; Gi, M.; Hanjra, M.A.; Parameswaran, P.; Stoltzfus, J.; Westerho, P.; Rittmann, B.E. Total Value of Phosphorus Recovery. *Environ. Sci. Technol.* **2016**, *2016*, 6606–6620. [CrossRef]
6. De Boer, M.A.; Wolzak, L.; Slootweg, J.C. Phosphorus: Reserves, Production, and Applications. In *Phosphorus Recovery and Recycling*; Springer: Singapore, 2019; pp. 75–100; ISBN 9789811080319.
7. Hamilton, H.A.; Brod, E.; Hanserud, O.; Müller, D.B.; Brattebø, H.; Haraldsen, T.K. Recycling potential of secondary phosphorus resources as assessed by integrating substance flow analysis and plant-availability. *Sci. Total Environ.* **2017**, *575*, 1546–1555. [CrossRef]
8. Cieślik, B.; Konieczka, P. A review of phosphorus recovery methods at various steps of wastewater treatment and sewage sludge management. The concept of "no solid waste generation" and analytical methods. *J. Clean. Prod.* **2017**, *142*, 1728–1740. [CrossRef]
9. Harder, R.; Wielemaker, R.; Larsen, T.A.; Zeeman, G.; Öberg, G. Recycling nutrients contained in human excreta to agriculture: Pathways, processes, and products. *Crit. Rev. Environ. Sci. Technol.* **2019**, *49*, 695–743. [CrossRef]
10. Hgm, S.; Sancho, D. *Technical Proposals for Selected New Fertilising Materials under the Fertilising Products Regulation (Regulation (EU) 2019/1009)*; European Commission: Luxembourg, 2019.
11. Smol, M. *Phosphorus Extraction and Sludge Dissolution*; Elsevier Inc.: Amsterdam, The Netherlands, 2019; ISBN 9780128159071.
12. Eurostat Sewage Sludge Production and Disposal. Available online: http://appsso.eurostat.europa.eu/nui/show.do?lang=en&dataset=env_ww_spd (accessed on 8 July 2021).
13. Desmidt, E.; Ghyselbrecht, K.; Zhang, Y.; Pinoy, L.; Van Der Bruggen, B.; Verstraete, W.; Rabaey, K.; Meesschaert, B. Global phosphorus scarcity and full-scale P-recovery techniques: A review. *Crit. Rev. Environ. Sci. Technol.* **2015**, *45*, 336–384. [CrossRef]
14. Melia, P.M.; Cundy, A.B.; Sohi, S.P.; Hooda, P.S.; Busquets, R. Trends in the recovery of phosphorus in bioavailable forms from wastewater. *Chemosphere* **2017**, *186*, 381–395. [CrossRef]
15. Horttanainen, M.; Kaikko, J.; Bergman, R.; Pasila-Lehtinen, M.; Nerg, J. Performance analysis of power generating sludge combustion plant and comparison against other sludge treatment technologies. *Appl. Therm. Eng.* **2010**, *30*, 110–118. [CrossRef]
16. Kacprzak, M.; Neczaj, E.; Fijałkowski, K.; Grobelak, A.; Grosser, A.; Worwag, M.; Rorat, A.; Brattebø, H.; Almås, Å.; Singh, B.R. Sewage sludge disposal strategies for sustainable development. *Environ. Res.* **2017**, *156*, 39–46. [CrossRef]
17. Donatello, S.; Cheeseman, C.R. Recycling and recovery routes for incinerated sewage sludge ash (ISSA): A review. *Waste Manag.* **2013**, *33*, 2328–2340. [CrossRef] [PubMed]
18. Egle, L.; Rechberger, H.; Krampe, J.; Zessner, M. Science of the Total Environment Phosphorus recovery from municipal wastewater: An integrated comparative technological, environmental and economic assessment of P recovery technologies. *Sci. Total Environ.* **2016**, *571*, 522–542. [CrossRef] [PubMed]
19. Chrispim, M.C.; Scholz, M.; Nolasco, M.A. Phosphorus recovery from municipal wastewater treatment: Critical review of challenges and opportunities for developing countries. *J. Environ. Manage.* **2019**, *248*, 109268. [CrossRef]
20. Kehrein, P.; Van Loosdrecht, M.; Osseweijer, P.; Garfí, M.; Dewulf, J.; Posada, J. A critical review of resource recovery from municipal wastewater treatment plants-market supply potentials, technologies and bottlenecks. *Environ. Sci. Water Res. Technol.* **2020**, *6*, 877–910. [CrossRef]
21. Tan, Z.; Lagerkvist, A. Phosphorus recovery from the biomass ash: A review. *Renew. Sustain. Energy Rev.* **2011**, *15*, 3588–3602. [CrossRef]
22. Kratz, S.; Vogel, C.; Adam, C. Agronomic performance of P recycling fertilizers and methods to predict it: A review. *Nutr. Cycl. Agroecosystems* **2019**, *115*, 1–39. [CrossRef]
23. Shiba, N.C.; Ntuli, F. Extraction and precipitation of phosphorus from sewage sludge. *Waste Manag.* **2017**, *60*, 191–200. [CrossRef]
24. Hannl, T.K.; Sefidari, H.; Kuba, M.; Skoglund, N.; Öhman, M. Thermochemical equilibrium study of ash transformation during combustion and gasification of sewage sludge mixtures with agricultural residues with focus on the phosphorus speciation. *Biomass Convers. Biorefinery* **2020**, *11*(1), 57–68. [CrossRef]
25. Stemann, J.; Peplinski, B.; Adam, C. Thermochemical treatment of sewage sludge ash with sodium salt additives for phosphorus fertilizer production - Analysis of underlying chemical reactions. *Waste Manag.* **2015**, *45*, 385–390. [CrossRef]
26. Ottosen, L.M.; Kirkelund, G.M.; Jensen, P.E. Extracting phosphorous from incinerated sewage sludge ash rich in iron or aluminum. *Chemosphere* **2013**, *91*, 963–969. [CrossRef] [PubMed]
27. Herzel, H.; Krüger, O.; Hermann, L.; Adam, C. Sewage sludge ash - A promising secondary phosphorus source for fertilizer production. *Sci. Total Environ.* **2016**, *542*, 1136–1143. [CrossRef] [PubMed]
28. Kabbe, C. *Sustainable Sewage Sludge Management Fostering Phosphorus Recovery and Energy Efficiency*; European Commission: Luxembourg, 2015.
29. Hermann, L.; Schaaf, T. Outotec (AshDec®) process for P fertilizers from sludge ash. In *Phosphorus Recovery and Recycling*; Springer: Singapore, 2018; pp. 221–233. ISBN 9789811080319.
30. Häggström, G.; Hannl, T.K.; Hedayati, A.; Kuba, M.; Skoglund, N.; Öhman, M. Single Pellet Combustion of Sewage Sludge and Agricultural Residues with a Focus on Phosphorus. *Energy Fuels* **2021**, *35*, 10009–10022. [CrossRef]
31. Grimm, A.; Skoglund, N.; Boström, D.; Boman, C.; Öhman, M. Influence of phosphorus on alkali distribution during combustion of logging residues and wheat straw in a bench-scale fluidized bed. *Energy Fuels* **2012**, *26*, 3012–3023. [CrossRef]

32. Falk, J.; Skoglund, N.; Grimm, A.; Öhman, M. Fate of Phosphorus in Fixed Bed Combustion of Biomass and Sewage Sludge. *Energy Fuels* **2020**, *34*, 4587–4594. [CrossRef]
33. Shaddel, S.; Bakhtiary-Davijany, H.; Kabbe, C.; Dadgar, F.; Østerhus, S.W. Sustainable sewage sludge management: From current practices to emerging nutrient recovery technologies. *Sustainability* **2019**, *11*, 3435. [CrossRef]
34. Schnell, M.; Horst, T.; Quicker, P. Thermal treatment of sewage sludge in Germany: A review. *J. Environ. Manage.* **2020**, *263*, 110367. [CrossRef]
35. Zhang, Q.; Hu, J.; Lee, D.J.; Chang, Y.; Lee, Y.J. Sludge treatment: Current research trends. *Bioresour. Technol.* **2017**, *243*, 1159–1172. [CrossRef]
36. Molinos-Senante, M.; Hernández-Sancho, F.; Sala-Garrido, R.; Garrido-Baserba, M. Economic feasibility study for phosphorus recovery processes. *Ambio* **2011**, *40*, 408–416. [CrossRef]
37. Daneshgar, S.; Buttafava, A.; Callegari, A.; Capodaglio, A.G. Economic and energetic assessment of different phosphorus recovery options from aerobic sludge. *J. Clean. Prod.* **2019**, *223*, 729–738. [CrossRef]
38. Oladejo, J.; Shi, K.; Luo, X.; Yang, G.; Wu, T. A review of sludge-to-energy recovery methods. *Energies* **2019**, *12*, 60. [CrossRef]
39. Syed-Hassan, S.S.A.; Wang, Y.; Hu, S.; Su, S.; Xiang, J. Thermochemical processing of sewage sludge to energy and fuel: Fundamentals, challenges and considerations. *Renew. Sustain. Energy Rev.* **2017**, *80*, 888–913. [CrossRef]
40. Widell, H. *Industrial-Scale Biomass Combustion Plants: Engineering Issues and Operation*; Woodhead Publishing Limited: Sawston, UK, 2013; ISBN 9780857091314.
41. Makarichi, L.; Jutidamrongphan, W.; Techato, K.A. The evolution of waste-to-energy incineration: A review. *Renew. Sustain. Energy Rev.* **2018**, *91*, 812–821. [CrossRef]
42. Leckner, B.; Lind, F. Combustion of municipal solid waste in fluidized bed or on grate—A comparison. *Waste Manag.* **2020**, *109*, 94–108. [CrossRef] [PubMed]
43. Lanzerstorfer, C. Grate-Fired Biomass Combustion Plants Using Forest Residues as Fuel: Enrichment Factors for Components in the Fly Ash. *Waste Biomass Valorization* **2017**, *8*, 235–240. [CrossRef] [PubMed]
44. Gao, N.; Kamran, K.; Quan, C.; Williams, P.T. Thermochemical conversion of sewage sludge: A critical review. *Prog. Energy Combust. Sci.* **2020**, *79*, 100843. [CrossRef]
45. Jurczyk, M.; Mikus, M.; Dziedzic, K. Flue Gas Cleaning in Municipal Waste-To-Energy Plants-Part I. *Infrastruct. Ecol. Rural Areas* **2016**, 1179–1193. [CrossRef]
46. Singh, R.; Shukla, A. A review on methods of flue gas cleaning from combustion of biomass. *Renew. Sustain. Energy Rev.* **2014**, *29*, 854–864. [CrossRef]
47. Chanaka Udayanga, W.D.; Veksha, A.; Giannis, A.; Lisak, G.; Chang, V.W.C.; Lim, T.T. Fate and distribution of heavy metals during thermal processing of sewage sludge. *Fuel* **2018**, *226*, 721–744. [CrossRef]
48. Kudra, T.; Gawrzynski, Z.; Glaser, R.; Stanislawski, J.; Poirier, M. Drying of pulp and paper sludge in a pulsed fluid bed dryer. *Dry. Technol.* **2002**, *20*, 917–933. [CrossRef]
49. Escala, M.; Zumbühl, T.; Koller, C.; Junge, R.; Krebs, R. Hydrothermal carbonization as an energy-efficient alternative to established drying technologies for sewage sludge: A feasibility study on a laboratory scale. *Energy Fuels* **2013**, *27*, 454–460. [CrossRef]
50. Kamizela, T.; Kowalczyk, M. *Sludge Dewatering: Processes for Enhanced Performance*; Butterworth-Heinemann: Oxford, UK, 2019; ISBN 9780128159071.
51. Mulchandani, A.; Westerhoff, P. Recovery opportunities for metals and energy from sewage sludges. *Bioresour. Technol.* **2016**, *215*, 215–226. [CrossRef]
52. Kelessidis, A.; Stasinakis, A.S. Comparative study of the methods used for treatment and final disposal of sewage sludge in European countries. *Waste Manag.* **2012**, *32*, 1186–1195. [CrossRef] [PubMed]
53. Vassilev, S.V.; Baxter, D.; Andersen, L.K.; Vassileva, C.G. An overview of the chemical composition of biomass. *Fuel* **2010**, *89*, 913–933. [CrossRef]
54. Phyllis2. Database for the Physico-Chemical Composition of (Treated) Lignocellulosic Biomass, Micro- and Macroalgae, Various Feedstocks for Biogas Production and Biochar. Available online: https://phyllis.nl/ (accessed on 9 September 2021).
55. Birgitta Strömberg, S.H.S. *The Fuel Handbook*; Värmeforsk: Stockholm, Sweden, 2012.
56. Raheem, A.; Sikarwar, V.S.; He, J.; Dastyar, W.; Dionysiou, D.D.; Wang, W.; Zhao, M. Opportunities and challenges in sustainable treatment and resource reuse of sewage sludge: A review. *Chem. Eng. J.* **2018**, *337*, 616–641. [CrossRef]
57. Gorazda, K.; Tarko, B.; Wzorek, Z.; Kominko, H.; Nowak, A.K.; Kulczycka, J.; Henclik, A.; Smol, M. Fertilisers production from ashes after sewage sludge combustion—A strategy towards sustainable development. *Environ. Res.* **2017**, *154*, 171–180. [CrossRef]
58. Smol, M.; Kulczycka, J.; Henclik, A.; Gorazda, K.; Wzorek, Z. The possible use of sewage sludge ash (SSA) in the construction industry as a way towards a circular economy. *J. Clean. Prod.* **2015**, *95*, 45–54. [CrossRef]
59. Salman, C.A.; Naqvi, M.; Thorin, E.; Yan, J. Gasification process integration with existing combined heat and power plants for polygeneration of dimethyl ether or methanol: A detailed profitability analysis. *Appl. Energy* **2018**, *226*, 116–128. [CrossRef]
60. Matthew Wittenstein, G.R. *Projected Cost of Generating Electricity*; NEA: Paris, France, 2015.
61. Doyle, G.; Borkowski, K.; Vantsiotis, G.; Dodds, J.; Critten, S. *Costs of Low-Carbon Generation Technologies*; Committee on Climate Change: Bristol, UK, 2011.

62. The World Bank Group. *Commodity Markets Outlook, January*; The World Bank Group: Washington, DC, USA, 2017; Volume 1, pp. 1689–1699.
63. European Union. *Study on Energy Prices, Costs and Their Impact on Industry and Households*; European Union: Brussels, Belgium, 2020.
64. European Union. *Report from the Commission to the European Parliament, the Council, the European Economic and Social Committee and the Committee of the Regions*; European Union: Brussels, Belgium, 2020.
65. Lo, S.L.Y.; How, B.S.; Leong, W.D.; Teng, S.Y.; Rhamdhani, M.A.; Sunarso, J. Techno-economic analysis for biomass supply chain: A state-of-the-art review. *Renew. Sustain. Energy Rev.* **2021**, *135*, 110164. [CrossRef]
66. Eurostat. Natural Gas Price Statistics—Statistics Explained. Available online: https://ec.europa.eu/eurostat/statistics-explained/index.php?title=Natural_gas_price_statistics#Electricity_prices_for_household_consumers (accessed on 8 July 2021).
67. Dal Pozzo, A.; Guglielmi, D.; Antonioni, G.; Tugnoli, A. Environmental and economic performance assessment of alternative acid gas removal technologies for waste-to-energy plants. *Sustain. Prod. Consum.* **2018**, *16*, 202–215. [CrossRef]
68. Savage, G.M. *Sludge Treatment and Disposal*; IWA Publishing: London, UK, 2003; Volume 23, ISBN 8790402057.
69. International Energy Agency (IEA). *Electricity Market Report—December 2020*; International Energy Agency (IEA): Paris, France, 2020; Volume 1.
70. Skoglund, N.; Grimm, A.; Öhman, M.; Boström, D. Effects on ash chemistry when co-firing municipal sewage sludge and wheat straw in a fluidized bed: Influence on the ash chemistry by fuel mixing. *Energy Fuels* **2013**, *27*, 5725–5732. [CrossRef]
71. Eurostat Statistics—Eurostat. Available online: https://ec.europa.eu/eurostat/web/products-datasets/product?code=env_ww_spd (accessed on 8 July 2021).
72. Holmgren, K.M. *Investment Cost Estimates for Gasification-Based Biofuel Production Systems*; IVL: Stockholm, Sweden, 2015.
73. Obernberger, I.; Hammerschmid, A. *Techno-Economic Evaluation of Selected Decentralised CHP Applications Based on Biomass Combustion with Steam Turbine and ORC Processes*; IEA Bioenergy: Graz, Austria, 2015.
74. Schneider, D.R.; Lončar, D.; Bogdan, Ž. Cost Analysis of Waste-to-Energy Plant. *Stroj. Časopis za Teoriju i Praksu u Strojarstvu* **2010**, *52*, 369–378.

Article

Biogenic Nanosilica Synthesis Employing Agro-Waste Rice Straw and Its Application Study in Photocatalytic Degradation of Cationic Dye

Garima Singh [1], Hossein Beidaghy Dizaji [2,3], Hariprasad Puttuswamy [1] and Satyawati Sharma [1,*]

[1] Centre for Rural Development and Technology, Indian Institute of Technology, New Delhi 110016, India; dpgarima@gmail.com (G.S.); phari@iitd.ac.in (H.P.)
[2] DBFZ Deutsches Biomasseforschungszentrum Gemeinnützige GmbH, Torgauer Straße 116, 04347 Leipzig, Germany; hossein.beidaghy@dbfz.de
[3] Institute of Chemical Technology, Leipzig University, Linnéstr. 3, 04103 Leipzig, Germany
* Correspondence: satyawatis@hotmail.com

Abstract: The current study aims towards a holistic utilization of agro-waste rice straw (RS) to synthesize nanosilica (SiNPs) employing the sol–gel method. The effect of ashing temperature was evaluated for the synthesis process. X-ray diffraction demonstrated a broad spectrum at 21.22° for SiNPs obtained using RSA-600, signifying its amorphous nature, whereas crystalline SiNPs were synthesized using RSA-900. The EDX and FTIR spectra confirmed the significant peaks of Si and O for amorphous SiNPs, confirming their purity over crystalline SiNPs. FE-SEM and TEM micrographs indicated the spheroid morphology of the SiNPs with an average size of 27.47 nm (amorphous SiNPs) and 52.79 nm (crystalline SiNPs). Amorphous SiNPs possessed a high surface area of 226.11 m^2/g over crystalline SiNPs (84.45 m^2/g). The results obtained attest that the amorphous SiNPs possessed better attributes than crystalline SiNPs, omitting the need to incorporate high temperature. Photocatalytic degradation of methylene blue using SiNPs reflected that 66.26% of the dye was degraded in the first 10 min. The degradation study showed first-order kinetics with a half-life of 6.79 min. The cost-effective and environmentally friendly process offers a sustainable route to meet the increasing demand for SiNPs in industrial sectors. The study proposes a sustainable solution to stubble burning, intending towards zero waste generation, bioeconomy, and achieving the Sustainable Development Goals (SDGs), namely SDG 13(Climate Action), SDG 3(Good health and well-being), SDG 7(use of crop residues in industrial sectors) and SDG 8 (employment generation).

Keywords: rice straw; ash; nanosilica; methylene blue; zero waste generation; decolorization; SDGs

Citation: Singh, G.; Dizaji, H.B.; Puttuswamy, H.; Sharma, S. Biogenic Nanosilica Synthesis Employing Agro-Waste Rice Straw and Its Application Study in Photocatalytic Degradation of Cationic Dye. *Sustainability* 2022, 14, 539. https://doi.org/10.3390/su14010539

Academic Editors: Marc A. Rosen and Antoni Sánchez

Received: 25 November 2021
Accepted: 28 December 2021
Published: 4 January 2022

Publisher's Note: MDPI stays neutral with regard to jurisdictional claims in published maps and institutional affiliations.

Copyright: © 2022 by the authors. Licensee MDPI, Basel, Switzerland. This article is an open access article distributed under the terms and conditions of the Creative Commons Attribution (CC BY) license (https://creativecommons.org/licenses/by/4.0/).

1. Introduction

Rice straw (RS), lignocellulosic biomass, is a very common agro-waste generated in the agriculture system after the post-harvesting of rice. Incorporating the crop and harvesting method, approximately 40–60% of residual biomass comprises RS [1,2]. RS is a stiff, voluminous lignocellulosic biomass with significant silica (SiO$_2$) deposits, for which the level of biogenic silica can reach up to 82% on a dry weight basis. The complex structure limits the usability of RS. Being a quick, easy and cheap process, most farmers opt for open field-burning as the most preferred approach to dispose of RS in agricultural fields [3,4]. The burning of farm waste causes the ghastly pollution of soil and water at the regional scale. This practice also adversely reduces the nutrient composition in the soil. The elemental carbon, nitrogen and sulphur become completely burnt and subsequently emit hazardous gases such as methane, nitrogen oxide and ammonia, causing austere atmospheric pollution. These gases also contribute and further add up to the existing ozone pollution. Burning releases fine particles which are known to aggravate chronic heart and lung diseases [3].

Rice plants accumulate Si by polymerizing water-soluble silicic acid (H_4SiO_4) absorbed from the soil into insoluble polysilicic acids, precipitated as amorphous silica and deposited on the plant cell's exterior [5,6]. Si is deposited in plants primarily as phytoliths, which consist of amorphous hydrated silica. These beneficial attributes and the rich Si content of RS make it a suitable alternate source of biogenic nano-silica [7]. The Si can be extracted from RS by ashing at a temperature beyond 400 °C. However, temperature above 700 °C leads to the production of crystalline Si such as cristobalite and tridymite, with limited applications, possessing higher risks of silicosis [8–10].

In recent years, many efforts have been made to synthesize silica nanoparticles (SiNPs) from various preparatory materials, precisely chemical and natural sources. Different approaches such as the sol–gel process, chemical precipitation method, microemulsion processing, plasma synthesis, chemical vapor deposition, combustion in a diffusion flame and hydrothermal treatment have been employed for preparing Si-NPs [11,12]. Among these, the sol–gel process, also known as the "Stöber method" is a relatively modest and low-cost process [13,14]. It is worth highlighting that the chemical route is not only expensive, but it also adds to the list of pollutants, therefore, adapting a green route is the need of the hour for a healthy and safe environment [15].

The scientific community has successfully utilized the potential of nanotechnology to develop different products and materials at the nanoscale for societal welfare [12,16]. Silica is an important inorganic material with a panoramic range of applications in the textile industry, automobile industry, biology, medicine, adsorbents, drug delivery system, etc. [11,12,17]; additionally, it also holds an advantage over conventional precursors owing to its abundancy as well as being a cheap substrate [18].

SiNPs holds potential application in various sectors due to its high surface area and reactivity in broad areas [12,13,19,20]. These versatile properties enhance their potentials for developing biosensors and biomarkers, holding application in the detection of platelet-derived microparticles and the identification of leukemia cells [21]. The combination of SiNPs with super absorbent polymers helps in assuaging the plastic shrinkage. Additionally, the unique ability of SiNPs to exhibit the nucleation effect and pozzolanic activity leads to a decrease in the setting time and the mitigation of calcium leaching losses for cement-based materials [22,23]. The suitability of SiNPs as fillers in nanosilica composites has also been investigated by Salimian et al. [24]. The unique ability of SiNPs also tends to enhance its catalytic and photocatalytic applicability for removing an organophosphate pesticide, elimination of heavy metals from wastewater, the treatment of textile effluents and dye decolorization [23,25].

Of the various pollutants reported, methylene blue is a prominent blue cationic thiazine dye used widely in textile, paper and wood industries. It is documented that intense exposure to this dye leads to release of aromatic amines with severe environmental and health hazards. In previous years, chemical-based products have been designed to degrade toxic pollutants effectively; yet again, the persistent and non-degradable nature with its known tendency to bioaccumulate serves well as another potential environmental health hazard, demanding an adequate replacement [26,27].

In this regard, nanomaterials have emerged as an influential factor in removing organic pollutants due to their excellent high surface area and adsorption capacity. Exploring the potential of naturally synthesized nanomaterials holds potential utility. With reference to the proposed methods, although silica is inert for many reactions, it shows noticeable catalytic activities under ultraviolet irradiation below ~390 nm, e.g., photo-oxidation of CO, photo-metathesis of propene, photo-epoxidation of propene and silica-based photocatalysts such as silica-alumina, silica-alumina-titania and gold-coated SiO_2 with practical photocatalytic activities possessing a significant utility in photodegradation of toxic products [27–29]. The utilization of agro-waste-derived SiNPs in the degradation of cationic dye serves as an excellent example of two birds with one stone, on one end promoting a natural route towards the treatment of the toxic effluents while on the other proposing an excellent alternative towards the minimization of chemical-based routes for the treatment of organic pollutants.

A dearth of literature highlights the difference observed in the characterization of SiNPs when RS ashing is performed at 900 °C vs. the ashing temperature of 550 to 600 °C. To this end, a comparative analysis of amorphous and crystalline SiNPs synthesized using the sol–gel method was conducted in the current study to empirically attest the amorphous SiNPs as a preferential choice over crystalline SiNPs. The optimal SiNPS were further explored for a cost-effective route for the removal of the toxic pollutant methylene blue. The study offers insights to researchers and stakeholders towards a sustainable route for the synthesis of SiNPs and its application in the degradation of the toxic cationic dye methylene blue. The increase in the utility of silica-rich RS can serve as an integral factor in avoidance of stubble burning and exploring its utility will provide new ventures towards a suitable replacement to chemical route adopted to eliminate the toxic dyes in industries, thus playing an integral role towards bioeconomy as well as to the safe and healthy environment. The study overall is an attempt to meet the Sustainable Development Goals (SDGs). An alternative to stubble burning will assist in targeting SDG 13 (Climate Action), SDG 3 (Good health and well-being), SDG 7 (Use of crop residues in industrial sectors) and SDG 8 (Employment generation), highlight the aim of the study undertaken [30].

2. Materials and Methods

2.1. Collection of Raw Material and Rice Straw Ash Preparation

RS was collected from the Khatauli village of Uttar Pradesh, India. Before any treatment, the piled agro-residue was washed thoroughly using distilled water and then dried at 105 °C. Dried RS was chopped into small pieces and pulverized in a supermass collider (Masuko Sangyo Co. Ltd., Kawaguchi, Japan). For uniform size powdered straw, the pulverized material was passed through a 20-mesh screen. The ground RS powder was re-washed to remove any dust particles and dried in a hot air oven at 60 °C. Finally, dried and cleaned straw powder was burned to ashes using a muffle furnace, maintaining the furnace temperature at 600 °C and 900 °C for 4 h leading to grey and white ash production, respectively [12]. The ashes obtained at two different temperatures are denoted hereafter as RSA-600 and RSA-900, used to synthesize SiNPs.

2.2. Nano Silica Extraction from Rice Straw Ash

A combined method for extracting nanosilica from RS was performed based on the methodology provided by Bahrami et al. [10] and Kalapathy et al. [31]. A detailed methodology followed is outlined in Figure 1.

2.3. Characterization of Rice Straw Ash and Nanosilica Powders

2.3.1. X-ray Diffraction (XRD)

The amorphous and crystalline nature of calcinated RSAs and synthesized SiNPs were determined using XRD X'Pert Pro (PANalytical The Netherlands). The samples were flattened in the sample container using a glass slide. Radical scans of scattering angle (2θ) vs. intensity scans were recorded from 5 to 100° with CuKα radiation of 1.54 Å.

2.3.2. Fourier Transform Infrared Spectroscopy (FTIR)

The functional bonds present in the RSA and SiNPs were studied using an FTIR spectrophotometer (Perkin-Elmer1600). The absorbance of the dried sample was measured in the spectral range of 4000–500 cm^{-1} for 128 scans at a speed of 16 cm s^{-1}. The spectral obtained was compared with the commercial nanosilica based on a literature review [4].

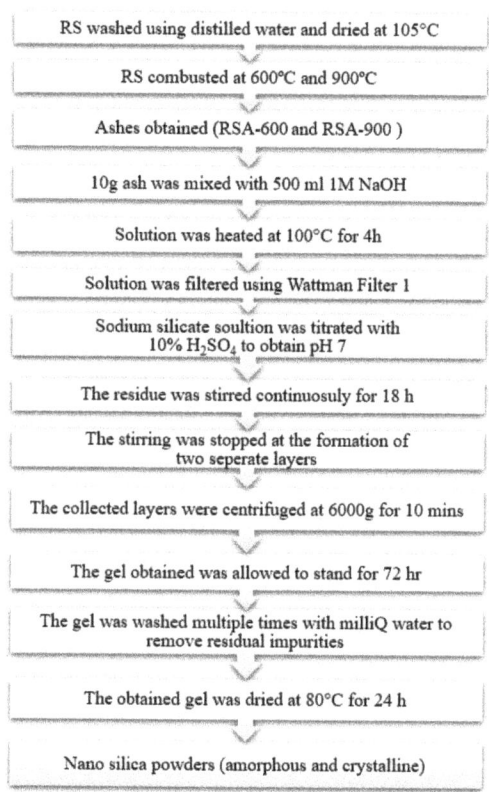

Figure 1. Synthesis of SiNPs using RSA combusted at 600 °C for amorphous particles and at 900 °C for crystalline particles.

2.3.3. Field Emission Scanning Electron Microscope-Energy Dispersive Spectroscopy (FE-SEM-EDX)

Microstructure and surface characteristics of the substrates (Raw RS, RSA, and SiNPs) were observed using FE-SEM-EDX (Field Emission Scanning Electron Microscope with Oxford-EDX system IE 250 X Max 80, The Netherlands). The dehydrated sample was mounted on the carbon tape. Gold sputtering was performed under vacuum 120 s with an acquisition time of 2 min, beam accelerating voltage of 10 kV at beam aperture (30 mm), with a working distance of 10 mm and probe current of 3×10^{-10} A. Mean EDX count rate was kept as 1600 ± 200 cps.

2.3.4. Transmission Electron Microscopy (TEM)

TEM analyses were performed by dissolving 5 mg of samples (RSA and SiNPs) in 50 mL of double distilled water and kept for 30 min ultrasonication. Next, 10 µL of the suspension (0.005% w/w) was mounted on carbon-coated copper grids. The shape and size of the samples prepared were characterized by TEM (JEOL JEM-1400) at an accelerating voltage of 100 kV. The diameter and size distribution of synthesized SiNPs was calculated using ImageJ software.

2.3.5. Surface Area and Porosity

A Brunauer–Emmett–Teller (BET) of the make (Micromeritics ASAP 2010, USA) was used to analyze the surface area of synthesized SiNPs at 77 K in N_2 atmosphere. The pore size distribution of the catalyst was calculated from the Barret–Joyner–Halenda (BJH) method using the adsorption data at relative pressure $P/P_0 - 0.990$.

2.3.6. Photocatalytic Degradation of Methylene Blue Dye

The photocatalytic effect of different concentrations of the optimum synthesized SiNPs (0.2–0.5 g/L) on methylene blue (100 ppm) at alkaline pH 11.0 was investigated following the protocols of Saleh and Dijaja [29] and Aly and Elhamid [26] with slight modifications. The photocatalytic experiment was conducted in a glass beaker equipped with continuous stirring under exposure to ultraviolet light (Philips 30 W, two tubes). Different suspensions were swirled in the dark for 30 min before irradiation to obtain a colloidal solution. The beakers were placed at a 15 cm distance from the light source. The samples were extracted every 2 h and centrifuged at 10,000 rpm for 5 min, and the study was conducted for 2 h. The absorbance of the solution was determined using a BioTek Epoch 2 microplate spectrophotometer at 630 nm (λmax), corresponding to the maximum absorption of methylene blue. The dye removal efficiency percentage was calculated as $(A_0-A_t)/A_0 \times 100$, where A_0 and A_t are the initial and final dye concentrations at time t, respectively.

3. Results

3.1. X-ray Diffraction Analysis

The XRD analysis of RSA-600 depicted a broad hump between 20.52°–22.71°, indicating the material to be amorphous (Figure 2A), possessing the ability for adsorption; whereas RSA-900 indicated sharp peaks between 20° and 30° (Figure 2B), documented to decrease the available surface area, thus restricting the adsorption potential [8].

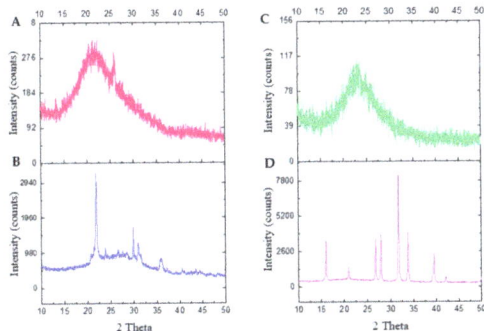

Figure 2. XRD pattern (**A**) RS ash obtained at 600 °C (RSA-600). (**B**) RS ash obtained at 900 °C (RSA-900). (**C**) SiNPs synthesized using RS ash obtained at 600 °C. (**D**) SiNPs synthesized using RS ash obtained at 900 °C.

The synthesis process of SiNPs carried out using RSA-600 depicted a broad peak centered at 22°; the absence of any other peaks confirmed the SiNPs to be of amorphous nature (Figure 2C). SiNPs synthesized from RSA-900 depicted the presence of crystalline components between 20°–30° with a crystallinity index of 65%, calculated according to the methodology of Mendes et al. [32]. However, various other peaks signified the impurity of the synthesized product, as shown in Figure 2D. The results are in coherence with the findings of several other researchers, where the diffraction peaks at 2θ angles between 20° and 30° is documented to be the characteristic peak of silica [6,33].

3.2. EDX Analysis

The EDX profiling of RSA-600 showed Si and O's presence and various other minor elements such as Mg, K, Na, Al, S, Fe and Ca, shown in Figure 3A. The profiling study of white RSA-900 also showed various elements along with Si and O, with a low K content, as shown in Figure 3B. The amorphous SiNPs confirmed the presence of Si and O and the absence of any other impurities, thus validating the purity of the product obtained as depicted (Figure 3C). The crystalline SiNPs denoted small peaks for Na and S, which could be due to traces of residual elements left in the process of washing (Figure 3D).

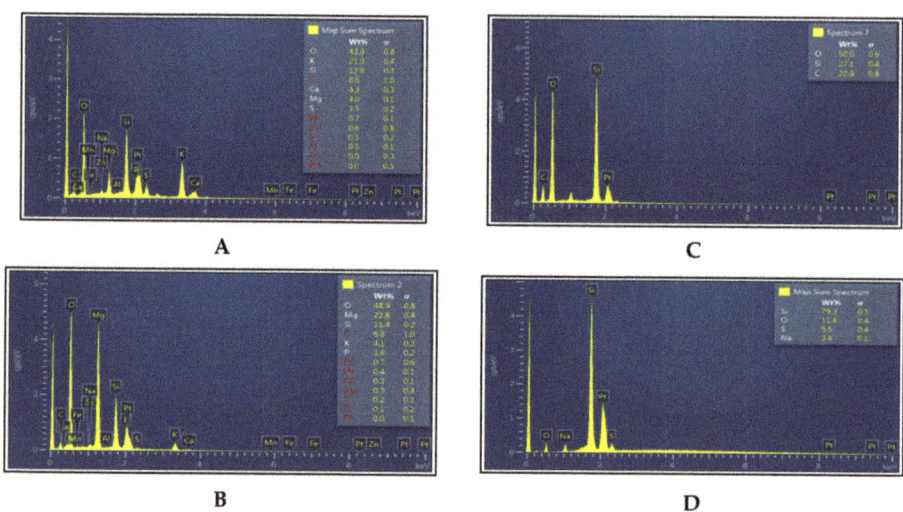

Figure 3. EDX analysis (**A**) RS ash obtained at 600 °C (RSA-600) (**B**) RS ash obtained at 900 °C (RSA-900) (**C**) SiNPs synthesized using RS ash obtained at 600 °C (**D**) SiNPs synthesized using RS ash obtained at 900 °C.

3.3. Fourier Transform Infrared Analysis

The Fourier transform IR analysis samples were recorded in the spectrum range of 4000–500 cm^{-1}. FTIR of RSAs, presented in Figure 4, showed significant bands at 794 cm^{-1}, 1109 cm^{-1} corresponding to the symmetric and asymmetric stretching vibration of the Si−O−Si bond [6,12]. The peak observed for RSA-600 at 991 cm^{-1} signified the Si−OH bond's bending vibration that diminished completely for RSA-900 [34]. The bonds 1629 cm^{-1} and 3448 cm^{-1} indicated the bending and stretching vibration of the H−OH bond. The results are indicated in (Figure 4A,B). FTIR analyses of synthesized SiNPs showed broadband ranged between 3000–3500 cm^{-1} that indicated the presence of silanol group (Si−OH) bonding. The little band in the region of 1633 cm^{-1} corresponded to the bending vibrations of H−O−H (water molecules). The dominant peak at 1097 cm^{-1} was due to the asymmetric vibration of the Si−O−Si bond. The band at 789 cm^{-1} corresponded to Si−O−Si bond stretching [13,35], as shown in Figure 4C,D. A comparison of the spectral values obtained, and literature reports are briefly outlined in Table 1.

Table 1. Infrared bands observed in ashes (RSA-600, RSA-900), SiNPs (amorphous and crystalline).

RSA-600	RSA-900	SiNPs (Amorphous)	SiNPs (Crystalline)	Literature	Functional Groups
459	470	462	452	460 [36]	Si−O
1091	11,081	1073	1101	1033 [12]	Si−O−Si
3474	3421	3423	NA	3000–3500 [13]	Si−OH
NA	NA	1663	1641	1600 [35]	H−OH
795	775	796	817	796 [6]	Si−O−Si

RSA 600—Rice straw ash obtained at 600 °C; RSA 900—Rice straw ash obtained at 900 °C; SiNPs—Silica nanoparticles.

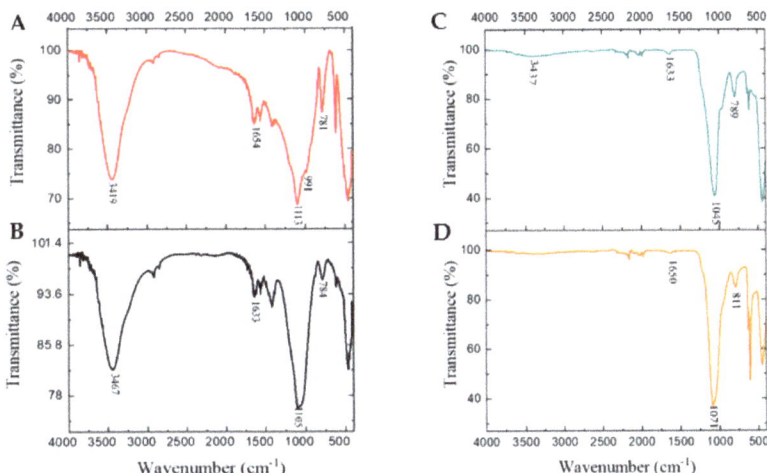

Figure 4. FTIR analysis (**A**) RS ash obtained at 600 °C (RSA-600). (**B**) RS ash obtained at 900 °C (RSA-900). (**C**) SiNPs synthesized using RS-600. (**D**) SiNPs synthesized using RS-900.

3.4. Morphology Studies

FE-SEM analysis of RS showed a stable, well-defined structure (Figure 5A). The microscopic analysis of RS ash (RSA-600 and RSA-900) showed a homogeneous distribution of dumbbell-shaped phytoliths commonly referred to as silica bodies over the entire surface (Figure 5B,C). The TEM analysis of RSAs showed the agglomeration of small circular bodies over the entire surface (Figure 5D,E).

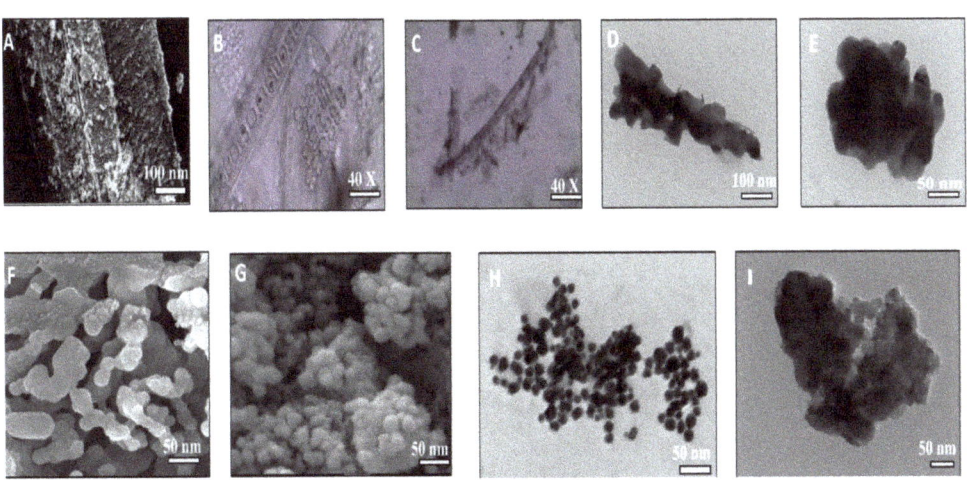

Figure 5. (**A**) FESEM micrographs of raw RS at 500 X. (**B**) Microscopic images of RSA-600 at 40X. (**C**) Microscopic images of RSA-900 D at 40 X. (**D**) TEM micrographs of RSA-600 at 120 kV. (**E**) TEM micrographs of RSA-900 at 120 kV. (**F**) FE-SEM micrographs of amorphous SiNPs at 10 kX. (**G**) FE-SEM micrographs of crystalline SiNPs at 10 kX. (**H**) TEM micrographs of amorphous SiNPs at 120 kV. (**I**) TEM micrographs of crystalline SiNPs at 120 kV.

FE-SEM analysis of amorphous SiNPs synthesized by RSA-600 depicted the particles of spheroid morphology with loose aggregates compared to crystalline SiNPs synthesized by RSA-900 (Figure 5F,G). The formation of aggregates could be attributed to the gel-like property of the hydrated silica and its high surface area [7,11,12]. TEM analysis revealed the amorphous SiNPs to be spherical, with the average particle size of 27.47 nm possessing little agglomeration. In contrast, crystalline SiNPs possessed average particle size of 52.79 nm with high agglomeration (Figure 5H,I). The results obtained can be well attested by the findings of Bahrami et al. [10] and Lu et al. [6] where it was postulated that high temperature leads to the formation of crystalline silica, keeping the Si bonds intact.

3.5. Surface Area and Porosity Studies

BET analysis revealed that amorphous SiNPs possessed a specific surface area (SSA) of 226.811 m^2/g with an average pore volume of 1.144 cm^3/g, whereas the crystalline SiNPs had a BET SSA of 84.45 m^2/g with a pore volume of 0.497 cm^3/g. Nitrogen adsorption-desorption isotherm for the amorphous and crystalline SiNPs are shown in Figure 6A,B. The results were in line with the findings of Beidaghy Dizaji et al. [37] who documented that the porosity of silica-rich ashes diminish once the crystallinity fraction is higher than 10 wt.%. Didamony et al. [38] reported a surface area of 160 m^2/g from SiNPs extracted using sodium silicate solution, while Yuvakumar et al. [39] reported the amorphous synthesized SiNPs with a surface area of 274 m^2/g and an average pore diameter of 1.46 nm.

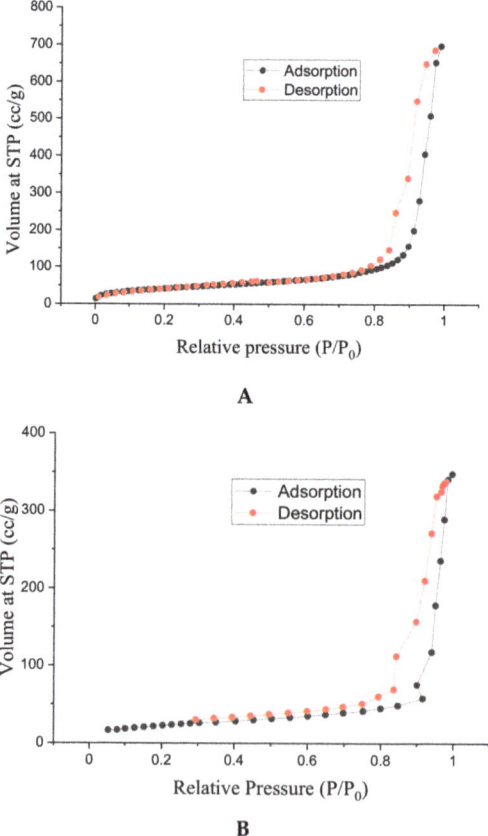

Figure 6. Nitrogen adsorption-desorption isotherms of (**A**) amorphous SiNPs (**B**) crystalline SiNPs.

The results attested the findings that the amorphous SiNPs had significantly better attributes when compared to the crystalline SiNPs, thus increasing its utility in industrial sectors.

3.6. Decolorization of Cationic Dye Methylene Blue Using Amorphous SiNPs

The photocatalytic degradation studies tested the effect of different concentrations of SiNPs on a constant concentration of methylene blue dye (100 ppm). The studies reflected that the dye was efficiently degraded by 66.26% within the first 10 min by the SiNPs at 50 ppm concentration. Degradation of the dye at the lowest SiNPs concentration of 10 ppm did not reveal any observable degradation pattern. It showed a similar trend as to the degradation process without SiNPs (Figure 7A). This could be possibly explained owing to the availability of fewer adsorption sites of SiNPs at a relatively high concentration of dye. Decolorization of ~100% was achieved within the initial 30 min of the study for the dye treated with SiNPs.

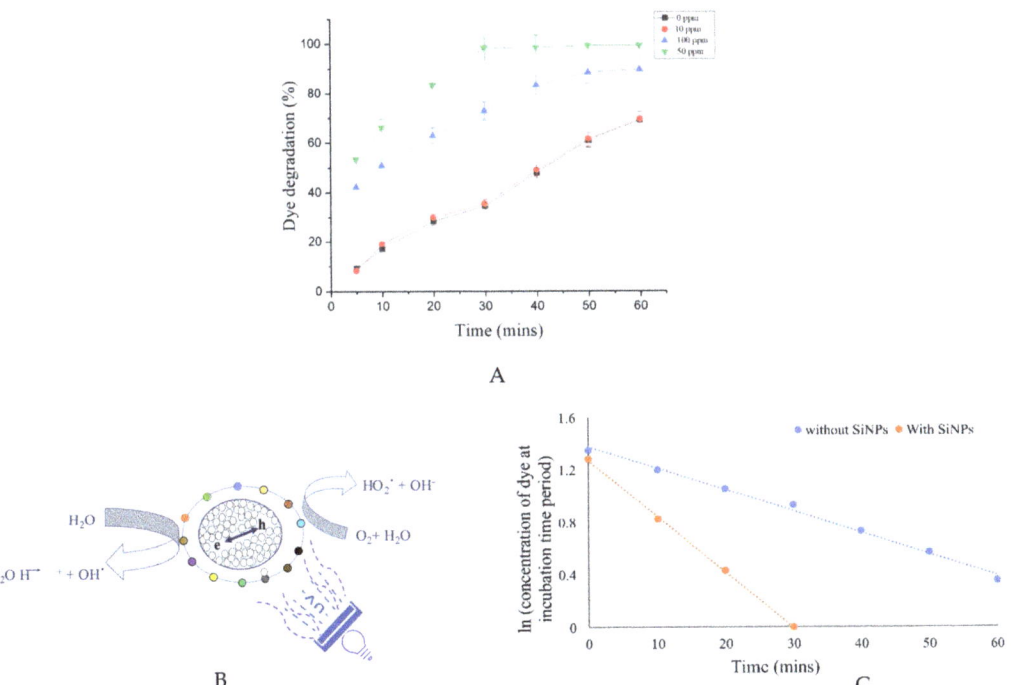

Figure 7. (**A**) Decolorization of methylene blue (MB) under the effect of UVC irradiation and different concentrations of SiNPs (**B**) Schematic representation of photocatalytic mechanism of methylene blue degradation by SiNPs (**C**) First-order kinetics plot of dye degradation, denoted by (....) with SiNPs and by (——) without SiNPs.

The result obtained could be well correlated to the effect of UV-C irradiation that assisted in the induction of the direct photolysis on the dye. Moreover, as the UV-C photons have a shorter penetration potential through photocatalyst particles, the possibility of electron-hole recombination is minimized due to shorter travel distances, leading to higher photocatalytic activity [27,29,40]. The high surface area of amorphous SiNPs accelerated the degradation process. A comprehensive detail of dye decolorization documented by numerous researchers using SiNPs is highlighted in Table 2.

Table 2. Nanosilica in photocatalytic degradation of textile dye effluents.

Dye	Concentration (Dye) and pH	Source of SiNPs	Concentration (Nanosilica)	Source	Degradation (%)	Time	References
Crystal Violet	50 mg/L; pH 7	Rice husk	1 g/L	Ultrasound	80%	60 min	Peres et al. [41]
Methyl red	0.05 Mm/100 mL; pH 7	Rice husk	1 g/100 mL	Sunlight	95%	120 min	Vinoda et al. [42]
Methylene blue	50 ppm/50 mL; pH-not reported	Rice husk	50 mg/100 mL	Ceramic material based on amorphous SiNPs	80%	30 min	Tolba et al. [25]
Methylene blue	100 mg/L; pH 11	Yellow sand	10 g/L	Ultraviolet light	100%	90 s	Aly and Elhamid [26]
Methylene blue	50 ppm; pH 7	ZnO/SiO_2 xerogel	0.075 g/L	Visible light	100%	30 min	Stanley [27]
Acridine orange	1×10^{-5} M	3-aminopropyl-functionalized silica NPs	10 mg/3 mL	Ultraviolet light	58%	50 min	Selvaggi et al. [43]
Acid ornage	150 mg/L	Sugar cane ash	1 g/L	Not reported	80%	30 min	Rovani et al. [44]
Malachite green	20 mg/L; pH 9	Fe_3O_4@SiO_2-COOH NPs	0.5 g/L	Not reported	97.5%	120 min	Galangash et al. [45]

3.6.1. Mechanism of Photocatalytic Degradation of Methylene Blue Using SiNPs

The mechanism of photocatalytic activity can be explained owing to the ability of SiNPs to be photoexcited under UV irradiation. This phenomenon can be supported by the charge transfer from Si−O bonding orbital to 2p non-bonding orbital of non-bridging oxygen. Interestingly, as observed in FTIR studies, the presence of Si−O and Si−OH groups imparts a negative charge on the silica surface, thus offering the SiNPs to serve as an excellent medium for adsorbent cationic dyes [43,46].

On striking the surface of SiO_2 by UV light, an electron transfer occurs from the valence band to the conduction band, generating a positive hole in the valence band (v_b) and a negative hole in the conduction band (c_b), leading to the formation of active photocatalytic centers on the surface of SiNPs (Equation (1)). The v_b hole further interacts with chemisorbed H_2O molecules to form OH radicals that successively attacks dye molecules (Equation (2)). The generation of heat in this process could be ascribed due to the combination of e_{cb}^- and h_{vb}^+ on the particle's surface. A plausible cause for dye decolorization can be attributed to the hydroxyl attack and conduction of the experiment at high pH, increasing OH^- groups on the silica surface, leading to an acceleration of dye degradation process [25,47].

A diagrammatic sketch of the mechanism is shown in Figure 7B.

$$SiO_2 + hv \rightarrow e_{cb}^- + h_{vb}^+ \quad (1)$$

$$HO_2 + OH^- + h_{vb}^- \rightarrow OH \quad (2)$$

The e_{cb}^- and h_{vb}^+ recombine on the particle's surface within nanoseconds, and the generated energy becomes dissipated in the form of heat. e_{cb}^- further reacts with the acceptor dissolved O_2 and is transformed to a super oxide radical anion ($O_2^{\cdot -}$), leading to the further growth of O_2H molecules (3):

$$O_2 + e_{cb}^- \rightarrow O_2^{\cdot -} + (H^+ + {}^-OH) \rightarrow HO_2 + OH^- \quad (3)$$

h_{vb}^+ interacts with the donor ^-OH and $\cdot O_2H$ forming $\cdot OH$ radical that attacks the MB in the following manner:

$$HO_2\cdot + OH^- + h_{vb}^- \rightarrow \cdot OH \quad (4)$$

The governing factor monitoring the efficiency of SiNPs is the amount of $\cdot OH$ radicals generated. Since the hydroxyl groups on the SiNPs surface are attached to the silicon atom, they are termed as silanols. The OH groups present in the silanols can preferentially complex particular chemicals or metal ions, imparting functionality to SiNPs [48]. It is worth highlighting that shifting the pH towards the alkaline range led to a dramatic boost in dye degradation. Henceforth, varying the pH value can significantly impact the interactions of various compounds with silanols. Subsequently, any factor that contributes to the generation of $\cdot OH$ radicals lead to an enhancement in the photocatalytic degradation process of methylene blue.

3.6.2. Kinetic Study of Dye Degradation

Kinetic studies were performed for the optimal concentration of SiNPs (50 ppm) that aided in the complete degradation of the methylene blue dye compared to the dye treated under UVC. The effect of SiNPs on dye decolorization indicated first-order kinetics. The linear form of the first-order rate equation is denoted by Equation (5), and the half-life ($t_{0.5}$) of dye decolorization was calculated using Equation (6) [47]:

$$\ln D_{Ab} = -kt + \ln D_{A0} \quad (5)$$

$$t_{0.5} = 0.693/k \quad (6)$$

where D_{Ab} is the dye absorbance at different incubation times, k is decolorization rate constant, D_{A0} is the initial absorbance of the dye, and $t_{0.5}$ is the time required to decolorize 50% dye.

The logarithm plot of dye concentration vs. treatment time with SiNPs exhibited a rate constant (k) of 0.102 min^{-1} and $t_{0.5}$ of 6.79 min. In contrast, the untreated dye exhibited a rate constant of 0.044 min^{-1} and $t_{0.5}$ of 15.75 min, reported for the 30 min study. The results obtained thus positively attest to the enhanced effect in dye degradation due to incorporation of the SiNPs as shown in Figure 7C.

3.6.3. FTIR Studies of Methylene Blue Decolorization

The methylene blue molecular structure transformation was further evaluated by the FTIR analysis of the treated and untreated samples. UV-Vis spectra illustrated the presence of aromatics along with conjugates of N–S heterocycle group and phenothiazine structure [29]. The results in Figure 8 showed that the broad peak at 3451 cm^{-1} was determined as the O–H stretching vibration of water molecules. It could be seen that the peak corresponding to C–H absorption of benzene ring occurred at 2975 cm^{-1}. The peak at 2105 cm^{-1} denoted the stretching vibration peak of the methyl group. The C=C framework corresponding to benzene ring vibration and the C=N stretching vibration was found at 1646 cm^{-1}. The absorption peak at 1420 cm^{-1} was related to another typical vibration in methyl bending. Other prominent peaks were in the vibration of C=O and -C-C at 1206 cm^{-1} and 991 cm^{-1}. It was noticed that the intensity of absorption peaks for C=N and O–H had significant decrease in C=C aromatic stretch in the treated sample, indicating a change in chemical composition. This could be possibly attributed to the breakdown of the N–S heterocyclic compound during the degradation process [25,26,49]. The characteristic peak of benzene and C–H bending vibration of aromatic C–H declined substantially, indicating the variation in chemical compositions of phenyl groups.

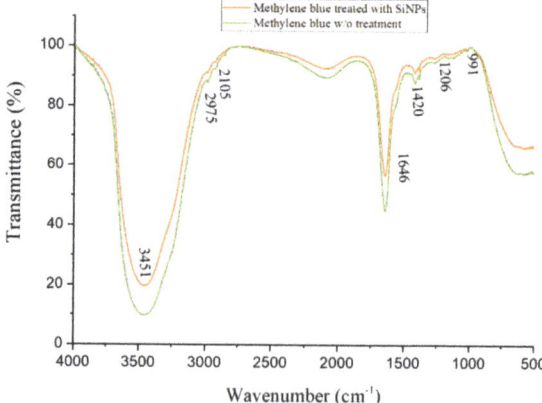

Figure 8. FTIR spectrum of methylene blue about its treatment with SiNPs.

It henceforth can be deduced from the spectra that the adsorption of methylene blue on the surface of SiNPs led to a remarkable change in infrared bands intensities, retaining its positions.

Stanley [26], Salimi [40] and Singh [43] hypothesized that the conjugate structure of N–S heterocyclic underwent variations in chemical composition, and consequently the aromatic ring was oxidized to open the ring leading to degradation of the dye molecules during the photocatalysis reaction. The result obtained provides new insight into potential waste usage and utilizing the low-cost synthesized SiNPs for dye degradation.

4. Conclusions

The study revealed that RS combusted at 600 °C served as an ideal condition for synthesizing amorphous SiNPs. The findings and characterization postulated that the amorphous SiNPs served better attributes when compared to crystalline SiNPs, establishing that all the researchers/stakeholders working in this domain can note that lower temperature offers a more sustainable route in product synthesis. This first-hand analysis provided the particulars pertaining to temperature's crucial role and effect on the characterization and properties of SiNPS. The efficacy of the green synthetic route for amorphous SiNPs holding potential applications in different sectors was accessed in the field of wastewater and textile effluents for degradation of the toxic dye methylene blue. Additionally, incorporating microwave, sonication, conjugation of substrates as laccase, surface modification of SiNPs can further be attempted towards the dye degradation process. However, the study has much to offer in terms of optimizing the synthesis process via integrating surfactants, catalyst, residence time, etc., to enhance the surface area and pore size of SiNPs; that will further assist in the degradation process with a minimal dose of SiNPs and enhancing its utility as an adsorbent in different industrial sectors. Contemplating the abundance of agro-waste RS worldwide, the study establishes a background in converting the waste to a value-added product, providing a comprehensive and viable sustainable resolution towards stubble burning to achieve the SDGs 3, 7, 8 and 13.

Author Contributions: Conceptualization, S.S. and H.B.D.; methodology, G.S. and H.B.D.; software, G.S.; validation, H.B.D. and S.S.; formal analysis, G.S. and H.B.D.; investigation, G.S. and H.B.D.; resources, H.B.D. and S.S.; data curation, G.S.; writing—original draft preparation, G.S.; writing—review and editing, H.B.D. and S.S.; visualization, H.P. and S.S.; supervision H.P. and S.S.; project administration, H.B.D. and S.S.; funding acquisition, H.B.D. and S.S. All authors have read and agreed to the published version of the manuscript.

Funding: The APC was funded by the Thermochemical Conversion department of DBFZ Deutsches Biomasseforschungszentrum gemeinnützige GmbH.

Institutional Review Board Statement: Not applicable.

Informed Consent Statement: Not applicable.

Data Availability Statement: The data that support the findings of this study are available upon request from the authors.

Acknowledgments: The authors are thankful to the Indian Institute of Technology, New Delhi, and DBFZ Deutsches Biomasse for schungszentrumgemeinnützige GmbH, Leipzig, Germany, for providing the research funds and financial support.

Conflicts of Interest: The authors declare no conflict of interest.

References

1. Beidaghy Dizaji, H.; Zeng, T.; Hartmann, I.; Enke, D.; Schliermann, T.; Lenz, V.; Bidabadi, M. Generation of high-quality biogenic silica by combustion of rice husk and rice straw combined with pre- and post-treatment strategies—A review. *Appl. Sci.* **2019**, *9*, 1083. [CrossRef]
2. Kumar, S.; D'Silva, T.C.; Chandra, R.; Malik, A.; Vijay, V.K.; Misra, A. Strategies for boosting biomethane production from rice straw: A systematic review. *Bioresour. Technol. Rep.* **2021**, *15*, 100813. [CrossRef]
3. Tipayarom, A.; Oanh, N.T.K. Influence of rice straw open burning on levels and profiles of semi-volatile organic compounds in ambient air. *Chemosphere* **2020**, *243*, 125379. [CrossRef]
4. Singh, G.; Arya, S.K. A review on management of rice straw by use of cleaner technologies: Abundant opportunities and expectations for Indian farming. *J. Clean. Prod.* **2020**, *291*, 125278. [CrossRef]
5. Marxen, A.; Klotzbücher, T.; Jahn, R.; Kaiser, K.; Nguyen, V.S.; Schmidt, A.; Schädler, M.; Vetterlein, D. Interaction between silicon cycling and straw decomposition in a silicon deficient rice production system. *Plant Soil* **2016**, *398*, 153–163. [CrossRef]
6. Lu, P.; Hsieh, Y.L. Highly pure amorphous silica nano-disks from rice straw. *Powder Technol.* **2012**, *225*, 149–155. [CrossRef]
7. Singh, G.; Tiwari, A.; Gupta, A.; Kumar, A.; Hariprasad, P.; Sharma, S. Bioformulation development via valorizing silica-rich spent mushroom substrate with *Trichoderma asperellum* for plant nutrient and disease management. *J. Environ. Manag.* **2021**, *297*, 113278. [CrossRef]

8. Bahrami, A.; Pech-Canul, M.I.; Gutierrez, C.A.; Soltani, N. Effect of rice-husk ash on properties of laminated and functionally graded Al/SiC composites by one-step pressureless infiltration. *J. Alloys Compd.* **2015**, *644*, 256–266. [CrossRef]
9. Beidaghy Dizaji, H.; Zeng, T.; Hölzig, H.; Bauer, J.; Klöß, G.; Enke, D. Ash transformation mechanism during combustion of rice husk and rice straw. *Fuel* **2022**, *307*, 121768. [CrossRef]
10. Bahrami, A.; Simon, U.; Soltani, N.; Zavareh, S.; Schmidt, J.; Pech-Canul, M.I.; Gurlo, A. Eco-Fabrication of hierarchical porous silica monoliths by ice-templating of rice husk ash. *Green Chem.* **2017**, *19*, 188–195. [CrossRef]
11. Bhattacharya, M.; Mandal, M.K. Synthesis of rice straw extracted nano-silica-composite membrane for CO_2 separation. *J. Clean. Prod.* **2018**, *186*, 241–252. [CrossRef]
12. Mor, S.; Manchanda, C.K.; Kansal, S.K.; Ravindra, K. Nanosilica extraction from processed agricultural residue using green technology. *J. Clean. Prod.* **2017**, *143*, 1284–1290. [CrossRef]
13. Kauldhar, B.S.; Yadav, S.K. Turning waste to wealth: A direct process for recovery of nano-silica and lignin from paddy straw agro-waste. *J. Clean. Prod.* **2018**, *194*, 158–166. [CrossRef]
14. Sachan, D.; Ramesh, A.; Das, G. Green synthesis of silica nanoparticles from leaf biomass and its application to remove heavy metals from synthetic wastewater: A comparative analysis. *Environ. Nanotechnol. Monit. Manag.* **2021**, *16*, 100467. [CrossRef]
15. Hassan, A.F.; Abdelghny, A.M.; Elhadidy, H.; Youssef, A.M. Synthesis and characterization of high surface area nanosilica from rice husk ash by surfactant-free Sol–Gel method. *J. Sol-Gel Sci. Technol.* **2014**, *69*, 465–472. [CrossRef]
16. Handojo, L.; Pramudita, D.; Mangindaan, D.; Indarto, A. Application of nanoparticles in environmental cleanup: Production, potential risks and solutions. *Emerg. Eco-Friendly Green Technol. Wastewater Treat.* **2020**, *18*, 45–76.
17. Liu, X.; Wang, Y.; Zhang, T.C.; Xiang, G.; Wang, X.; Yuan, S. One-Pot Synthesis of a Magnetic TiO_2/PTh/γ-Fe_2O_3 Heterojunction Nanocomposite for Removing Trace Arsenite via Simultaneous Photocatalytic Oxidation and Adsorption. *Ind. Eng. Chem. Res.* **2020**, *60*, 528–540. [CrossRef]
18. Krawczyk, A.; Domagała-Świątkiewicz, I.; Lis-Krzyścin, A.; Daraż, M. Waste Silica as a Valuable Component of Extensive Green-Roof Substrates. *Polish J. Environ. Stud.* **2017**, *26*, 643–653. [CrossRef]
19. Liu, S.; Wang, Y.; Liao, C.; Wang, Y.; He, J.; Fu, C.; Yang, K.; Bai, Z.; Zhang, F. Nano silica diaphragm in-fiber cavity for gas pressure measurement. *Sci. Rep.* **2017**, *7*, 787. [CrossRef]
20. Le, T.M.; Tran, U.P.N.; Duong, Y.H.P.; Nguyen, Q.D.; Tran, V.T.; Mai, P.T.; Le, P.K. Sustainable bioethanol and value-added chemicals production from paddy residues at pilot scale. *Clean Technol. Environ. Policy* **2021**, 1–13. [CrossRef]
21. Singh, P.; Srivastava, S.; Singh, S.K. Nanosilica: Recent progress in synthesis, functionalization, biocompatibility, and biomedical applications. *ACS Biomater. Sci. Eng.* **2019**, *5*, 4882–4898. [CrossRef]
22. Olivier, G.; Combrinck, R.; Kayondo, M.; Boshoff, W.P. Combined effect of nano-silica, super absorbent polymers, and synthetic fibres on plastic shrinkage cracking in concrete. *Constr. Build. Mater.* **2018**, *192*, 85–98. [CrossRef]
23. Zhang, H.; Goeppert, A.; Olah, G.A.; Prakash, G.K.S. Remarkable effect of moisture on the CO_2 adsorption of nano-silica supported linear and branched polyethylenimine. *J. CO_2 Util.* **2017**, *19*, 91–99. [CrossRef]
24. Salimian, S.; Zadhoush, A.; Naeimirad, M.; Kotek, R.; Ramakrishna, S. A review on aerogel: 3D nanoporous structured fillers in polymer-based nanocomposites. *Polym. Compos.* **2018**, *39*, 3383–3408. [CrossRef]
25. Tolba, G.M.K.; Barakat, N.A.M.; Bastaweesy, A.M.; Ashour, E.A.; Abdelmoez, W.; El-Newehy, M.H.; Al-Deyab, S.S.; Kim, H.Y. Effective and highly recyclable nanosilica produced from the rice husk for effective removal of organic dyes. *J. Ind. Eng. Chem.* **2015**, *29*, 134–145. [CrossRef]
26. Aly, H.F.; Abd-Elhamid, A.I. Photocatalytic degradation of methylene blue dye using silica oxide nanoparticles as a catalyst. *Water Environ. Res.* **2018**, *90*, 807–817. [CrossRef]
27. Stanley, R. Enhanced sunlight photocatalytic degradation of methylene blue by rod-like ZnO-SiO_2 nanocomposite. *Optik* **2019**, *180*, 134–143. [CrossRef]
28. Kusdianto, K.; Widiyastuti, W.; Shimada, M.; Qomariyah, L.; Winardi, S. Fabrication of ZnO-SiO_2 nanocomposite materials prepared by a spray pyrolysis for the photocatalytic activity under UV and sunlight irradiations. In Proceedings of the IOP Conference Series: Materials Science and Engineering, Borovets, Bulgaria, 26–29 November 2020; IOP Publishing: Bristol, UK, 2020; Volume 778, p. 12105. [CrossRef]
29. Saleh, R.; Djaja, N.F. UV light photocatalytic degradation of organic dyes with Fe-doped ZnO nanoparticles. *Superlattices Microstruct.* **2014**, *74*, 217–233. [CrossRef]
30. Venkatramanan, V.; Shah, S.; Rai, A.K.; Prasad, R. Nexus between Crop Residue Burning, Bioeconomy and Sustainable Development Goals over North-Western India. *Front. Energy Res.* **2021**, *8*, 614212. [CrossRef]
31. Kalapathy, U.; Proctor, A.; Shultz, J. A simple method for production of pure silica from rice hull ash. *Bioresour. Technol.* **2000**, *73*, 257–262. [CrossRef]
32. Mendes, C.; Adnet, F.; Leite, M.; Furtado, C.R.G.; Sousa, A. Chemical, physical, mechanical, thermal and morphological characterization of corn husk residue. *Cellul. Chem. Technol.* **2015**, *49*, 727–735.
33. Sarkar, P.; Moyez, S.A.; Dey, A.; Roy, S.; Das, S.K. Experimental investigation of photocatalytic and photovoltaic activity of titania/rice husk crystalline nano-silica hybrid composite. *Sol. Energy Mater. Sol. Cells* **2017**, *172*, 93–98. [CrossRef]
34. Yan, F.; Jiang, J.; Chen, X.; Tian, S.; Li, K. Synthesis and characterization of silica nanoparticles preparing by low-temperature vapor-phase hydrolysis of $SiCl_4$. *Ind. Eng. Chem. Res.* **2014**, *53*, 11884–11890. [CrossRef]

35. Zamani, A.; Marjani, A.P.; Mousavi, Z. Agricultural waste biomass-assisted nanostructures: Synthesis and application. *Green Process. Synth.* **2019**, *8*, 421–429. [CrossRef]
36. Ferreira, C.S.; Santos, P.L.; Bonacin, J.A.; Passos, R.R.; Pocrifka, L.A. Rice Husk Reuse in the Preparation of SnO_2/SiO_2 Nanocomposite. *Mater. Res.* **2015**, *18*, 639–643. [CrossRef]
37. Beidaghy Dizaji, H.; Zeng, T.; Enke, D. New fuel indexes to predict ash behavior for biogenic silica production. *Fuel* **2022**, *310B*, 122345. [CrossRef]
38. El-Didamony, H.; El-Fadaly, E.; Amer, A.A.; Abazeed, I.H. Synthesis and characterization of low cost nanosilica from sodium silicate solution and their applications in ceramic engobes. *Bol. Soc. Esp. Cerám. Vidr.* **2020**, *59*, 31–43. [CrossRef]
39. Yuvakkumar, R.; Elango, V.; Rajendran, V.; Kannan, N. High-Purity nano silica powder from rice husk using a simple chemical method. *J. Exp. Nanosci.* **2014**, *9*, 272–281. [CrossRef]
40. Kong, X.; Liu, X.; Li, J.; Yang, Y. Advances in pharmacological research of eugenol. *Curr. Opin. Complement. Altern. Med.* **2014**, *1*, 8–11.
41. Peres, E.C.; Favarin, N.; Slaviero, J.; Almeida, A.R.F.; Enders, M.P.; Muller, E.I.; Dotto, G.L. Bio-nanosilica obtained from rice husk using ultrasound and its potential for dye removal. *Mater. Lett.* **2018**, *231*, 72–75. [CrossRef]
42. Vinoda, B.M.; Vinuth, M.; Bodke, Y.; Manjanna, J. Photocatalytic degradation of toxic methyl red dye using silica nanoparticles synthesized from rice husk ash. *J. Env. Anal. Toxicol.* **2015**, *5*, 525–2161. [CrossRef]
43. Selvaggi, R.; Tarpani, L.; Santuari, A.; Giovagnoli, S.; Latterini, L. Silica nanoparticles assisted photodegradation of acridine orange in aqueous suspensions. *Appl. Catal. B Environ.* **2015**, *168*, 363–369. [CrossRef]
44. Rovani, S.; Santos, J.J.; Corio, P.; Fungaro, D.A. Highly pure silica nanoparticles with high adsorption capacity obtained from sugarcane waste ash. *ACS Omega* **2018**, *3*, 2618–2627. [CrossRef]
45. Mohammadi Galangash, M.; Mohaghegh Montazeri, M.; Ghavidast, A.; Shirzad-Siboni, M. Synthesis of carboxyl-functionalized magnetic nanoparticles for adsorption of malachite green from water: Kinetics and thermodynamics studies. *J. Chin. Chem. Soc.* **2018**, *65*, 940–950. [CrossRef]
46. Salimi, F.; Tahmasobi, K.; Karami, C.; Jahangiri, A. Preparation of modified nano-SiO_2 by bismuth and iron as a novel remover of methylene blue from water solution. *J. Mex. Chem. Soc.* **2017**, *61*, 250–259. [CrossRef]
47. Badr, Y.; Abd El-Wahed, M.G.; Mahmoud, M.A. Photocatalytic degradation of methyl red dye by silica nanoparticles. *J. Hazard. Mater.* **2008**, *154*, 245–253. [CrossRef]
48. Jadhav, S.A.; Garud, H.B.; Patil, A.H.; Patil, G.D.; Patil, C.R.; Dongale, T.D.; Patil, P.S. Recent advancements in silica nanoparticles based technologies for removal of dyes from water. *Colloid Interface Sci. Commun.* **2019**, *30*, 100181. [CrossRef]
49. Wang, X.; Han, S.; Zhang, Q.; Zhang, N.; Zhao, D. Photocatalytic oxidation degradation mechanism study of methylene blue dye waste water with GR/iTO_2. In Proceedings of the MATEC Web of Conferences, Warsaw, Poland, 5–7 October 2017; EDP Sciences: Les Ulis, France, 2018; Volume 238, p. 3006.

Article

Characteristics of Smoldering on Moist Rice Husk for Silica Production

Shengtai Yan [1], Dezheng Yin [1], Fang He [1,*], Junmeng Cai [2], Thomas Schliermann [3] and Frank Behrendt [1,4]

[1] School of Transportation and Vehicle Engineering, Shandong University of Technology, Zibo 255049, China; Shengtai_Yan@163.com (S.Y.); ydz2332398336@163.com (D.Y.); frank.behrendt@tu-berlin.de (F.B.)
[2] School of Agriculture and Biology, Shanghai Jiao Tong University, Shanghai 200240, China; jmcai@sjtu.edu.cn
[3] DBFZ Deutsches Biomasseforschungszentrum Gemeinnützige GmbH, 04347 Leipzig, Germany; thomas.schliermann@dbfz.de
[4] Institute of Energy Engineering, Technische Universität Berlin, 10623 Berlin, Germany
* Correspondence: fanghe916@daad-alumni.de

Citation: Yan, S.; Yin, D.; He, F.; Cai, J.; Schliermann, T.; Behrendt, F. Characteristics of Smoldering on Moist Rice Husk for Silica Production. *Sustainability* **2022**, *14*, 317. https://doi.org/10.3390/su14010317

Academic Editor: Alberto-Jesus Perea-Moreno

Received: 2 December 2021
Accepted: 24 December 2021
Published: 29 December 2021

Publisher's Note: MDPI stays neutral with regard to jurisdictional claims in published maps and institutional affiliations.

Copyright: © 2021 by the authors. Licensee MDPI, Basel, Switzerland. This article is an open access article distributed under the terms and conditions of the Creative Commons Attribution (CC BY) license (https://creativecommons.org/licenses/by/4.0/).

Abstract: In order to assess the possibility of silica production via smoldering of moist rice husk, experiments of washed (moist) rice husk (7 kg with moisture content of 51%) in a newly designed smoldering apparatus was performed. The temperature inside the fuel bed during smoldering was recorded, and characteristics of ash were analyzed. Results showed that the highest temperature in the middle of the naturally piled fuel bed was about 560.0 °C, lower than those in most of combustors. Some volatiles from the lower part of the fuel bed adhere to its upper ash during piled smoldering. Silica content and specific surface area of ash from smoldering of washed (moist) rice husk were 86.4% and 84.9 m^2/g, respectively. Compared to our experiments, they are close to smoldering of unwashed rice husk (89.0%, 67.7 m^2/g); different from muffle furnace burning (600 °C, 2 h) of washed (93.4%, 164.9 m^2/g) and un-washed (90.2%, 45.7 m^2/g) rice husk. The specific surface area is higher than those from most industrial methods (from 11.4 to 39.3 m^2/g). After some improvements, the smoldering process has great potential in mass product of high quality silica directly from moist rice husk.

Keywords: smoldering; rice husk; high moisture content; silica; specific surface area

1. Introduction

Smoldering is slow, low-temperature, and flameless burning of porous fuels, which is an important and complex phenomenon [1,2]. The application of it in the field of waste-to-energy conversion such as sludge treatment [3], recovery of resources from waste streams [4], and biomass energy conversion [5] has attracted lots of attention in recent years. The main advantages are its low temperature of the solid phase [6] and self-sustainability in a fuel bed with high moisture content (75–80 wt.%) [7]. From an environmental point of view, these characteristics avoid the ash-related slagging/corrosion [8], making nutrients recovery easy via recycling of ash directly to farms [9] and reducing the pollution of solid waste. As to energy consumption, it makes the complete burning of moist solid waste possible [10], reducing the energy consumption for drying fuel.

Rice husk is a typical biomass waste [11], accounting for 14–25% of the grain's overall mass [12]. In 2021, approximately 150 million tons of rice husk were produced around the world, with China contributing approximately 40 million tons. Nowadays most rice husk is directly buried or open burned [13] due to its low nutritive value for humans compared with rice grain and rice bran [14]. Direct burying results in soil pollution, because of its slow decomposition owing to its hard surface resulting from its high silicon and high lignin content [15]. Open burning leads to air pollution because of the release of fine dust and incomplete combustion gases of CO, NO_x, CH_4, poly-cyclic hydrocarbons (PAH) and soot [16].

There is great potential in producing silica from rice husk due to its high content of amorphous silica (around 18–23% [17]) and ash (around 85–95% [18]). Silica is an important inorganic material and is widely used in various fields such as fertilizer, insulator, adsorbent and catalyst [19]. It is characterized by high mechanical strength, good chemical stability, high-temperature resistance, easy dispersion in solvents, etc [20,21]. With the widespread application of silica, a variety of methods have been adopted to produce it, such as precipitation, plasma synthesis, chemical vapor deposition, micro emulsion processing, combustion synthesis and hydrothermal technique [22,23]. At present, most popular mass producing methods are precipitation from alkaline silicates and hydrothermal treatment of sand with lye [24]. However, both methods are expensive, intensive energy input, and environmentally harmful due to the production of dust, nitrogen and sulfur oxides, etc in the process of obtaining silica [25,26].

Producing silica from thermochemical conversion of rice husk has received considerable attention due to its economic and environmental advantages [27]. As to lab-scale production, Dizaji et al. [28] prepared silica by burning raw rice husk and pretreated rice husk (water washing at 50 °C for 2 h) in a muffle furnace at 600 °C for 4 h. The specific surface area was around 45.0 and 240.0 m^2/g, respectively. Abu Bakar et al. [29] prepared silica by burning rice husk (unleached/acid-leached) in a muffle furnace (600 °C for 2 h). The purity of silica from unleached and acid-leached rice husk was 95.8 and 99.6 wt.% (XRF results), respectively, and specific surface area was 116.0 and 218.0 m^2/g, respectively. Almeida et al. [30] prepared a mixture of silica and carbon by pyrolysis of raw rice husk in a tubular furnace. The obtained silica was black, in a mixture of amorphous and crystalline, with purity of 81.6 wt.% and specific surface area of 114.0 m^2/g. Schliermann et al. [31] obtained ashes produced from water washed (50 °C for 2 h) rice husk using ÖKOTHERM® furnaces. The ashes are post-treated with acid and then thermally treated at 650 °C using a muffle furnace. The specific surface area of silica is about 150–200 m^2/g. As to industrial production, Fernandes et al. [32] investigated characteristics of ash from burning rice husk in a grate furnace, a fluidized bed, and a suspension/entrained combustor. The silica content in these three types of ash was 90.0, 96.7, 93.6 wt.%, and the specific surface area was 39.3, 11.4, 26.7 m^2/g, respectively. The specific surface area of silica from mass production tends to be lower than that prepared in a laboratory, which may be related to none pretreatment of rice husk and the high burning temperature in industrial combustors. It is recorded that high combustion temperature results in the transformation of amorphous silica to crystalline material [33].

Pretreatment of rice husk is an effective way to increase the purity and specific surface area of silica and typical pretreatments for rice husk are acid-leaching and water-washing [29,34,35]. Moisture content of the treated rice husk is normally high. According to our pre-experiments, moisture content of rice husk is around 50% after washing. The moist rice husk is not suitable to be burned directly in a normal combustor (the moisture content in a fluidized bed combustor needs to be <35% [36], for a suspension burner <15% [37]). Smoldering might be a good choice for thermochemical conversion of moist rice husk directly to silica. Yet, to the best of our knowledge there are no experiments using smoldering in literature.

The objective of this study is to assess the possibility of silica production via smoldering of rice husk with high moisture content omitting a drying step before thermochemical conversion. A smoldering apparatus was designed and smoldering experiment of rice husk with high moisture content was performed. Temperature inside a fuel bed was recorded and characteristics of ash (silica content, specific surface area and mass loss characteristic) were analyzed.

2. Materials and Methods

2.1. Material

Rice husk was collected from rural southern China in 2021. Two types of rice husk (washed/unwashed) were chosen as raw material in the experiment. The washing process is the following: 3.5 kg of rice husk was put into a bucket (ϕ42 cm diameter × 40 cm height), then the bucket was filled with tap water (mass ration of water to rice husk: 10:1). The mixture was stirred for 10 min at ambient temperature. After being immersed for 12 h, rice husk was taken out and leaked above a screen in air for 1 h. The proximate and ultimate analysis of washed and unwashed rice husk were performed three times and the results were shown in Table 1.

Table 1. Proximate and elemental analysis of the rice husk. Oxygen is calculated by difference (C + H + O + N + S + Ash = 100 wt.%, dry basis).

	Proximate Analysis (Arrival Basis, wt.%)				Elemental Analysis (Dry Basis, wt.%)				
	Moisture	Volatile	Ash	Fixed Carbon	C	H	O	N	S
Washed	51.0 ± 0.2	32.0 ± 1.5	9.3 ± 0.3	7.7 ± 0.8	40.2 ± 0.3	5.5 ± 0.1	35.0 ± 0.2	0.3 ± 0.1	<0.1
Unwashed	8.1 ± 0.1	59.0 ± 1.4	18.4 ± 0.1	14.5 ± 0.7	38.7 ± 0.3	5.4 ± 0.1	35.5 ± 0.2	0.2 ± 0.1	<0.1

2.2. Experimental Set-Up

A batch smoldering apparatus consisting of three parts (a smoldering chamber, a gas burning chamber and a heat exchanger) was designed, as shown in Figure 1. The smoldering chamber is rectangular (inner size 60 × 30 × 30 cm) with an insulation layer (4 cm) outside of its inner wall. An air inlet (ϕ5 cm) and a flue gas outlet (ϕ5 cm) are on the left and the right wall, respectively. A hole (ϕ30 cm) with door on the top wall is used for material feeding and ash removal. The gas burning chamber is a square cavity (30 × 30 × 30 cm) with an insulation layer (4 cm) outside of its inner wall. Its inlet (ϕ5 cm) on the left wall is connected with the flue gas outlet of the smoldering chamber. A quartz glass tube inside the burner is used to introduce the incomplete combustion flue gas from the smoldering chamber to the bottom of the burner, then the gas is ignited using an igniter. The heat exchanger is a cylinder with a diameter of 30 and a height of 100 cm. The heat is transferred from high temperature flue gas through three tubes (diameter of 5 and a height of 40 cm) to the water outside them. Rice husk is smoldered in the smoldering chamber, smoke is burned out in the gas burning chamber, and flue gas is discharged into the air after cooled in the heat exchanger.

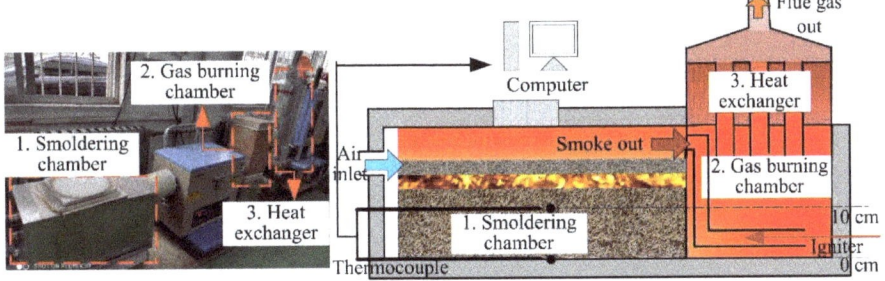

Figure 1. Photo and schematic of the smoldering apparatus.

2.3. Ash Preparation and Treatment

2.3.1. Ash Preparation from Rice Husk

For comparison, four types of ash—from washed/unwashed rice husk via smoldering and muffle furnace burning—were prepared. The washed and unwashed rice husk (3.5 kg of raw materials) were put naturally piled into the smoldering chamber. The height of the fuel beds was around 24 and 20 cm, and the bulk density were about 160 and 100 kg/m³, respectively. Then, it was ignited with a block of solid alcohol at the air inlet. Two thermocouples (ϕ1 mm, KMTXL-040-G) were put into the bottom and middle (10 cm from the bottom) layer of the fuel bed to monitor the change of temperature inside the fuel bed, and data were recorded every 60 s. It was found that a thin layer of upper ash was black due to the low temperature and the other part of ash was homogeneous after extinguishing and cooling of the fuel bed. A vacuum cleaner was used to remove the black ash on the upper layer and 200 g rice husk ash was taken out from the center of the piled residue every time.

About 3 g of washed and unwashed rice husk were put into an ash tray (6 × 3 × 2 cm) respectively, heated in a muffle furnace from ambient temperature to 600 °C at 10 °C/min and holding for 2 h. Air entered the muffle furnace through a 2–3 cm gap between the furnace door and the wall. Then the ash was taken out and cooled to ambient temperature in a desiccator for later use.

2.3.2. Grinding and Drying of Ash

For homogeneity in the subsequent measurements, the four types of ash were separately crushed to powder with a particle size <120 mesh using an agate mortar. After that, these powders were dried in an oven at 105 °C for 24 h.

2.4. Thermal and Physical Characterization of Ash

Mass loss characteristic was analyzed using a simultaneous thermogravimetric analysis (TGA DSC1, Mettler TOLEDO). To avoid corrosion to the instrument, a pair of crucibles (inner: alumina 50 µL, outer: platinum 70 µL) were used in experiments. About 8 mg of ash was put into the inner crucible and heated from 50 to 950 °C at a heating rate of 10 °C/min. An air flow rate of 200 mL/min was used as reactive gas and 20 mL/min of N2 was used as protective gas. Each experiment was repeated three times to check reproducibility.

The contents of silica and other elements in ash were triplicate and measured by Wavelength Dispersive X-Ray Fluorescence Spectrometer (ZXS100e, Rigaku Corporation) at room temperature. It should be noted that contents are only reliable for elements with atomic weight ≥23 [38]. Because of the data overflow of the results of carbon and boron, the results of them were deleted before calculation of the oxides' content.

Specific surface area was evaluated according to the Brunnauer, Emmett and Teller (BET) method and based on the nitrogen adsorption of the material at 77K. It was determined in the pressure range of p/p_0 = 0.05–0.3 [28], where p is the system pressure, and p_0 is the initial pressure (1 bar in this experiment). The measurement was conducted in a surface area analyzer (ASAP 2460, Micromeritics).

3. Results and Discussion

3.1. Characteristics of Smoldering Process

3.1.1. Temperature inside Fuel Bed

Temperature inside the washed (moist) and unwashed fuel bed is shown in Figure 2. The temperature history at one spot of the batch fuel bed can be divided into two stages—drying and oxidation. At the drying stage, the temperature first increases and then stabilizes at a temperature of about 60 °C. This is similar to the temperature of smoldering pine bark particle [39] and sewage sludge [4], which is different from the temperature (around 100 °C) of smoldering corn stalk powder [40] and corn flour [6]. Supplement experiments show this temperature is always around 60 °C in natural piled rice husk. In our experiments of smoldering branches, there is even no obvious plat temperature at the preheating period, which is similar to smoldering of unwashed rice husk. It implies temperature in the fuel bed at drying stage might be related to the porosity inside the fuel bed, materials, particle size, air flow, etc. The detailed analysis of this will be left for future work. At the oxidation stage, the temperature first increases rapidly and then increases at a stable rate. It drops quickly at the end stage of oxidation. For both cases the highest temperatures are around 560.0 °C, being much lower than those in most combustors (>700 °C) [41,42]. The low temperature is favorable to maintain the amorphous state of the silica [43].

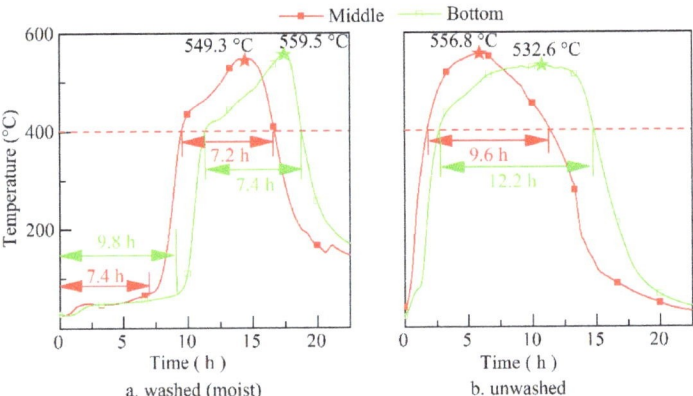

Figure 2. History of temperature of middle and bottom layer of washed (**a**) and unwashed (**b**) fuel bed (red and green pentacles for the maximum temperatures of middle and bottom layers, respectively).

Comparing the temperature development of the washed (moist) fuel bed with the unwashed fuel bed, the former has a longer duration of drying stage. The lower the part of the zone in the fuel bed, the longer the drying stage lasts. At oxidation stage, the duration of temperature > 400 °C of the moist fuel bed is shorter than for the unwashed fuel bed. This happens due to more heat generated in the process of smoldering is used to dry the moist rice husk and heat transfer rate to dry fuel is bigger than those for unwashed rice husk. The oxidation duration is also affected by bulk density of the fuel bed. In our supplementary experiments, smoldering of unwashed rice husk with bulk density of 170 kg/m^3 was performed in a small apparatus. It was found that the maximum temperature becomes higher (around 600 °C) than for naturally piled rice husk. The bigger bulk density decreases the porosity inside the fuel bed, reducing thermal dispersions [39]. Besides, another possible reason is that the dwell time of gaseous species is extended resulting in longer duration.

3.1.2. Absorption of Volatiles by the Upper Ash

A one-dimensional simplified illustration on temperature field of the whole fuel bed after formation of a thin layer of ash at top surface is shown in Figure 3. It was drawn according to the history of temperature inside the fuel bed, our previous experiments [40] and the characteristic temperature profile in a forward smoldering system in the literature [4]. Due to the longer drying time of the moist fuel bed than an unwashed one shown in Figure 2, the former has a thinner layer of high temperature area (reaction zone) than the latter after a short time, as illustrated in the curve of Figure 3. The amount of the unreacted rice husk (without pyrolysis) is proportional to the marked area of the left side. During their devolatilization, part of the volatiles can be absorbed by the upper ash due to its low temperature. As a result, the ash of the moist fuel bed has a higher tendency to absorb volatiles from its lower part than the ash of fuel with less moisture.

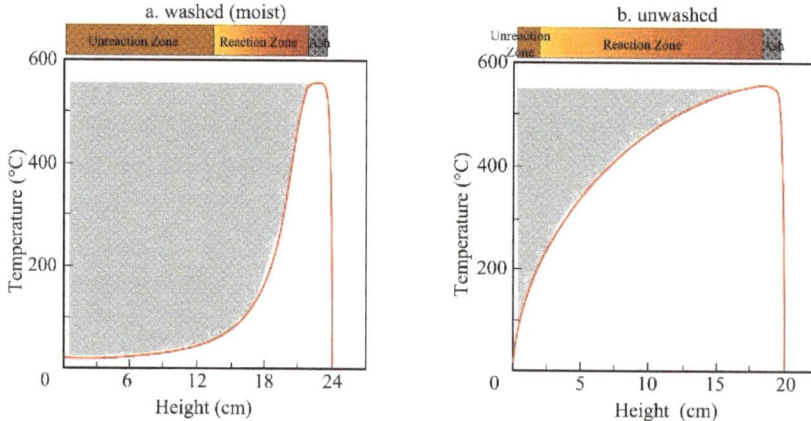

Figure 3. Schematic of spatial temperature distributions of washed (**a**) and unwashed (**b**) fuel bed.

3.2. Physical Properties and Mass Loss Characteristic of Rice Husk Ash

3.2.1. Physical Properties of Ash

Photos of ash before and after grinding from washed and unwashed smoldering as well as washed and unwashed burning are shown in Figure 4. These ashes are gray, soft, and almost retains the shape of rice husk itself. The whitest ash stems from washed burning, followed by unwashed burning. As for the other two ashes, the difference in the whiteness is negligible. The main reason is that temperature in smoldering (560 °C) is significantly lower than in muffle furnace (600 °C) and the duration of this temperature in smoldering is shorter. Other possible reasons are the removal of impurities like dust by washing or absorption of volatiles by the upper ash. The higher the whiteness, the higher silica content in ash [44]. It is worth mentioning that there are always some black particles in the ash. These might be the rice husk with incomplete combustion. Bridge forming in the fuel bed that makes the cooling of the related particle faster than the dense piled should be the reason.

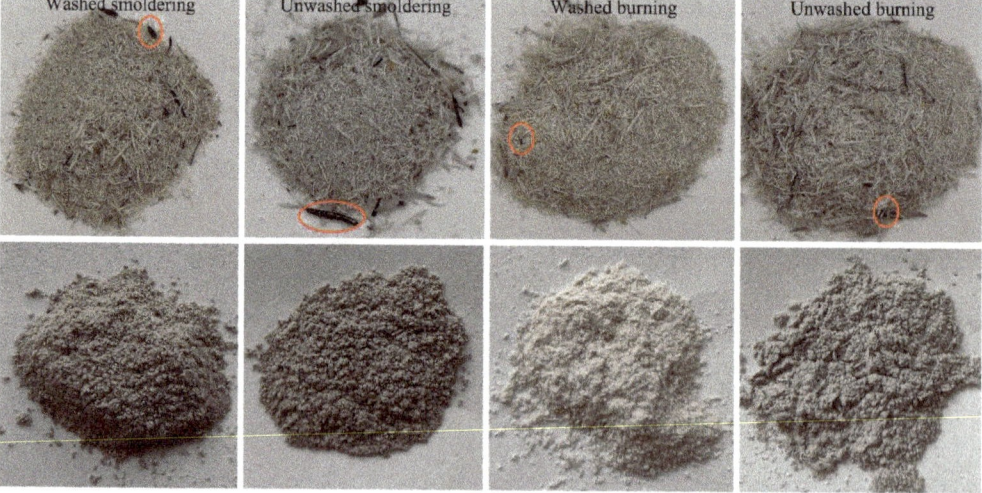

Figure 4. Photos of four types of rice husk ash (circles show black particles in ash).

3.2.2. Mass Loss Characteristic of Rice Husk Ash

Thermogravimetric (TG) and derivative thermogravimetric (DTG) curves of four types of ash (washed and unwashed smoldering, washed and unwashed burning) are shown in Figure 5. It is seen from TG data that total mass-loss of four ashes is <5%. The mass loss from 50 to 950 °C for the above four types of ash are 4.2 wt.%, 3.1 wt.%, 2.5 wt.%, and 2.3 wt.%, respectively. The lower the combustion temperature and oxidation duration, the higher total mass loss.

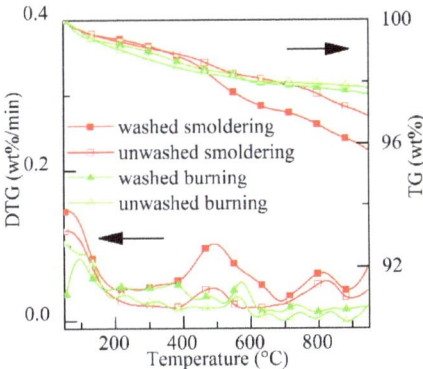

Figure 5. TG & DTG curves of four types of ash.

As shown in the DTG curves, there are three stages of mass loss: drying, oxidation, and combustion of residual carbon. Mass loss at each stage is shown in Table 2. At the drying stage (<200 °C), more water and longer drying time are there for burning ash than those for smoldering ash. This implies that the absorption of condensed materials stated in Section 3.1.2 decreases the capability of moisture absorption of ash. At the oxidation stage (200–700 °C), more mass is lost at lower temperatures for the two smoldering ashes than those for the burning ashes. For smoldering ashes, especially for smoldering of moist rice husk, mass loss occurs in the range 400–560 °C. In theory, the mass in this range is burned out during smoldering due to the maximum temperature inside the fuel bed is around 560 °C. The mass loss in this range indicates the oxidation of volatiles and carbon. At the stage of combustion of residual carbon (>700 °C), the mass loss of ash from smoldering is higher compared with burning ash due to more carbon in ash. The reason for the formation of residual carbon is that the melted silica obstructs the transport of oxygen to carbon [45]. At a higher temperature, the residual carbon can be burned out. Besides, the mass loss at this stage might also relate to the evaporation of KCl [38], the decomposition of carbonates [42].

Table 2. Mass loss of different stages (wt.%).

	Drying	Oxidation	Combustion of Resdiual Carbon
Washed smoldering	0.6 ± 0.1	2.4 ± 0.2	1.2 ± 0.1
Unwashed smoldering	0.7 ± 0.1	1.4 ± 0.1	1.0 ± 0.1
Washed burning	0.9 ± 0.1	1.3 ± 0.1	0.3 ± 0.1
Unwashed burning	0.8 ± 0.2	1.3 ± 0.1	0.2 ± 0.1

3.3. Silica Content in Rice Husk Ash

3.3.1. Reproducibility and Reliability of XRF Measurement

Main compositions (>0.5%) from triplicate XRF measurements of ash produced by smoldering of washed rice husk are shown in Table 3. It is seen that the reproducibility is good (relative error < 10%). As to reliability, if there are elements with atomic weights <23, the absolute contents of element are not reliable as pointed in Section 2.4. For smoldering ash, residue carbon and absorbed volatiles affect this measured elemental content.

Table 3. Contents of main elements in ash from triplicate measurements (wt.%).

NO.	SiO_2	K_2O	CaO	SO_3	P_2O_5	MgO	Cl	Fe_2O_3	Al_2O_3
1	86.6	4.16	2.61	1.25	1.15	1.08	0.78	0.73	0.72
2	86.4	4.20	2.70	1.26	1.23	1.14	0.81	0.71	0.70
3	86.2	4.20	2.76	1.26	1.22	1.15	0.90	0.70	0.69
Ave	86.4 ± 0.20	4.19 ± 0.03	2.69 ± 0.08	1.26 ± 0.01	1.20 ± 0.05	1.12 ± 0.04	0.83 ± 0.07	0.71 ± 0.02	0.70 ± 0.02

3.3.2. Content of Silica and Other Main Compositions in Rice Husk Ash

The contents of SiO_2 and others main compositions in the four types of ash are listed in Table 4. The main component is SiO_2 and the contents of it in all ashes is >85%. The SiO_2 content in descending order is washed burning (93.4%), unwashed burning (90.2%), unwashed smoldering (89.0%), and washed smoldering (86.4%). Contents of other compositions in descending order of the content are K_2O, CaO, SO_3, P_2O_5, MgO, Cl, Fe_2O_3, Al_2O_3 in smoldering ash, and this order holds for most elements in other ashes.

Table 4. Main compositions in 4 types of ash.

Types	SiO_2	K_2O	CaO	SO_3	P_2O_5	MgO	Cl	Fe_2O_3	Al_2O_3
Washed smoldering	86.4	4.19	2.69	1.26	1.20	1.12	0.83	0.71	0.70
Unwashed smoldering	89.0	4.46	1.34	0.75	1.17	0.91	0.75	0.49	0.41
Washed burning	93.4	1.07	2.08	0.58	0.50	0.70	0.30	0.39	0.21
Unwashed burning	90.2	4.12	1.22	0.82	0.94	0.83	0.61	0.36	0.19

3.3.3. Effect of Production Method on Silica Content

The purity of silica in ash is affected by three main factors: absorption of volatiles by upper ash, combustion temperature and pretreatment of rice husk. Volatiles adhering to the surface of upper-ash smoldering decreases the SiO_2 content. Incomplete burn out of solid organics at low combustion temperature also decreases SiO_2 content in ash. The pretreatment way of washing can remove some water soluble inorganics, such as K, Cl and dust [46,47]. The removal of water soluble inorganics can increase the silica content [35]. According to Table 4, the measured silica content in ash of washed smoldering is similar to or lower than that in unwashed smoldering, while the content of water-soluble ions (Ca, S, P, Mg, Cl, Fe, Al) is higher than that in unwashed smoldering. The lower silica content for washed smoldering is related to the shorter oxidation duration in the moist fuel bed, which results in incomplete combustion of rice husk. The higher content of water-soluble ions might relate to the inaccuracy of XRF measurement as mentioned in Section 3.3.1. The removing of carbon results in an increase of those water-soluble ions contents as a percentage of the whole ash.

3.4. BET Specific Surface Area

3.4.1. Specific Surface Area of the Four Types of Rice Husk Ash

The BET specific surface area of ash produced in this study (washed and unwashed smoldering, washed and unwashed burning) and in literature is shown from Figure 6. Two characteristics can be seen from this data of this study: (1) The specific surface area of ash prepared from washed rice husk is higher than that from unwashed rice husk; (2) for the ash of prepared from washed rice husk, the specific surface area is lower when smoldering is used compared with burning in the muffle furnace. However, the situation is opposite for the ash prepared from unwashed rice husk.

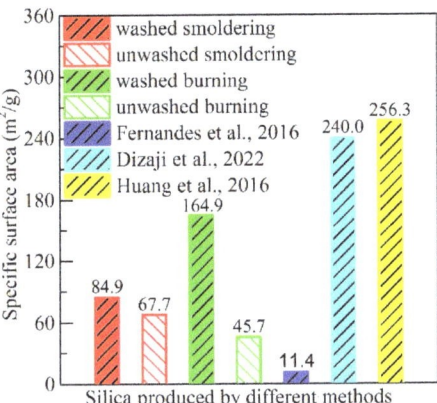

Figure 6. Specific surface area of silica produced by different methods. ([32]: no-pretreatment rice husk burned in a fluidized bed combustor; [28]: washed rice husk (50 °C tap water for 2 h) burned in a muffle furnace (600 °C for 4 h); [48]: citric acid-leached rice husk (1 wt.% at 80 °C for 3 h) burned in a muffle furnace (700 °C for 2 h)).

3.4.2. Factors of Specific Surface Area

It can be seen from the above characteristics that the specific surface area is affected by pretreatment conditions and combustion temperature. Pretreatment, such as washing, removes part of potassium. This decreases the formation of eutectic from interaction of K and Si. The decrease in amount of the eutectic partly avoids the transformation from amorphous silica to crystalline via melting in eutectic and condensing in cooling stage, and consequently increases the specific surface area [49]. Pretreatment of washing also removes soil particles which normally have lower specific surface area than amorphous silica.

As to temperature, low temperature avoids sintering/eutectic melting of the mixed components in ash of rice husk and is beneficial for silica to maintain its amorphous state and high specific surface area [50]. There is a combination effect of the two factors. It is very hard to get high-specific-surface-area ash from burning original rice husk at a high temperature (>700 °C).

3.4.3. Comparison of Specific Surface Area in This Study with Those of Silica Prepared Using Methods in Literature

The specific surface area of ash produced by smoldering of washed rice husk is 84.9 m^2/g, which is lower than those prepared in the laboratory (99.2–293.9 m^2/g) [13,29,51], but higher than those produced in the industry (11.4–39.3 m^2/g) [32,52], as shown in Figure 6. In the laboratory, rice husk is normally washed or leached using water and acid to remove alkali and alkaline earth metals, such as the experiments performed by Dizaji [28] and Huang [48]. In the industry, no-pretreatment rice husk is burned directly in combustors. The high temperature (>700 °C) of most combustor is not suitable to produce silica with high specific surface area.

3.5. Supplementary Experiment of Smoldering Air-Dried Rice Husk after Washing

To decrease the absorption of volatiles by upper ash, a supplementary smoldering experiment using air dried rice husk after washing was performed in the experimental set-up. It was found that the silica content is 93.5%, and specific surface area is 145.9 m^2/g. Both the purity and specific surface are significantly higher than those (86.4%, 84.9 m^2/g) from moist rice husk. It shows that drying before smoldering does increase both silica content and specific surface area.

4. Potential of Mass Product of Silica from Smoldering of Rice Husk

4.1. Measures to Increase Silica Content and Specific Surface Area

As mentioned before, the purity and specific surface area of silica produced from smoldering of rice husk are affected by pretreatment of the raw material, the volatile absorption of ash in reactor, and the solid temperature and dwell time in the fuel bed. According to the above experience in smoldering, two measures are proposed here to improve silica properties. The first is to develop a lateral continuously smoldering involving a dry stage apparatus. The second is to supply a little amount of air to improve fuel bed temperature slightly.

4.1.1. Lateral Continuously Smoldering Involving a Dry Stage of Rice Husk

A continuously lateral propagation smoldering is proposed here as shown in Figure 7. The smoldering of rice husk can be divided into three stages using a grate: drying, pyrolysis and oxidation. Due to its lateral propagation, the volatiles generated in the pyrolysis stage are discharged into the gas burning chamber directly and burned out, which cannot adhere to the ash in oxidation stage. The heat generated by oxidation can be used for the drying of rice husk. The condensate water generated in the process of drying can be collected and used for the washing of rice husk.

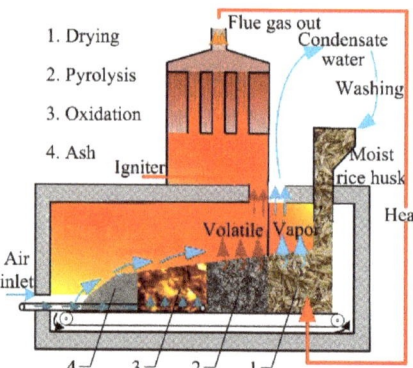

Figure 7. Schematic diagram of lateral continuously smoldering.

4.1.2. The Supply of Air at the Oxidation Stage

To increase the combustion temperature, a small amount of air can be provided in the stage of oxidation. As mentioned before, temperature of 700 °C can increase the purity and specific surface area. The highest temperature of the current piled smoldering is around 600 °C, which can be increased by supplying a small amount of air.

4.2. Feasibility of Mass Product of Silica from Smoldering of Rice Husk

Although silica prepared in the laboratory from pretreated rice husk has a high silica content and high specific surface area, considerable energy is necessary for removal of the water from moist rice husk. Besides, a lot of waste acid and lye is produced during pretreatment and post-treatment, which is harmful to the environment. As to a traditional combustor, the high temperature of the solid during combustion is disadvantageous in maintaining a high specific surface area of silica. Besides, most combustors are not suitable for burning rice husk with high moisture content.

Smoldering is characterized by a low temperature in the fuel bed and is self-sustained with moist content. Our experiments show that the highest silica content and specific surface area of ash is from smoldering of air-dried-washed rice husk. It indicates that after some improvement of our experimental set-up, it has great potential in the mass production of high-quality silica directly from moist rice husk.

5. Conclusions

Smoldering experiments with moist rice husk were performed in a self-designed apparatus to check its possibility for silica production. Temperature inside the fuel bed and silica content/specific surface area/mass-loss-characteristics of ash were analyzed. The main conclusions are:

1. Smoldering is a novel method for producing silica from rice husk. In our experimental set-up with a smoldering chamber, a gas burning chamber and a heat exchanger, the smoldering was self-sustained for naturally-piled rice husk with moisture content of 51.0%. Incomplete-combustion smoldering gas can be burned out in a gas burning chamber, and the heat generated during smoldering can be used for heating.
2. The highest temperature inside the fuel bed is around 560 °C, which was lower than those in many combustors. In the piled smoldering process of rice husk, part of the volatiles will be absorbed by the upper ash, which can decrease the silica content and specific surface area of ash. It should be avoided in the future.
3. Specific surface area of ash prepared from washed (moist) rice husk by smoldering was lower than those prepared in the laboratory, but it is higher than those produced from most industrial methods. It was greatly improved by air drying of moist rice husk before smoldering.
4. In future, a lateral continuously smoldering scheme involving a drying stage and small amount of air supply can be used for silica production. Smoldering of rice husk shows great potential for the industrial production of high-quality silica.

Author Contributions: Conceptualization, F.H.; methodology, F.H. and J.C.; software, S.Y. and F.H.; investigation, T.S.; data curation, T.S.; writing—original draft preparation, T.S.; writing—review and editing, D.Y., F.H., J.C., T.S. and F.B.; supervision, F.H.; funding acquisition, F.H. and F.B. All authors have read and agreed to the published version of the manuscript.

Funding: This research was funded by [National Natural Science Foundation of China] grant number [51676115] and [Sino-German Center for Research Promotion] grant number [M-0183].

Institutional Review Board Statement: Exclude this statement.

Informed Consent Statement: Not applicable.

Data Availability Statement: Data are available upon request.

Conflicts of Interest: The authors declare no conflict of interest.

References

1. Torero, J.L.; Gerhard, J.I.; Martins, M.F.; Zanoni, M.A.B.; Rashwan, T.L.; Brown, J.K. Processes defining smoldering combustion: Integrated review and synthesis. *Prog. Energy Combust. Sci.* **2020**, *81*, 100869. [CrossRef]
2. Lin, S.; Chow, T.H.; Huang, X. Smoldering propagation and blow-off on consolidated fuel under external airflow. *Combust. Flame* **2021**, *234*, 111685. [CrossRef]
3. Zhao, C.; Li, Y.; Gan, Z.; Nie, M. Method of smoldering combustion for refinery oil sludge treatment. *J. Hazard. Mater.* **2021**, *409*, 124995. [CrossRef] [PubMed]
4. Rashwan, T.L.; Fournie, T.; Torero, J.L.; Grant, G.P.; Gerhard, J.I. Scaling up self-sustained smoldering of sewage sludge for waste-to-energy. *Waste Manag.* **2021**, *135*, 298–308. [CrossRef]
5. Rashwan, T.L.; Torero, J.L.; Gerhard, J.I. The improved energy efficiency of applied smoldering systems with increasing scale. *Int. J. Heat Mass Transf.* **2021**, *177*, 121548. [CrossRef]
6. Rosa, A.; Hammad, A.W.A.; Qualharini, E.; Vazquez, E.; Haddad, A. Smoldering fire propagation in corn grain: An experimental study. *Results Eng.* **2020**, *7*, 100151. [CrossRef]
7. Wyn, H.K.; Konarova, M.; Beltramini, J.; Perkins, G.; Yermán, L. Self-sustaining smoldering combustion of waste: A review on applications, key parameters and potential resource recovery. *Fuel Process. Technol.* **2020**, *205*, 106425. [CrossRef]
8. Mitchell, E.J.S.; Gudka, B.; Whittaker, C.; Shield, I.; Price-Allison, A.; Maxwell, D.; Jones, J.M.; Williams, A. The use of agricultural residues, wood briquettes and logs for small-scale domestic heating. *Fuel Process. Technol.* **2020**, *210*, 106552. [CrossRef]
9. Stávková, J.; Maroušek, J. Novel sorbent shows promising financial results on P recovery from sludge water. *Chemosphere* **2021**, *276*, 130097. [CrossRef]
10. Feng, C.; Huang, J.; Yang, C.; Li, C.; Luo, X.; Gao, X.; Qiao, Y. Smoldering combustion of sewage sludge: Volumetric scale-up, product characterization, and economic analysis. *Fuel* **2021**, *305*, 121485. [CrossRef]

11. Silva, L.A.; Santos, I.F.S.d.; Machado, G.d.O.; Tiago Filho, G.L.; Barros, R.M. Rice husk energy production in Brazil: An economic and energy extensive analysis. *J. Clean. Prod.* **2021**, *290*, 125188. [CrossRef]
12. Quispe, I.; Navia, R.; Kahhat, R. Energy potential from rice husk through direct combustion and fast pyrolysis: A review. *Waste Manag.* **2017**, *59*, 200–210. [CrossRef] [PubMed]
13. Santana Costa, J.A.; Paranhos, C.M. Systematic evaluation of amorphous silica production from rice husk ashes. *J. Clean. Prod.* **2018**, *192*, 688–697. [CrossRef]
14. Azat, S.; Korobeinyk, A.V.; Moustakas, K.; Inglezakis, V.J. Sustainable production of pure silica from rice husk waste in Kazakhstan. *J. Clean. Prod.* **2019**, *217*, 352–359. [CrossRef]
15. Zhang, H.; Ding, X.; Chen, X.; Ma, Y.; Wang, Z.; Zhao, X. A new method of utilizing rice husk: Consecutively preparing D-xylose, organosolv lignin, ethanol and amorphous superfine silica. *J. Hazard. Mater.* **2015**, *291*, 65–73. [CrossRef]
16. Luu, L.Q.; Halog, A. Rice Husk Based Bioelectricity vs. Coal-fired Electricity: Life Cycle Sustainability Assessment Case Study in Vietnam. *Procedia CIRP* **2016**, *40*, 73–78. [CrossRef]
17. Téllez, J.F.; Silva, M.P.; Simister, R.; Gomez, L.D.; Fuertes, V.C.; De Paoli, J.M.; Moyano, E.L. Fast pyrolysis of rice husk under vacuum conditions to produce levoglucosan. *J. Anal. Appl. Pyrolysis* **2021**, *156*, 105105. [CrossRef]
18. Siddika, A.; Mamun, M.A.A.; Alyousef, R.; Mohammadhosseini, H. State-of-the-art-review on rice husk ash: A supplementary cementitious material in concrete. *J. King Saud Univ.-Eng. Sci.* **2020**, *33*, 294–307. [CrossRef]
19. Adam, F.; Appaturi, J.N.; Iqbal, A. The utilization of rice husk silica as a catalyst: Review and recent progress. *Catal. Today* **2012**, *190*, 2–14. [CrossRef]
20. Ma, M.; Li, H.; Xiong, Y.; Dong, F. Rational design, synthesis, and application of silica/graphene-based nanocomposite: A review. *Mater. Des.* **2021**, *198*, 109367. [CrossRef]
21. Lei, Q.; Guo, J.; Noureddine, A.; Wang, A.; Wuttke, S.; Brinker, C.J.; Zhu, W. Sol-Gel-Based Advanced Porous Silica Materials for Biomedical Applications. *Adv. Funct. Mater.* **2020**, *30*, 1909539. [CrossRef]
22. Soltani, N.; Bahrami, A.; Pech-Canul, M.I.; González, L.A. Review on the physicochemical treatments of rice husk for production of advanced materials. *Chem. Eng. J.* **2015**, *264*, 899–935. [CrossRef]
23. Goodman, B.A. Utilization of waste straw and husks from rice production: A review. *J. Bioresour. Bioprod.* **2020**, *5*, 143–162. [CrossRef]
24. Beidaghy Dizaji, H.; Zeng, T.; Hartmann, I.; Enke, D.; Schliermann, T.; Lenz, V.; Bidabadi, M. Generation of High Quality Biogenic Silica by Combustion of Rice Husk and Rice Straw Combined with Pre- and Post-Treatment Strategies—A Review. *Appl. Sci.* **2019**, *9*, 1083. [CrossRef]
25. Tchakouté, H.K.; Rüscher, C.H.; Kong, S.; Ranjbar, N. Synthesis of sodium waterglass from white rice husk ash as an activator to produce metakaolin-based geopolymer cements. *J. Build. Eng.* **2016**, *6*, 252–261. [CrossRef]
26. Pfeiffer, T.; Enke, D.; Roth, R.; Roggendorf, H. Hydrothermal Dissolution of Opal in Sodium Hydroxide Lyes for the Synthesis of Water Glass. *Adv. Chem. Eng. Sci.* **2017**, *7*, 76–90. [CrossRef]
27. Schneider, D.; Wassersleben, S.; Weiß, M.; Denecke, R.; Stark, A.; Enke, D. A Generalized Procedure for the Production of High-Grade, Porous Biogenic Silica. *Waste Biomass Valorization* **2018**, *11*, 1–15. [CrossRef]
28. Beidaghy Dizaji, H.; Zeng, T.; Hölzig, H.; Bauer, J.; Klöß, G.; Enke, D. Ash transformation mechanism during combustion of rice husk and rice straw. *Fuel* **2022**, *307*, 121768. [CrossRef]
29. Bakar, R.A.; Yahya, R.; Gan, S.N. Production of High Purity Amorphous Silica from Rice Husk. *Procedia Chem.* **2016**, *19*, 189–195. [CrossRef]
30. Almeida, S.R.; Elicker, C.; Vieira, B.M.; Cabral, T.H.; Silva, A.F.; Sanches Filho, P.J.; Raubach, C.W.; Hartwig, C.A.; Mesko, M.F.; Moreira, M.L.; et al. Black SiO_2 nanoparticles obtained by pyrolysis of rice husk. *Dye. Pigment.* **2019**, *164*, 272–278. [CrossRef]
31. Schliermann, T.; Hartmann, I.; Dizaji, H.B.; Zeng, T.; Schneider, D.; Wassersleben, S.; Enke, D.; Jobst, T.; Lange, A.; Roelofs, F.; et al. High quality biogenic silica from combined energetic and material utilization of agricultural residues. In Proceedings of the 7th International Symposium of Energy from Biomass and Waste, Venice, Italy, 15–18 October 2018.
32. Fernandes, I.J.; Calheiro, D.; Kieling, A.G.; Moraes, C.A.M.; Rocha, T.L.A.C.; Brehm, F.A.; Modolo, R.C.E. Characterization of rice husk ash produced using different biomass combustion techniques for energy. *Fuel* **2016**, *165*, 351–359. [CrossRef]
33. Aprianti, E.; Shafigh, P.; Bahri, S.; Farahani, J.N. Supplementary cementitious materials origin from agricultural wastes—A review. *Constr. Build. Mater.* **2015**, *74*, 176–187. [CrossRef]
34. Fernandes, I.J.; Calheiro, D.; Sánchez, F.A.L.; Camacho, A.L.D.; Rocha, T.L.A.d.C.; Moraes, C.A.M.; Sousa, V.C.d. Characterization of Silica Produced from Rice Husk Ash: Comparison of Purification and Processing Methods. *Mater. Res.* **2017**, *20*, 512–518. [CrossRef]
35. Pa, F.C.; Kein, W.K. Removal of iron in rice husk via oxalic acid leaching process. *IOP Conf. Ser. Mater. Sci. Eng.* **2019**, *701*, 012021. [CrossRef]
36. Kuprianov, V.I.; Kaewklum, R.; Sirisomboon, K.; Arromdee, P.; Chakritthakul, S. Combustion and emission characteristics of a swirling fluidized-bed combustor burning moisturized rice husk. *Appl. Energy* **2010**, *87*, 2899–2906. [CrossRef]
37. Werther, J.; Saenger, M.; Hartge, E.-U.; Ogada, T.; Siagi, Z. Combustion of agricultural residues. *Prog. Energy Combust. Sci.* **2000**, *26*, 1–27. [CrossRef]
38. Li, X.; He, F.; Behrendt, F.; Gao, Z.; Shi, J.; Li, C. Inhibition of K_2SO_4 on evaporation of KCl in combustion of herbaceous biomass. *Fuel* **2021**, *289*, 119754. [CrossRef]

39. Gianfelice, G.; Della Zassa, M.; Biasin, A.; Canu, P. Onset and propagation of smouldering in pine bark controlled by addition of inert solids. *Renew. Energy* **2019**, *132*, 596–614. [CrossRef]
40. He, F.; Yi, W.; Li, Y.; Zha, J.; Luo, B. Effects of fuel properties on the natural downward smoldering of piled biomass powder: Experimental investigation. *Biomass Bioenergy* **2014**, *67*, 288–296. [CrossRef]
41. Faé Gomes, G.M.; Philipssen, C.; Bard, E.K.; Zen, L.D.; de Souza, G. Rice husk bubbling fluidized bed combustion for amorphous silica synthesis. *J. Environ. Chem. Eng.* **2016**, *4*, 2278–2290. [CrossRef]
42. Modolo, R.C.E.; Silva, T.; Senff, L.; Tarelho, L.A.C.; Labrincha, J.A.; Ferreira, V.M.; Silva, L. Bottom ash from biomass combustion in BFB and its use in adhesive-mortars. *Fuel Process. Technol.* **2015**, *129*, 192–202. [CrossRef]
43. Mlonka-Mędrala, A.; Magdziarz, A.; Gajek, M.; Nowińska, K.; Nowak, W. Alkali metals association in biomass and their impact on ash melting behaviour. *Fuel* **2020**, *261*, 116421. [CrossRef]
44. Gu, S.; Zhou, J.; Yu, C.; Luo, Z.; Wang, Q.; Shi, Z. A novel two-staged thermal synthesis method of generating nanosilica from rice husk via pre-pyrolysis combined with calcination. *Ind. Crops Prod.* **2015**, *65*, 1–6. [CrossRef]
45. Krishnarao, R.V.; Subrahmanyam, J.; Kumar, T.J. Studies on the formation of black particles in rice husk silica ash. *J. Eur. Ceram. Soc.* **2001**, *21*, 99–104. [CrossRef]
46. Bandara, Y.W.; Gamage, P.; Gunarathne, D.S. Hot water washing of rice husk for ash removal: The effect of washing temperature, washing time and particle size. *Renew. Energy* **2020**, *153*, 646–652. [CrossRef]
47. He, F.; Li, X.; Behrendt, F.; Schliermann, T.; Shi, J.; Liu, Y. Critical changes of inorganics during combustion of herbaceous biomass displayed in its water soluble fractions. *Fuel Process. Technol.* **2020**, *198*, 106231. [CrossRef]
48. Huang, M.; Cao, J.; Meng, X.; Liu, Y.; Ke, W.; Wang, J.; Sun, L. Preparation of SiO_2 nanowires from rice husks by hydrothermal method and the RNA purification performance. *Chem. Phys. Lett.* **2016**, *662*, 42–46. [CrossRef]
49. Mirmohamadsadeghi, S.; Karimi, K. Recovery of silica from rice straw and husk. In *Current Developments in Biotechnology and Bioengineering*; Elsevier: Amsterdam, The Netherlands, 2020; pp. 411–433. [CrossRef]
50. Prasara-A, J.; Gheewala, S.H. Sustainable utilization of rice husk ash from power plants: A review. *J. Clean. Prod.* **2017**, *167*, 1020–1028. [CrossRef]
51. Lee, J.H.; Kwon, J.H.; Lee, J.-W.; Lee, H.-S.; Chang, J.H.; Sang, B.-I. Preparation of high purity silica originated from rice husks by chemically removing metallic impurities. *J. Ind. Eng. Chem.* **2017**, *50*, 79–85. [CrossRef]
52. Blissett, R.; Sommerville, R.; Rowson, N.; Jones, J.; Laughlin, B. Valorisation of rice husks using a TORBED® combustion process. *Fuel Process. Technol.* **2017**, *159*, 247–255. [CrossRef]

Article

Biomethane from Manure, Agricultural Residues and Biowaste—GHG Mitigation Potential from Residue-Based Biomethane in the European Transport Sector

Katja Oehmichen [1,*], Stefan Majer [1] and Daniela Thrän [2]

[1] Deutsches Biomasseforschungszentrum gGmbH (DBFZ), 04347 Leipzig, Germany; stefan.maje@dbfz.de
[2] Helmholtz Centre for Environmental Research, 04347 Leipzig, Germany; daniela.thraen@ufz.de
* Correspondence: katja.oehmichen@dbfz.de; Tel.: +49-341-2343-717

Abstract: Biomethane from manure, agricultural residues, and biowaste has been prioritized by many energy strategies as a sustainable way to decrease greenhouse gas (GHG) emissions in the transport sector. The technology is regarded as mature; however, its implementation is still at an early stage. At EU level, there are currently two major instruments relevant for promoting the production of biomethane from waste and residues and which are likely to contribute to unlocking unused GHG mitigation potentials: the Renewable Energy Directive 2018/2001 (RED II) and the European Emission Trading System (EU ETS). Our study analyzes the effects of these two instruments on the competitiveness of biomethane as an advanced transport fuel in relation to different policy scenarios within the RED II framework and under EU ETS conditions. Within the RED II market framework for advanced biofuels, biomethane concepts that use manure as a substrate or as a cosubstrate show significantly lower GHG mitigation costs compared to advanced biofuels. With respect to the current EU ETS conditions for bioenergy, it is helpful to consider the GHG reduction potential from the non-ETS agricultural sector as a way to unlock unused potential for reducing GHG emissions.

Keywords: biogas; lifecycle assessment; greenhouse gas emissions; mitigation potential; GHG mitigation costs; manure; biomethane; RED II; EU ETS

Citation: Oehmichen, K.; Majer, S.; Thrän, D. Biomethane from Manure, Agricultural Residues and Biowaste—GHG Mitigation Potential from Residue-Based Biomethane in the European Transport Sector. *Sustainability* **2021**, *13*, 14007. https://doi.org/10.3390/su132414007

Academic Editor: Mohammad Aslam Khan Khalil

Received: 27 October 2021
Accepted: 10 December 2021
Published: 18 December 2021

Publisher's Note: MDPI stays neutral with regard to jurisdictional claims in published maps and institutional affiliations.

Copyright: © 2021 by the authors. Licensee MDPI, Basel, Switzerland. This article is an open access article distributed under the terms and conditions of the Creative Commons Attribution (CC BY) license (https://creativecommons.org/licenses/by/4.0/).

1. Introduction

Driven by ambitious GHG reduction targets, the European Union has created different political instruments to promote renewable energy [1–3]. Biomethane from animal manure, agricultural residues, such as straw, and biowaste is regarded by different policy strategies as a sustainable way to decrease greenhouse gas (GHG) emissions in the energy and transport sectors [4,5]. In particular, the use of animal excrement in anaerobic digestion plants is thought to contribute significantly to reducing GHG emissions because of improved manure management [6]. When manure is conventionally stored in open storage systems on farms, it inevitably produces GHG emissions [7]. By introducing manure as a substrate in a biogas plant early on, these GHG emissions can largely be avoided [8].

1.1. Background

In addition to replacing fossil fuels, the production of biomethane from animal manure achieves additional savings that are increasing in importance [9]. Nevertheless, there remains a vast amount of unlocked GHG mitigation potential by using animal manure [10–12].

There are currently two major instruments at EU level that are (or can be) relevant for the promotion of biomethane from waste and residues.

Firstly, the share of renewables in the EU transport sector is (and will be) mainly driven by the targets defined in the Renewable Energy Directive 2009/28/EC (EU RED) and the follow-up Directive 2018/2001 (RED II). RED II defines the market size for renewables in the EU transport sector, in particular for advanced fuels such as biomethane from waste and residues. Because of discussions on the sustainability of bioenergy and the resulting introduction of several sustainability criteria as part of the directive, the GHG mitigation potential of biofuels and bioenergy carriers has increased in significance and has become an important criterion for a biofuel's success on the market. With the introduction of a cap on biofuels from food and feed crops and the introduction of a subtarget for advanced biofuels, the 2015 version of the RED and the follow-up directive RED II have triggered a shift towards the increased use of biofuels from residues and waste materials.

The maximum proportion of biofuels produced from food and feed crops will be frozen at 2020 consumption levels, plus an additional 1% (with a maximum cap of 7%) for road and rail transport fuel. At the same time, advanced fuels must supply a minimum of 0.2% of transport energy in 2022, 1% in 2025, and at least 3.5% (double-counted) by 2030. The revised Renewable Energy Directive (revised RED II), which results from the changes to the proposal of the European Commission and the Council [13], increases the overall target for the share of renewable energies in gross final energy consumption in the European Union from 32% to 40% in 2030. Furthermore, it proposes to increase the share of sustainable advanced biofuels from at least 0.2% in 2022 to 0.5% in 2025 and 2.2% in 2030 (compared to 1.75% without double-counting in the current RED II).

Assuming that the RED II (and Red II revision) subtarget will create a "protected market" for advanced fuels, several technological pathways suitable for converting the feedstocks listed in Annex IX, Part A of RED II will compete within this market as described above. Thus, the first step in evaluating the competitiveness of biomethane as an advanced biofuel is to assess potential competitors within the advanced fuel subtarget market. With respect to the characteristics of the qualified feedstock, technology options such as biomethane or the production of synthetic fuels seem to be appropriate. Consequently, we consider in our analysis two alternative, market-ready advanced fuels as potential competitors to biomethane.

1.2. Scope of the Study

The aim of the second instrument, the EU Emission Trading System (EU ETS), is to increase the cost of using carbon-intense energy carriers. In theory, raising the cost of using carbon-intense energy carriers would bring these costs more in line with the production costs for renewable energy carriers (which are typically associated with higher costs and lower GHG emissions). The low price for CO_2 certificates means that certificate trading is currently not an effective way to increase the cost of CO_2-intensive energy carriers and thus to create a balance between the provision of energy from biomass and fossil energy supply. Additionally, the EU ETS is not applied to all sectors and, in particular, the provision of biomethane using manure as a feedstock relies on a trans-sectoral production chain.

The GHG reduction potential of using manure-based biomethane affects the energy sector by replacing fossil fuels and influences the agricultural sector by improving manure management. It could be an appropriate way to allocate the GHG emission reduction potential to the two related sectors—energy and agriculture—and to identify policy fields in which additional instruments should be implemented so that unused GHG mitigation potentials can be unlocked [9,11]. In Germany, for example, only around 50% of the mobilizable technical biomass potential of animal manure with a value of 37 PJ [12,14] is used for energy purposes. At the same time, including non-ETS emissions (or reductions) is an interesting way to take into account cross-sectoral GHG mitigation in the framework of EU ETS.

Consequently, for the future utilization of this potential, one of the most important questions is: How competitive is biomethane as an advanced biofuel in the European road transport sector?

While GHG emissions and mitigation costs for biogas and biomethane have been addressed in many studies, for instance by coupling economic and GHG accounting models [15], by assessing the economics of biogas GHG mitigation potential [16], and by calculating the environmental impact of biomethane production, considering different types of technologies and substrates [17]. The effects of the different support schemes on biomethane have been rarely compared and are mostly within the framework of the EU RED [18,19] and RED II [20]. At the same time, the market for biomethane and its competitiveness with other advanced fuels have not been taken into account [21].

To answer the scoping question, we assessed the market for biomethane and the competitiveness of biomethane within this study. For that purpose, we considered different perspectives. On the one hand, we assessed the competitiveness of biomethane by comparing GHG emissions and the production costs of biomethane with values of other advanced biofuels as competitors in the advanced fuels market within the RED II framework. Therefore, we consider different biomethane concepts based on animal manure, straw, and biowaste, and two other advanced biofuels: ethanol based on straw and FT diesel based on waste wood. In doing so, we use the definition of advanced biofuels according to Annex IX of RED II. On the other hand, we examined under which conditions the EU ETS is an effective instrument to decrease the GHG mitigation costs of biomethane pathways.

2. Materials and Methods

In order to assess the GHG mitigation potential of biomethane in the European transport sector, we (1) defined different model concepts for biomethane and two other competing advanced fuels, (2) built three scenarios for the assessment framework, and (3) calculated the GHG emission mitigation costs for the different fuels.

2.1. Model Concepts

2.1.1. Biomethane Concepts

Three model pathways for biomethane production were assessed with regard to their specific GHG emissions and GHG mitigation costs. The three biomethane value chains investigated were biomethane from animal manure (Figure 1A), biomethane from a mixture of animal manure and straw (Figure 1B), and biomethane from biowaste (Figure 1C). For our assessment, the biomethane concepts had an assumed annual biomethane production of approximately 1,700,000 m^3 with a feedstock requirement of 40,000 t of animal manure per year (plant A), 26,000 t of animal manure and 6500 t of straw per year (plant B), and 26,000 t of biowaste per year (plant C) based on KTBL data [22]. After biogas was produced via digestion, a pressurized water-washing technology was used in the upgrading step with a capacity of 350 Nm^3 biogas per hour. In addition to carbon dioxide separation, the upgrading process also includes biogas pretreatment processes such as desulfurization and drying. The plant is running for 8300 h per year. Heat for the biomethane facility was supplied internally. In this case, part of the biogas produced was converted in a biogas boiler. The electricity for the biogas plant and the upgrading plant was taken from the public electricity grid. Methane emissions from biogas production and upgrading were included, assuming a methane loss of 1% during biogas production [17] and 0.2% during the upgrading process [23]. Depending on the retention time of the substrates in the fermenter, post-fermentation processes of the fermentation residues (digestate) in the digestate storage tanks can result in high residual methane emissions. With an uncovered storage facility, climate-relevant emissions between 19 [24] and 69 g CO_2-eq.*MJ^{-1} [25] can be assumed. Considering that in modern biogas plants the digestate storage is gas-tight covered, and that a gas-tight cover will be mandatory in the future, we assume that the digestate is stored in closed tanks. For this reason, no GHG emissions from biogas digestate were taken into account. Due to the declaration of the used biogas substrates (manure, straw, and biowaste) as waste and residues [3], the assessed lifecycle of the investigated biomethane concepts starts with the collection and transport processes of the substrates (e.g., straw baling, straw transport from field to biogas plant). Greenhouse gas emissions from upstream processes associated with, e.g., livestock breeding such as, for example, emissions caused by crop cultivation for animal feed, or in the case of straw use, the emissions from cereals cultivation were not included in the assessment. Also not considered in this GHG calculation are two aspects, related to the handling of the digestate. The first aspect concerns GHG emissions from the transport and the field application of the digestate. In particular, nitrous oxide emissions from spreading the digestate can have an influence on the GHG balance. The second aspect is related to the fertilizing effect of the digestate and the associated substitution of, in particular, synthetic nitrogen fertilizer. Climate-relevant emissions resulting from the energy-intensive production of synthetic nitrogen fertilizers can be avoided in this manner. In this sense, digestate based on residual and waste materials is not related to cultivation areas directly. As upstream processes are not considered, the nutrients remaining in the digestate (which could replace synthetic fertilizers) cannot be credited to the system. The same also applies to the GHG emissions from digestate field applications.

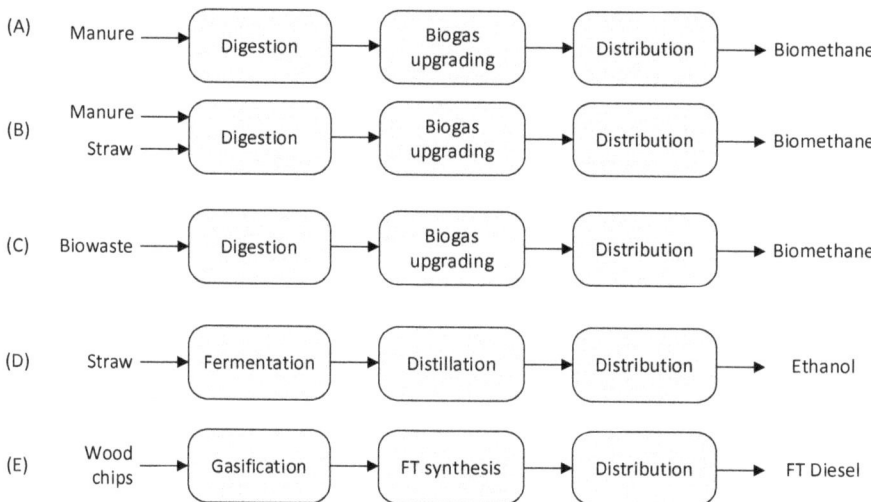

Figure 1. Main process steps of the advanced biofuel pathways under consideration: (**A**) biomethane from manure, (**B**) biomethane from manure and straw, (**C**) biomethane from biowaste, (**D**) ethanol from straw, and (**E**) FT diesel from woodchips.

2.1.2. Bioethanol from Straw

In our assessment, we included a straw-to-ethanol plant concept (Figure 1D) with an annual production capacity of approximately 50,000 t of bioethanol and a feedstock requirement of 275,000 t of wheat straw per year [26]. First, the straw is crushed and then broken down into its basic constituents (cellulose, hemicellulose, and lignin) using steam and pressure. In a subsequent liquefaction step, hemicellulose is dissolved into C5 sugars and cellulose is dissolved into C6 sugars, with the aid of enzymes. The sugars are subsequently fermented into bioethanol. The produced bioethanol is then concentrated through multiple distillation and rectification steps and through subsequent dehydration. Biomass byproducts supply the auxiliary energy for process heat and electricity.

2.1.3. Biomass-to-Liquid (BtL) from Woodchips

For the biomass-to-liquid pathway (shown in Figure 1E) considered in this study, we assumed an annual Fischer–Tropsch fuel (FT diesel) production of approximately 188,000 t with an annual feedstock requirement of 1,641,921 t of woodchips from short-rotation coppice [26]. The fuel production is based on the multistage gasification of dried woodchips. The gasification process is followed by gas scrubbing and by conditioning to syngas. This subsequently undergoes FT synthesis in a fixed-bed reactor and is converted to FT crude products using catalysts. After product separation, the wax fractions are converted to FT diesel through hydrocracking using an H_2 feed. The necessary process heat is supplied via natural gas. The saturated steam, which is not used in the process, is superheated and used in a steam turbine to generate surplus electricity.

2.2. Scenarios for GHG Mitigation Costs

The biomethane concepts were assessed in relation to various support instruments: on the one hand, in comparison to other advanced biofuels as competitors within the subtarget set out in RED II, and on the other hand in comparison to natural gas in the framework of the EU Emission Trading System (EU ETS). The assessment was carried out using a three-scenario approach based on the different support instruments:

- In scenario 1, the GHG mitigation costs for the biomethane concepts are assessed and compared with the values for the competing advanced biofuels as defined by RED II. In this case, the GHG emission values for the biomethane under consideration and other advanced biofuels include the complete lifecycle emissions calculated according to the methodology set out in the directive.
- In scenario 2, we assessed the GHG mitigation costs for biomethane concepts in the context of the EU ETS and calculated the CO_2 certificate price at which the advanced provision costs are competitive with the natural gas price. A crucial factor in this scenario is that the EU ETS defines GHG emissions from bioenergy as zero. Therefore, the differences between the GHG mitigation costs for the various biomethane value chains are the result of the different production costs.
- Scenario 3 aims to investigate the potential impact on the GHG mitigation costs for biomethane when the GHG mitigation effects from improved animal manure management (non-ETS agricultural sector) and the overall GHG emissions are taken into account. Here, we assumed that emission savings in the agricultural sector from biomethane production could be monetized within the EU ETS. Consequently, the corresponding GHG emission savings associated with the production of biomethane from manure is incorporated in the GHG emission values relevant for calculating the GHG mitigation costs.

2.3. Calculating the GHG Mitigation Costs

GHG mitigation costs express the price for the mitigation of a specific amount of greenhouse gas emissions by the use of an energy carrier with relatively higher costs and lower emissions compared to the reference fuel. The following Equation (1) was used to calculate GHG mitigation for the reduction of 1 ton of CO_2-eq. by substituting fossil fuels with biofuels:

$$C_{\text{GHG mitigation}} = \frac{C_{\text{Biofuel}} - C_{\text{Fossil fuel}}}{GHG_{\text{Fossil fuel}} - GHG_{\text{Biofuel}}} = \frac{\Delta C}{\Delta GHG} \quad (1)$$

where the additional cost of biofuels (ΔC) results from the difference between the specific production cost of biofuels (C_{Biofuel}) and the specific production cost of fossil fuel ($C_{\text{Fossil fuel}}$), whereas the net GHG emissions avoided by replacing fossil fuels with biofuels (ΔGHG) is the difference between the GHG emissions of the fossil fuel ($GHG_{\text{Fossil fuel}}$) and the GHG emissions of the biofuel (GHG_{Biofuel}). According to this methodology, the mitigation costs are sensitive to two factors: on the one hand, the cost difference between biomethane and the reference, and on the other hand, the difference between the specific GHG emissions of the fuels. As costs for biofuels and fossil fuels are both volatile, we present our results in the following sections as the function of GHG mitigation costs in relation to the cost/price difference between biofuels and fossil comparators.

2.4. Database

The specific greenhouse gas emissions of the advanced biofuel concepts considered here were calculated based on the guidelines set out in RED II [3]. The RED II methodology clearly defines the basic framework for the calculation of the specific GHG emissions. The methodology used in the evaluation of GHG emissions is described in the following section. The system boundaries, which define the framework in which the calculation takes place and determine which energy material flows are taken into account in the assessment, are set as well-to-wheel. The greenhouse gases relevant for GHG calculation are carbon dioxide (CO_2), methane (CH_4), and nitrous oxide (N_2O). The global warming potential of greenhouse gases is expressed in kg of carbon dioxide equivalents (CO_2-eq.). To convert a specific methane mass to kg CO_2-eq., the methane weight is multiplied by 25 and the nitrous oxide mass is multiplied by 298 (based on a period of 100 years, according to IPCC 2007). The CO_2 emissions from biofuels use are not included in GHG calculation; according to the IPCC, biogenic CO_2 emissions are considered to be offset by the CO_2 sequestration during plant growth. The functional unit which defines the quantification of the product, and which shall provide a reference to which the inputs and outputs are related, has been set for biofuels as 1 MJ biofuel. According to the GHG calculation approach of the RED II, the option for the consideration of coproducts is allocation according to the lower heating value. Allocation means that emissions shall be divided between the main and the coproducts in proportion to their energy content (lower heating value according to RED II methodology). Furthermore, The RED II framework allows to include credits for the use of manure/slurry as biogas substrate. The conventional storage of manure can lead to significant emissions of methane. These emissions can be reduced in case manure is used as a substrate for biogas production. The RED II recognizes this benefit by a credit of 45 g CO_2-eq. per MJ manure used (i.e., 54 kg CO_2-eq. per t fresh matter). According to the RED II methodology, GHG emissions associated with the construction and the demolition of the biofuel production plants were not considered. The specific overall greenhouse gas emissions calculated according to the described methodology for the investigated advanced biofuels are shown in Table 1. While the values for biomethane from biowaste, bioethanol from straw, and BtL from woodchips are of a similar order of magnitude, the provision of biomethane based on animal manure and biomethane based on a mixture of manure and straw cause significantly fewer emissions. This is primarily due to the credits for the improved manure management. The difference in overall GHG emissions between the two concepts (100% manure) shows how highly sensitive the credit is to the GHG emissions.

As we did not calculate specific costs for the production of the advanced biofuel options, literature values and references for advanced biofuel production were used. Due to the high volatility of costs and nonharmonized assumptions and framework conditions, we used bandwidths for the comparison of advanced biofuels with each other in scenario 1 shown in Table 1. The GHG mitigation potential in scenario 1 was calculated according to the fossil reference defined in RED II, with a value of 94 g CO_2-eq.*MJ^{-1} [3].

Table 1. Basic assumptions for GHG mitigation cost calculations.

	Advanced Biofuels				
	Biomethane from Manure	Biomethane from Manure (80%) and Straw (20%)	Biomethane from Biowaste	Bioethanol from Straw	BtL from Woodchips
Overall GHG emissions in g CO_2-eq./MJ (for biofuels: calculated according the methodology set out in RED II [3])	−76 (own calculations based on [22,27,28])	−10 (own calculations based on [22,27,28])	13 (own calculations based on [22,27,28])	13.7 [26]	13.7 [26]
GHG emissions from improved manure management in g CO_2-eq./MJ	−88 (own calculations based on [22,27,28])	−19 (own calculations based on [22,27,28])	-	-	-
Costs in EUR ct/kWh	6.7 [27]–11 [29]	8.8 [27]–12.85 [29]	6.2 [27]–9.2 [28]	6.3 [30]–14 [31]	7.5 [32]–12, 45 [33]

3. Results of GHG Mitigation Costs

3.1. Scenario 1: Comparison of the GHG Mitigation Costs for Biomethane and Other Advanced Biofuels as Competitors in the Advanced Fuels Market within the RED II Framework

The introduction of the subtarget for advanced biofuels in RED II has resulted in the development of a defined market for advanced biofuels and has meant that the GHG mitigation costs for bioenergy carriers have become a factor in distinguishing between the various bioenergy options. The calculations in this scenario include the total lifecycle greenhouse gas emissions of the advanced biofuel pathways for biomethane and other advanced biofuel competitors.

Based on the figures in Figure 2 for overall GHG emissions and the bandwidths of the costs, the specific GHG mitigation potential for the advanced biofuel concepts considered here is calculated and compared to the GHG emissions of the fossil references defined in RED II. The respective cost bandwidths are shown in Figure 2. Based on the assumptions made, biomethane value chains that use animal manure show the highest GHG mitigation potential compared to the fossil fuel reference, due to the given credit for improved manure management, and the production costs are lower than for the other advanced biofuels under consideration: bioethanol from straw and BtL from woodchips. The production of biomethane from biowaste causes similar costs but higher GHG emissions, due to the higher demand on process energy (electricity from the grid) [28]. A higher share of renewable energies in the electricity production mix can lower the GHG emissions associated with the production of biowaste-based biomethane. It can be stated that all the biofuels considered have a high GHG reduction potential compared to the fossil reference (defined in RED II with a value of 94 g CO_2-eq.*MJ^{-1}).

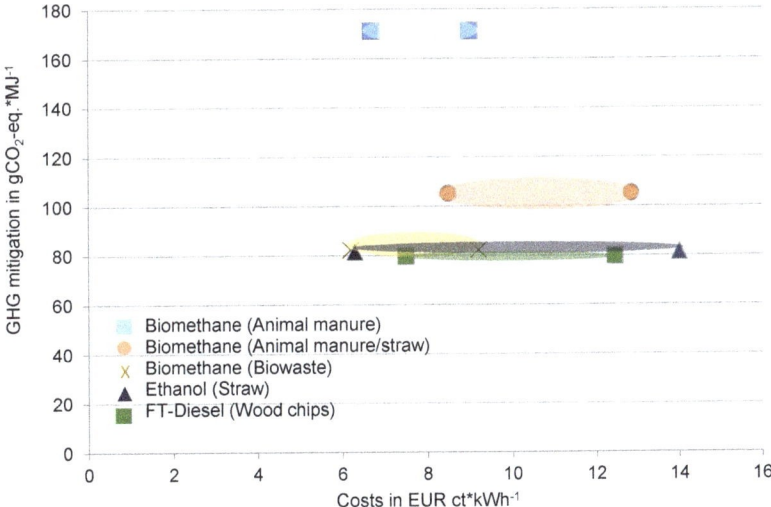

Figure 2. GHG mitigation potential and cost bandwidths for the biomethane concepts under consideration and other advanced biofuels.

As mentioned before, costs for biofuels and fossil fuels are both volatile. Therefore, we calculated the GHG mitigation cost (according to Equation (1)) based on GHG mitigation potential shown in Figure 2 as functions using cost/price differences between advanced biofuels and fossil reference fuels. This means that we calculated GHG mitigation costs based on the relative price difference and not based on absolute cost figures. Figure 3 shows the functions of GHG mitigation cost for each biofuel option in relation to the cost/price difference between the biofuel option and fossil fuel. Under the assumptions made, the results in Figure 3 show significantly lower GHG mitigation costs (and consequently more flat-curve characteristics) for biomethane concepts using animal manure as a biogas and biomethane substrate and as a cosubstrate in a mixture with straw compared to the other advanced biofuels under consideration. The values for the GHG mitigation cost shown in Figure 3 are highly sensitive to the GHG reduction potential, similar to the values shown in Figure 2.

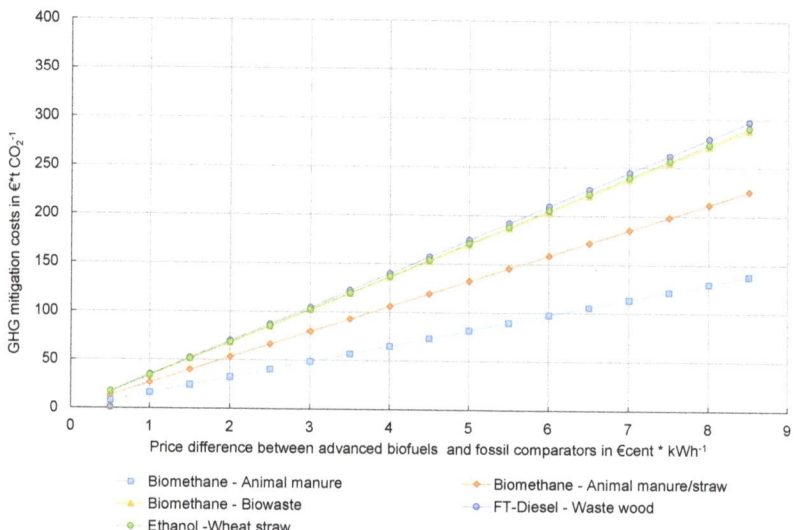

Figure 3. CO_2 mitigation costs functions based on cost/price differences for advanced biofuels, assuming that GHG emissions from biofuel production are calculated according to the RED II methodology (see Table 1 for emission values).

3.2. Scenario 2: GHG Mitigation Costs for Biomethane in the EU ETS

In contrast to scenario 1, the GHG mitigation costs for the biomethane concepts were calculated in this scenario based on the assumption that emissions from biofuel production are not accounted for within the EU Emission Trading System framework. In addition, in contrast to scenario 1, we calculated the GHG mitigation costs with actual cost values, in order to better illustrate the special consideration of bioenergy in the context of the EU ETS. For biomethane production we used values from [27]. For the fossil fuel we assumed a price of 3 EURct*kWh^{-1} [34] and emission values of 67.6 g CO_2-eq.*MJ^{-1} [35] for natural gas. As emissions from biofuels are considered to be zero, differences in the GHG mitigation costs only result from differences in production costs. Figure 4 shows the mitigation costs for the biomethane concepts (based on assumptions in Table 1) in relation to a cost difference between biomethane and natural gas.

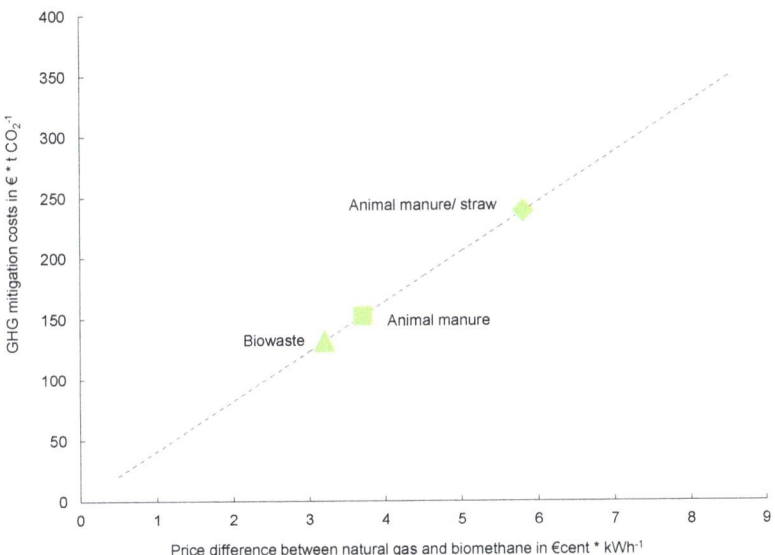

Figure 4. CO_2 mitigation costs in relation to cost/price differences for advanced biofuels based on biomethane pathways in comparison to natural gas, assuming that GHG emissions from biomethane are zero within the EU ETS.

The results show clear differences between the biomethane value chains under investigation. Because of the relatively lower production costs for biomethane from biowaste and animal manure, the mitigation costs for these value chains are comparably lower than for biomethane concepts using a mixture of animal manure and straw.

The GHG mitigation cost value chains range from 123 EUR*t CO_2-eq.$^{-1}$ to 247 EUR*t CO_2-eq.$^{-1}$. These values represent the CO_2 certificate price at which the costs for natural gas and the respective biomethane value chain reach a breakeven point (in fact, an increasing price for CO_2 certificates would increase the costs for the utilization of carbon intense energy carriers in the EU ETS). Figure 4 also shows the relation between GHG mitigation costs and the cost/price difference between the biomethane value chains and natural gas. Assuming a consistent difference of greenhouse gas emissions between natural gas and biomethane, the mitigation costs increase as the cost/price differences between both energy carriers rise.

3.3. Scenario 3: GHG Mitigation Costs for Biomethane in the EU ETS Including Non-ETS Sectors

Scenario 3 moves beyond the current EU ETS to include greenhouse gas emissions from the non-ETS agricultural sector, in this case in the form of GHG emissions from improved animal manure management. The values of the GHG emissions from improved manure management for the biomethane concepts (biomethane based on 100% animal manure and biomethane based on a mixture of animal manure and straw) are shown in Table 1. As a consequence of the inclusion, the GHG mitigation costs for the two biomethane value chains using animal manure as a substrate decrease significantly. Figure 5 shows the CO_2 certificate price at which the costs for natural gas and the respective biomethane value chain reach a breakeven point (as mentioned, an increasing price for CO_2 certificates would increase the costs for the utilization of carbon intense energy carriers in the EU ETS). The mitigation costs for biomethane based on animal manure decrease from 152 EUR*t CO_2-eq.$^{-1}$ to 66 EUR*t CO_2-eq.$^{-1}$. The mitigation costs for biomethane from animal manure and straw, compared to natural gas, decrease from 247 EUR*t CO_2-eq.$^{-1}$ to 185 EUR*t CO_2-eq.$^{-1}$.

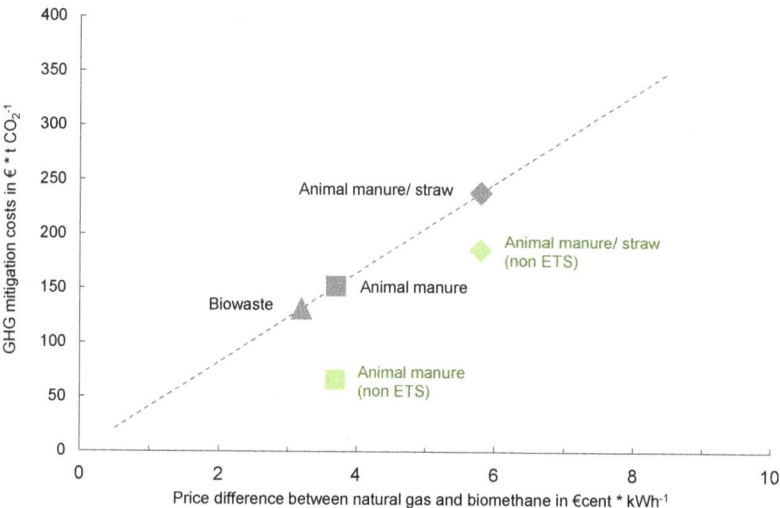

Figure 5. CO$_2$ mitigation costs in relation to cost/price differences for different biomethane pathways in comparison to natural gas, assuming that GHG emissions from biomethane are zero within the EU ETS. For biomethane from animal manure and animal manure + straw, the GHG benefit from avoiding methane emissions from manure storage (animal manure credit) is included here.

4. Discussion

4.1. RED II Perspective

Biomethane produced from animal manure, straw, or biowaste can achieve very high savings in greenhouse gas emissions of up to 200% by using animal manure as substrate over fossil fuels and can be considered very competitive compared to other advanced biofuels. As the specific production costs are in similar bandwidths for all advanced biofuels under consideration, the GHG mitigations costs are highly sensitive to the specific GHG mitigation potential. Due to the credit given for improved manure management according to RED II methodology (and highly dependent on it), the use for manure-based biomethane is associated with the highest potential for GHG mitigation. In addition, the use of straw and biowaste shows high GHG mitigation potential. The GHG mitigation potential of biomethane from biowaste will increase, due to the use of process energy associated with lower GHG emission.

Although the subtarget for advanced biofuels set out in RED II offers particular advantages for biomethane from waste and residues, there are some market restrictions. The subtarget of 1.75% in 2030 will develop a market of approximately 31 PJ [36]–39 PJ [37] for advanced fuels in the German transport sector and between 227 PJ [38] and 416 PJ [39] in the European transport sector. In the case of Germany, for instance, biomethane production capacities of 36 PJ [40] and a mobilizable technical biomass potential for animal manure in the amount of 37 PJ and for straw with 26 PJ are available for this [12].

Nevertheless, the current low rate of 5–6 PJ [41,42] gaseous fuels in the German transport sector and 84 PJ [43] in the European transport sector reveals the existing limitations for the use of biomethane, which is primarily due to the low share of gas-fueled vehicles in the passenger car and truck fleet. However, the advantages of biomethane listed here with regard to the GHG reduction potential and costs could, with the right incentives, lead to an increase in gas-fueled vehicles. Optimistic scenarios expect, for instance, that the number of gas-fueled trucks will increase from 9000 to 480,000 by 2030 in Europe [44].

4.2. EU ETS Perspective

In contrast to the RED II perspective, the EU ETS does not calculate greenhouse gas emissions from bioenergy based on a lifecycle assessment approach that includes comprehensive mass and energy balances of all process steps. Greenhouse gas emissions from bioenergy use are considered zero in the context of the EU ETS; in other words, biomethane users do not need emission certificates for the biomethane they use. When, in this context, greenhouse gas emissions from bioenergy are considered zero, there are no differences in the specific GHG performance of biomethane value chains. Consequently, differences in the GHG mitigation costs for the biomethane value chains calculated based on EU ETS conditions solely result from differences in their production cost. Particularly high GHG reduction effects are not taken into account and, therefore, have no competitive advantage. The GHG mitigation costs range from 123 EUR*t CO_2-eq.$^{-1}$ to 247 EUR*t CO_2-eq.$^{-1}$.

The use of animal manure in biomethane plants can result in substantial GHG reductions in different sectors. This is due to the substitutions in the transport and energy sector and improved manure management in the agricultural sector, which avoids climate-related GHG emissions from animal manure storage. With respect to emissions from the agricultural sector, the production of biomethane based on animal manure can result in a CO_2 certificate price of 75 EUR*t CO_2, which is much lower than other GHG mitigation options in the transport sector.

4.3. Limitations of the Study

In our study, we used typical model concepts to calculate GHG mitigation costs. GHG balances are individual balances that are highly dependent on process-specific characteristics, methodological assumptions, and data availability. The values shown in this study for the advanced fuels are therefore only valid under the specified framework conditions. Furthermore, the proposed consideration of the GHG mitigation potential in the non-EU ETS agricultural sector is an amalgamation between the entire value chain assessment according to RED II and current EU ETS conditions that set bioenergy at zero emissions. This approach is more a political measure to unlock unused GHG reduction potentials. However, if greenhouse gas emissions from manure storage are taken into account, the question arises as to whether downstream emissions also have to be considered, even though they can be associated with other EU ETS sectors.

Another aspect to consider is that GHG mitigation in intersectoral value chains is linked to allocation problems. There are two main challenges here. Firstly, determining bioenergy emissions within the framework of the EU ETS as zero and including non-EU ETS emissions and/or the entire upstream chain. Secondly, finding a way to promote GHG reductions in the agricultural and energy sector in order to unlock unused GHG reduction potential.

5. Conclusions

Biomethane from waste and residues can mitigate significant GHG emissions compared to fossil fuels, but the production costs of biomethane often exceed the costs of fossil fuels.

GHG mitigation costs are typically calculated from a specific point of view (for example, from that of political decision-makers). Our investigation focused on the conditions under which biomethane from animal manure, biowaste, and agricultural residues can be competitive as an advanced fuel in the European transport sector.

The first perspective addresses the policy instruments that promote renewable energy carriers under the RED II framework. Biomethane based on animal manure, on a mixture of animal manure and straw, and on biowaste can achieve a GHG mitigation potential of up to 200% compared to the fossil reference, and enjoys the advantages offered by the subtarget for advanced fuels set out in RED II. The perspective of the calculations then shifted towards the EU ETS mechanism. Depending on the actual GHG performance and cost of the specific biomethane pathway, CO_2 emission certificate prices of between ~120 EUR*t CO_2-eq.$^{-1}$ and ~250 EUR*t CO_2-eq.$^{-1}$ were calculated.

Based on the calculations, the following points can be concluded:

The RED II perspective

- In general, there are significant differences in the GHG mitigation costs for the various advanced biofuels pathways investigated. Because of lower GHG emissions per MJ, biomethane based on animal manure tends to have lower GHG mitigation costs compared to other advanced biofuels. This is a competitive advantage within the protected market for advanced fuels set out by RED II. Sufficient production capacities and biomass potentials are available to serve this defined market with biomethane from waste and residues. However, the current low rate of gaseous fuels in the German and European transport sectors reveals the existing limitations for the use of biomethane. An expansion of the gas-fueled fleet and the filling station infrastructure could be one measure to unlock the unused potential of GHG mitigation by using biomethane from waste and residues, in particular animal manure, in the transport sector.

The EU ETS perspective

- Under the current conditions, the EU ETS is not an effective instrument for increasing the costs of carbon-intensive energy carriers to reach a breakeven between the costs for natural gas and biomethane. Considering the current price for CO_2 certificates, we propose taking into account the GHG reduction potential of the non-ETS agricultural sector in order to unlock unused GHG mitigation potential for the reduction of GHG emissions. The mitigation costs for biomethane from animal manure would thus decrease from 164 EUR*t CO_2-eq.$^{-1}$ to 75 EUR*t CO_2-eq.$^{-1}$.
- Due to the flexibility in terms of the types of substrates used for biomethane production, this technology can, in theory, tap great potentials for GHG mitigation in sectors outside the EU ETS (e.g., agriculture, waste treatment, and disposal). Internalizing, and thus capitalizing on, these effects will significantly decrease the GHG mitigation costs for the respective biomethane pathways in the EU ETS.

Biomethane from waste and residues is an interesting and potentially powerful way to mitigate GHG emissions in the EU transport sector. However, the magnitude of the GHG mitigation effects from biomethane strongly depends on the type of substrate used in biomethane production processes, the greenhouse gas emissions from the fossil energy carrier substituted by biomethane, and the calculation method used by the different GHG mitigation instruments.

Author Contributions: Conceptualization, all authors; methodology, K.O. and S.M.; software, K.O.; validation, S.M. and D.T.; formal analysis, K.O.; investigation, K.O. and S.M.; resources, K.O. and S.M.; data curation, K.O. and S.M.; writing—original draft preparation, K.O. and D.T.; writing—review and editing, D.T. and S.M.; visualization, K.O.; supervision, D.T.; project administration, K.O.; funding acquisition, S.M. All authors have read and agreed to the published version of the manuscript.

Funding: The project is supported (was supported) by funds of the Federal Ministry of Food and Agriculture (BMEL) based on a decision of the Parliament of the Federal Republic of Germany and by funds of European Union's Horizon 2020 research and innovation programme under grant agreement No. 646533.

Institutional Review Board Statement: Not applicable.

Informed Consent Statement: Not applicable.

Data Availability Statement: Not applicable.

Conflicts of Interest: The authors declare no conflict of interest.

References

1. European Commission. *Directive 2001/77/EC of the European Parliament and of the Council of 27 September 2001 on the Promotion of Electricity Produced from Renewable Energy Sources in the Internal Electricity Market*; European Commission: Brussels, Belgium, 2001.
2. European Commission. *Directive 2009/30/EC of the European Parliament and of the Council of 23 April 2009 Amending Directive 98/70/EC as Regards the Specification of Petrol, Diesel and Gas-Oil and Introducing a Mechanism to Monitor and Reduce Greenhouse Gas Emissions and Amending Council Directive 1999/32/EC as Regards the Specification of Fuel Used by Inland Waterway Vessels and Repealing Directive 93/12/EEC*; European Commission: Brussels, Belgium, 2009.
3. European Commission. *Directive (EU) 2018/ 2001 of the European Parliament and of the Council—Of 11 December 2018—On the Promotion of the Use of Energy from Renewable Sources*; European Commission: Brussels, Belgium, 2018.
4. European Commission. *Directive (EU) 2015/1513 of the European Parliament and of the Council of 9 September 2015 Amending Directive 98/70/EC Relating to the Quality of Petrol and Diesel Fuels and Amending Directive 2009/28/EC on the Promotion of the Use of Energy from Renewable Sources September 2015*; European Commission: Brussels, Belgium, 2015.
5. Federal Government of Germany. *Act on Granting Priority to Renewable Energy Sources (Renewable Energy Sources Act-EEG)*; Federal Government of Germany: Berlin, Germany, 2012.
6. Giuntoli, J.; Agostini, A.; Edwards, R.; Marelli, L. *Solid and Gaseous Bioenergy Pathways: Input Values and GHG Emissions. Calculated According to the Methodology set in COM(2010) 11 and SWD(2014) 259, Version 1a*; Joint Research Center (JRC): Luxembourg, 2015. [CrossRef]
7. Haenel, H.-D.; Claus, R.; Ulrich, D.; Eike, P.; Annette, F.; Sebastian, W.; Brigitte, E.-M.; Helmut, D.; Carsten, S.; Beate, B.; et al. *Calculations of Gaseous and Particulate Emis-sions from German Agriculture 1990–2012*; Thünen Report 17; Thünen Institut: Braunschweig, Germany, 2014.
8. Dämmgen, U.; Webb, J. The development of the EMEP/CORINAIR Guidebook with respect to the emissions of different ni-trogen and carbon species from animal production. *Agric. Ecosyst. Environ. Mitig. Greenh. Gas Emiss. Livest. Prod.* **2006**, 2–3, 241–248.
9. Oehmichen, K.; Thrän, D. Fostering renewable energy provision from manure in Germany: Where to implement GHG emis-sion reduction incentives. *Energy Policy* **2017**, *110*, 471–477. [CrossRef]
10. Scheftelowitz, M.; Thrän, D. Unlocking the Energy Potential of Manure—An Assessment of the Biogas Production Potential at the Farm Level in Germany. *Agriculture* **2016**, *6*, 20. [CrossRef]
11. van Melle, T.; Peters, D.; Cherkasky, J.; Wessels, R.; Mir, G.; Hofsteenge, W. *Gas for Climate: How Gas Can Helpto Achievethe Paris Agreement Target in an Affordable Way*; European Biogas Association: Brussels, Belgium, 2018.
12. DBFZ. Resource Data Repository, Webapplicaton, Leipzig. 2021. Available online: https://webapp.dbfz.de/resource-database/?lang=en (accessed on 26 October 2021).
13. European Parliament and Council. *Proposal for a Directive of the European Parliament and of the Council Amending Directive (EU) 2018/2001, Regulation (EU) 2018/1999 and Directive 98/70/EC as Regards the Promotion of Energy from Renewable Sources, and Repealing Council Directive (EU) 2015/652*; European Commission: Brussels, Belgium, 2021.
14. Brosowski, A.; Krause, T.; Mantau, U.; Mahro, B.; Noke, A.; Richter, F.; Raussen, T.; Bischof, R.; Hering, T.; Blanke, C.; et al. How to measure the impact of biogenic residues, wastes and by-products: Development of a national resource monitoring based on the example of Germany. *Biomass Bioenergy* **2019**, *127*, 105275. [CrossRef]
15. Bartoli, A.; Hamelin, L.; Rozakis, S.; Borzęcka, M.; Brandão, M. Coupling economic and GHG emission accounting models to evaluate the sustainability of biogas policies. *Renew. Sustain. Energy Rev.* **2019**, *106*, 133–148. [CrossRef]
16. Agostini, A.; Battini, F.; Padella, M.; Giuntoli, J.; Baxter, D.; Marelli, L.; Amaducci, S. Economics of GHG emissions mitigation via biogas production from Sorghum, maize and dairy farm manure digestion in the Po valley. *Biomass Bioenergy* **2016**, *89*, 58–66. [CrossRef]
17. Vetter, A.; Arnold, K. *Klima- und Umwelteffekte von Biomethan, Anlagentechnik und Substratauswahl*; Wuppertal Institut: Wuppertal, Germany, 2010.
18. Rana, R.; Ingrao, C.; Lombardi, M.; Tricase, C. Greenhouse gas emissions of an agro-biogas energy system: Estimation under the Renewable Energy Directive. *Sci. Total Environ.* **2016**, *550*, 1182–1195. [CrossRef] [PubMed]
19. Manninen, K.; Koskela, S.; Nuppunen, A.; Sorvari, J.; Nevalainen, O.; Siitonen, S. The applicability of the renewable energy directive calculation to assess the sustainability of biogas production. *Energy Policy* **2013**, *56*, 549–557. [CrossRef]
20. Long, A.; Bose, A.; O'Shea, R.; Monaghan, R.; Murphy, J.D. Implications of European Union recast Renewable Energy Directive sustainability criteria for renewable heat and transport: Case study of willow biomethane in Ireland. *Renew. Sustain. Energy Rev.* **2021**, *150*, 111461. [CrossRef]
21. Kraussler, M.; Pontzen, F.; Müller-Hagedorn, M.; Nenning, L.; Luisser, M.; Hofbauer, H. Techno-economic assessment of biomass-based natural gas substitutes against the background of the EU 2018 renewable energy directive. *Biomass Conv. Bioref.* **2018**, *8*, 935–944. [CrossRef]
22. KTBL. Leistungs-Kostenrechnung Pflanzenbau; Web-Application. 2020. Available online: https://daten.ktbl.de/dslkrpflanze/postHV.html (accessed on 26 October 2021).
23. GasNZV. *Verordnung über den Zugang zu Gasversorgungsnetzen*; Deutscher Bundestag: Berlin, Germany, 2017.
24. Johann Heinrich von Thünen-Institut. *Biogas-Messprogramm II*; Johann Heinrich von Thünen-Institut: Gülzow, Germany, 2010.
25. BioGrace II Calculation Rules. Harmonised Greenhouse Gas Calculations for Electricity, Heating and Cooling from Biomass. Version 4a. Available online: www.biograce.net (accessed on 8 June 2021).

26. Edwards, R.; Padella, M.; Giuntoli, J.; Koeble, R.; O'Connell, A.; Bulgheroni, C.; Marelli, L.; Lonza, L. *Definition of Input Data to Assess GHG Default Emissions from Biofuels in EU Legislation: Version 1d—2019*; Publications Office of the European Union: Luxembourg, 2019. [CrossRef]
27. Majer, S.; Oehmichen, K. *Comprehensive Methodology on Calculating Entitlement to CO2 Certificates by Biomethane Producers: BIOSURF Deliverable D5.5*; DBFZ: Leipzig, Germany, 2017.
28. Daniel-Gromke, J.; Rensberg, N.; Denysenko, V.; Barchmann, T.; Oehmichen, K.; Beil, M.; Beyrich, W.; Krautkremer, B.; Trommler, M.; Reinholz, T.; et al. *Optionen für Bio-Gas-Bestandsanlagen bis 2030 aus Ökonomischer und Energiewirtschaftlicher Sicht: Abschlussbericht*; Umweltbundesamt: Dessau-Roßlau, Germany, 2019.
29. Majer, S.; Kornatz, P.; Daniel-Gromke, J.; Rensberg, N.; Brosowski, A.; Oehmichen, K.; Liebetrau, J. *Stand und Perspektiven der Bio-gaserzeugung aus Gülle*; DBFZ: Leipzig, Germany, 2019.
30. Meisel, K.; Millinger, M.; Naumann, K.; Majer, S.; Müller-Langer, F.; Thrän, D. *Untersuchungen zur Ausgestaltung der Biokraftstoffgesetzgebung, Abschlussbericht*; DBFZ Deutsches Biomasseforschungszentrum Gemeinnützige GmbH: Leipzig, Germany, 2019.
31. Zech Konstantin, M.; Meisel, K.; Brosowski, A.; Toft, L.V.; Müller-Langer, F. Environmental and economic assessment of the Inbicon lignocellulosic ethanol technology. *Appl. Energy* 2016, *171*, S347–S356. [CrossRef]
32. Millinger, M.; Ponitka, J.; Arendt, O.; Thrän, D. Competitiveness of advanced and conventional biofuels. Results from least-cost modelling of biofuel competition in Germany. *Energy Policy* 2017, *107*, S394–S402. [CrossRef]
33. Albrecht Friedemann, G.; König Daniel, H.; Baucks, N.; Dietrich, R. A standardized methodology for the techno-economic evaluation of alternative fuels—A case study. *Fuel* 2017, *194*, S11–S26. [CrossRef]
34. EUROSTAT. Natural Gas Price Values for Europe. Available online: https://ec.europa.eu/eurostat/statistics-explained/index.php?title=Natural_gas_price_statistics/de&oldid=363671#Erdgaspreise_f.C3.BCr_Industriekunden (accessed on 8 June 2021).
35. Giuntoli, J.; Agostini, A.; Edwards, R.; Marelli, L. *Solid and Gaseous Bioenergy Pathways: Input Values and GHG Emissions: Calculated According to Methodology Set in COM(2016) 767: Version 2*; Joint Research Center (JRC): Luxembourg, 2017. [CrossRef]
36. Purr, K.; Günther, J.; Lehmann, H.; Nuss, P. Wege in Eine Ressourcenschonende Treibhausgasneutralität. RESCUE-Studie. Hg. v. Umweltbundesamt. Dessau-Roßlau (CLIMATE CHANGE, 36/2019). Available online: https://www.umweltbundesamt.de/sites/default/files/medien/376/publikationen/rescue_studie_cc_36-2019_wege_in_eine_ressourcenschonende_treibhausgasneutralitaet.pdf (accessed on 6 October 2021).
37. Kemmler, A.; Kirchner, A.; Auf der Maur, A.; Ess, F.; Kreidelmeyer, S.; Pjégsa, A.; Spillmann, T.; Wünsch, M.; Ziegenhagen, I. *Energiewirtschaftliche Projektionen und Folgeabschätzungen 2030/2050: Dokumentation vonReferenzszenariound Szenario mit Klimaschutzprogramm*; Prognos: Berlin, Germany, 2020.
38. Concawe's Transport and Fuel Outlook towards EU 2030 Climate Targets. Concawe Brussels. April 2021. Available online: https://www.concawe.eu/wp-content/uploads/Rpt_21-2.pdf (accessed on 25 November 2021).
39. USDA Foreign Agricultural Service. EU-28: Biofuels Annual—EU Biofuels Annual. 2019. Available online: https://apps.fas.usda.gov/newgainapi/api/report/downloadreportbyfilename?filename=Biofuels%20Annual_The%20Hague_EU-28_7-15-2019.pdf (accessed on 26 October 2021).
40. Deutsche Energie-Agentur GmbH (Dena). *Dena Analyse Branchenbarometer Biomethan 2020*; Deutsche Energie-Agentur: Berlin, Germany, 2020.
41. Bundesministerium für Wirtschaft und Energie (BMWi). *Zeitreihen zur Entwicklung der Erneuerbaren Energien in Deutschland unter Verwendung der Daten der Arbeitsgruppe Erneuerbare Energien-Statistik (AGEE-Stat) Mit Stand Februar 2020*; Bundesministerium für Wirtschaft und Energie (BMWi): Berlin, Germany, 2020.
42. Radke, S. *Verkehr in Zahlen 2019/2020*; Bundesministerium für Verkehr und digitale Infrastruktur: Berlin, Germany, 2019.
43. EUROSTAT. Complete Energy Balances. NRG_BAL_C. Hg. v. Eurostat. 2021. Available online: https://ec.europa.eu/eurostat/databrowser/view/NRG_BAL_C__custom_1071932/default/table?lang=de (accessed on 26 October 2021).
44. Pääkkönen, A.; Aro, K.; Aalto, P.; Konttinen, J.; Kojo, M. The Potential of biomethane in replacing fossil fuels in heavy transport—A case study on Finland. *Sustainability* 2019, *11*, 4750. [CrossRef]

Article

Exhausted Grape Marc Derived Biochars: Effect of Pyrolysis Temperature on the Yield and Quality of Biochar for Soil Amendment

Kawthar Frikha [1], Lionel Limousy [1], Muhammad Bilal Arif [1], Nicolas Thevenin [2], Lionel Ruidavets [2], Mohamed Zbair [1] and Simona Bennici [1,*]

1. Institut de Science des Materiaux de Mulhouse, Université de Haute-Alsace, F-68057 Mulhouse, France; kawthar.frikha@uha.fr (K.F.); lionel.limousy@uha.fr (L.L.); muhammadbilal.arif@etu.toulouse-inp.fr (M.B.A.); mohamed.zbair@uha.fr (M.Z.)
2. Department of Agronomy, Rittmo Agroenvironment, F-68025 Colmar, France; nicolas.thevenin@rittmo.com (N.T.); lionel.ruidavets@rittmo.com (L.R.)
* Correspondence: simona.bennici@uha.fr; Tel.: +33-38-933-6729

Citation: Frikha, K.; Limousy, L.; Arif, M.B.; Thevenin, N.; Ruidavets, L.; Zbair, M.; Bennici, S. Exhausted Grape Marc Derived Biochars: Effect of Pyrolysis Temperature on the Yield and Quality of Biochar for Soil Amendment. *Sustainability* **2021**, *13*, 11187. https://doi.org/10.3390/su132011187

Academic Editors: Dirk Enke, Hossein Beidaghy Dizaji, Volker Lenz and Thomas Zeng

Received: 27 August 2021
Accepted: 3 October 2021
Published: 11 October 2021

Publisher's Note: MDPI stays neutral with regard to jurisdictional claims in published maps and institutional affiliations.

Copyright: © 2021 by the authors. Licensee MDPI, Basel, Switzerland. This article is an open access article distributed under the terms and conditions of the Creative Commons Attribution (CC BY) license (https://creativecommons.org/licenses/by/4.0/).

Abstract: The present study focuses on the valorisation of winery industry wastes through slow pyrolysis of exhausted grape marc (EGM). The optimal pyrolysis parameters were firstly identified by small scale experiments carried out using thermogravimetric analysis. Nine pyrolysis temperatures were tested and their influence on the decomposition of the EGM residue and biochar yield was evaluated. Then, biochar production was conducted in a pilot plant at three chosen temperatures (450, 500 and 550 °C) at which the biochar was shown to be stable. The effects of biochar application to soil with respect to plant (ryegrass) growth was also evaluated. Pyrolysis of EGM at the 450–550 °C temperature range has been shown to generate thermally stable and nutrient-rich biochars, but only the biochar produced at 450 °C showed a marked benefit effect of ryegrass growth.

Keywords: waste; exhausted grape marc; biochar; pyrolysis; soil amendment

1. Introduction

Grapes are among the most cultivated fruit crops worldwide [1]. One-third of the grape total production is intended for the wine industry [2–4]. France is among the leading producers of wine in the world. In 2020, France was ranked as the second largest wine-producing country, just after Italy, with an estimated volume of 46.6 million hectolitres [5,6]. Considering the high volumes produced every year, the wine industry can generate thousands of tons of solid residues, which raises serious environmental and economic issues [7]. Throughout the winemaking process, a large amount of solid residue, referred to as grape marc (pomace), is generated and it represents 10–30 wt. % of the grape fresh weight [3]. This residue generally consists of skins, seeds, stalks and moisture. In the red wine making process, the grape marc residue is produced after fermentation and pressing, and it contains a valuable alcoholic fraction [4]. As a traditional and reliable way to recover the alcoholic fraction (polyphenols), fermented grape marc undergoes a distillation process, which results in a final solid residue known as exhausted grape marc [8]. The chemical composition of grape marc (similarly to exhausted grape marc) is very diverse, consisting mainly of fibres (cellulose, hemicellulose and lignin), polyphenols, lipids, proteins, oligosaccharides and minerals [9]. Such a rich and complex composition holds great potential for further valorisation through value-added processing, including the thermochemical and biochemical processes [3]. Thermal conversion technologies for energy recovery from grape marc include combustion, gasification, hydrothermal carbonisation, torrefaction and pyrolysis [10]. Pyrolysis is a thermal decomposition process in oxygen-free or oxygen-poor conditions at temperatures of around 400–700 °C [11]. Pyrolysis yields a mix of gas, liquid (tar or biooil) and solid (biochar) products, depending on the operating conditions (heating rate, vapours/solid residence time, pyrolysis temperature, pyrolysis atmosphere) and the origin of the feedstocks [11]. Low pyrolysis temperatures and low heating rates are generally required when the biochar is the product of interest. The pyrolysis of exhausted grape marc could provide a sustainable approach to managing waste from the wine industry. From this standpoint, several studies were carried out to explore optimal thermal conversion pathways of winery wastes and evaluate their potential for the recovery of energy and high-added value compounds [12–24]. Encinar et al. [12] investigated the influence of the pyrolysis temperature and feedstocks' particle size on the quality of the biochars produced from grape marc, using a batch pyrolysis system with nitrogen as a carrier gas. They found that increasing the temperature yields an increase in the fixed carbon content, gases produced and ash content. Particle size has a negligible effect on the properties of the biochars. Demiral et al. [13] investigated changes in pyrolysis product yields by varying the operating conditions, including the temperature, heating rate and nitrogen gas flow. Encinar et al. also reported that an increase in the pyrolysis temperature led to higher liquid and gas yields. Marculescu and Ciuta [14] studied the optimal pyrolysis conditions for grape marc treatment using experimental and modelling techniques. The study revealed that the pyrolysis temperature had a significant impact on products distribution and energy content. An optimum treatment temperature for the maximum net energy content of the combined pyrolysis products (char, liquid and gas) was identified to be 550 °C. As mentioned before, the pyrolysis products range from solid to liquid to gaseous compounds, and all are considered as valuable bio-products. Some of these products can be used directly, as a feedstock for a further industrial process or can be converted into energy. Charcoal as a bio-product has potential for use in the energy, industrial and agricultural sectors [25–27]. A significant increase in the agricultural use of biochar has already been recorded since 2015 [28]. Agricultural applications include soil amendment, composting, carriers for fertilisers, silage additives and feed additives [28]. Biochar as a soil amendment has received considerable attention because of its carbon sequestration potential and ability to enhance soil productivity [29–33]. Compared to non-waste feedstocks (e.g., woody biomass), biochar production from exhausted grape marc for agronomic uses has not been widely explored and only few studies reported uses of biochar in soil amendment and connected applications.

The main motivations for this study were (1) to evaluate the potential of grape marc waste as a feedstock for biochar production, (2) to examine the effects of pyrolysis temperature on the physicochemical properties of the produced biochars and (3) to consider the potential agronomic implications of applying biochar to soil.

2. Materials and Methods

2.1. Feedstock

Exhausted Grape Marc (EGM), a distillery waste or an agro-industrial by-product, was collected from a distillery industry located in the Alsace region (France). An EGM sample was prepared following the NF EN ISO 14780 standard. An EGM lot (batch) was mixed by turning and piling it upside-down several times, then different subsamples were arbitrarily taken and gathered from different spots of the homogenised EGM lot. After mixing, one portion of the new subsample was dried at 60 °C for 24 h, ground using a laboratory batch mill (IKA Mod. A11 Basic analytical mill, IKA, Staufen im Breisgau, Germany), thoroughly mixed and sieved to obtain particle sizes in the range of 250 to 400 µm. The dried milled sample was stored in a sealed container. Before laboratory analysis, representative subsamples were weighed and stored into another set of sealed containers. The pictures of (a) raw and (c) dried milled EGM samples are shown in Figure 1.

2.2. Feedstock Analysis

Characterisation of EGM is important to assess its properties and quality as a feedstock for biochar production. A series of physicochemical characterisation were performed, including thermogravimetric analysis (TGA), elemental analysis (C, N, H, O and S), Higher Heating Value (HHV) measurements and mineral and chemical composition analysis.

Figure 1. Pictures of (**a**) raw and (**b**) dried milled (250–400 μm) EGM samples.

A non-isothermal pyrolysis experiment of EGM was carried out in a thermogravimetric analyser (thermo-balance, Mettler Toledo TGA 850). Prior to experiment, an EGM sample was dried overnight in a ventilated oven at 105 °C under air. A typical run was carried out as follows: about 15 mg of EGM sample was weighed into an open-type alumina crucible (70 μL) and introduced into the TGA furnace. The temperature was raised from room temperature up to 900 °C at a heating rate of 5 °C min^{-1} under 100 mL min^{-1} nitrogen and maintained at this temperature for 1h. Then, the atmosphere was switched to synthetic air (100 mL min^{-1}) for 60 min.

The hemicellulose, cellulose and lignin contents were determined, by Cirad laboratory, according to the Van Soest acid detergent fibre method [34].

The total carbon (C), nitrogen (N), Sulphur (S), hydrogen (H) and oxygen (O) contents measurements were analysed by Filab laboratory, according to NF EN ISO 16948 and NF EN ISO 16994 standards.

The inorganic composition (Mg, Na, K, Ca, Si, Al, P, Fe and Cu contents) was analysed using inductively coupled plasma atomic emission spectroscopy (ICP-AES), provided by Filab laboratory (Accreditation COFRAC ISO 17025).

The Higher Heating Value (HHV) (MJ kg^{-1}, dry basis) of EGM sample was measured on a dry basis in a calorimeter bomb according to the standard NF EN ISO 18125 by Eurofins laboratory. The Lower Heating Value LHV (MJ kg^{-1}, dry basis) was calculated based on the measured HHV (MJ kg^{-1}, dry basis) using Equation (1) [11].

$$\text{LHV} \left(\text{MJ kg}^{-1}, \text{dry basis}\right) = \text{HHV} \left(\text{MJ kg}^{-1}, \text{dry basis}\right) - 0.2122 \times \text{H (wt. \%, dry basis)} \quad (1)$$

2.3. Pyrolysis

Slow pyrolysis experiments were conducted in both a thermobalance and a pilot plant pyrolizer. In TGA experiments, a very low heating rate (5 °C min^{-1}) and a very small sample size (about 15 mg) were used to minimise the effects of secondary reactions and heat and mass transfer (optimal pyrolysis conditions) [35]. In the pilot plant experiments, a larger size sample was used (about 1.5 kg).

2.3.1. TGA Study

TGA experiments were conducted, based on ASTM D7582-15 protocol, in a TGA/DSC 1 LF1100 thermogravimetric analyser (Mettler-Toledo, Columbus, OH, USA), following two stages: in a first stage, a pyrolysis step was performed by heating under nitrogen from an ambient temperature up to the pyrolysis temperature (350, 400, 450, 475, 500, 525, 550, 575 and 600 °C), then holding for 90 min and finally heating from pyrolysis temperature to 900 °C. In a second stage, a combustion step was carried out by switching the gas environment to synthetic air and holding at 900 °C for 60 min. For the whole run, a constant heating rate of 5 °C min^{-1} and a constant gas flow rate of 100 mL min^{-1} were maintained. Prior to TGA experiments, the sample was dried overnight in a ventilated oven at 105 °C under air.

2.3.2. Pyrolysis Experiments

Pyrolysis experiments were performed by an external, accredited laboratory RAPSODEE-UMR CNRS 5302. Before each pyrolysis test, the raw EGM sample was dried at 105 °C for 24 h. Dried grape marc was deposited onto the different drawers and loaded into the reactor. The reactor was purged for 10 min with nitrogen to remove the air. Pyrolysis experiments were conducted in a crossed fixed bed reactor. As illustrated in Figure 2, the reactor consisted of several drawers, disposed inside an electric vertical furnace. The temperature profiles in each drawer were instantly recorded using a thermocouple. The inert atmosphere gas, N_2, was fed from the bottom at a flow rate of 10 L min^{-1}, heated along the lower part of the reactor and crossing the sample bed. This method would ensure a good thermal contact between the gas and the particles of the sample, while taking the gas products away from the bed. The temperature was raised at a constant heating rate of 10 °C min^{-1} to the required hold temperature (450, 500 and 550 °C), and was maintained for 1h before cooling. The evolved gases reached the condensing system carried by the nitrogen inert gas. The condensed gas fraction (biooil) was recovered in a liquid collector after moving down by gravity. The non-condensable gases were sent into a filter before reaching the atmosphere. After completing the pyrolysis process, the reactor was naturally cooled under N_2 flow. The solid residue fraction (biochar) was recovered at an ambient temperature. The biochar samples were immediately stored in sealed polyethylene containers. Liquid and solid yields were directly obtained by weight. The gas yield was calculated using the following balance: Gas yield = 100% − (biochar yield + biooil yield).

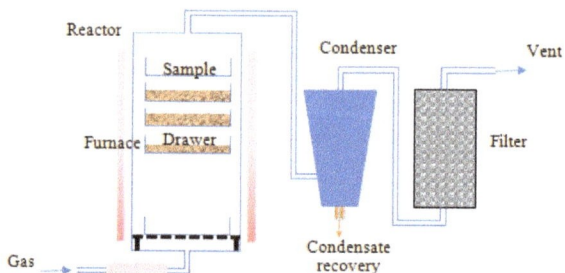

Figure 2. Pilot plant pyrolizer.

The biochar samples obtained at the three pyrolysis temperatures were labelled as EGM450, EGM500 and EGM550, respectively.

2.4. Biochar Characterisation

Biochars derived from pyrolysis were characterised using different techniques, including TGA, elemental analysis, X-ray diffraction analysis (XRD), Scanning Electron Microscopy (SEM), Energy Dispersive X-ray spectrometry (EDX), CO_2 adsorption analysis and pH measurement. Before analysis, representative samples of the different biochars were ground by hand using a mortar and pestle. Ground samples were sieved to a uniform size fraction of 250–400 µm and stored in sealed containers until use.

Proximate analysis was conducted in a TGA/DSC 1 LF1100 thermogravimetric analyser (Mettler-Toledo, Columbus, OH, USA), according to the ASTM D7582-15 standard, whereby samples were heated from ambient to 900 °C in a N_2 atmosphere, after which air was introduced. Weight loss up to 110 °C represents moisture content, and between 110 and 900 °C, volatile matter content. The residue remaining after combustion in air at 900 °C represents the ash fraction. Fixed carbon content was determined by difference. Before analysis, biochar samples were dried at 105 °C overnight.

Surface area and porosity measurements were carried out using Micromeritics ASAP2420 equipment (Norcross, GA, USA). Prior to analysis, the biochar samples were degassed at 250 °C under vacuum for 12 h to remove surface water and volatile organic species. Analysis was carried out using CO_2 adsorption at 0 °C, with temperature control being achieved with an ice-water bath. Adsorption was carried out up to a relative pressure (P/P°) of 0.03 to examine the micropore region. Pore size distribution for the biochar was determined using the Density Functional Theory (DFT) method. The mean pore size was estimated from the pore size distribution data. The micropore volume was determined according to the Dubinin–Radushkevich equation.

SEM analysis was carried out on a JSM 7900 JOEL microscope equipped with an energy dispersive X-ray analysis spectrometer. SEM images were acquired at 5 kV accelerating voltage, while SEM-EDX microanalysis were conducted at 15 kV accelerating voltage. Prior to the observations, the samples were adhered to a specimen stub with carbon-based adhesive tape, then carbon coating was applied by sputtering to ensure high electrical conductivity of the biochar sample.

X-ray diffraction analysis was conducted using a PANalytical MPD X'Pert Pro diffractometer (Eindhoven, The Netherlands) operating with Cu K α radiation, λ = 0.15406 nm at 40 mA and 45 kV. Data were recorded at room temperature, applying a 2-theta scanning range of 10–70°, and a step size of 0.017° with a scan step time of 220 s. Data processing was then carried out using High Score software. The different mineralogical phases were identified according to the Joint Committee on Powder Diffraction Standards (JCPDS) database.

Biochar pH was measured by mixing 1 g of biochar sample with 10 mL of deionised water. After 1 h of stirring, samples were allowed to stand for 60 min and then pH was measured using a calomel electrode-glass electrode system. Before pH measurements, the pH meter was calibrated using buffers of pH 4, 7 and 10.

Quantification of total C, N and H elements was carried out, by Eurofins laboratory, according to the NF EN ISO 17294-2 standard. S content was determined separately according to NF EN ISO 16994 standard. O content was calculated by difference, using Equation (2) [11].

$$O \ (wt \%, dry \ basis) = 100 - (ash + C + H + N + S) \qquad (2)$$

where Ash, C, H, N and S are, respectively ash, carbon, hydrogen, nitrogen and sulfur content (wt. %, dry basis).

Analysis of Al, Ca, Fe, K, Mg, Na, P, Si, As and Cr elements was conducted, by Eurofins laboratory, according to DIN EN ISO 17294-2 (E29): 2017-0 standard.

2.5. Plant Growth Study

The influence of biochar on plant growth was assessed using EGM450, EGM500 and EGM550 biochars. The microculture technique was applied as described by Lombaert [36]. The Italian ryegrass, used as a model plant, was cultivated in pots containing 1.18 kg of dry soil with a seeding density of 1.2 g pot^{-1} (Table 1). A loamy-sand soil, with a sand content of 50% and a pH value of 7.8, was used. Regulation ensured a daytime temperature of 25 °C and a night temperature of 20 °C. During cultivation, the soil humidity was maintained between 70 and 80%. Plant growth experiments were conducted using the following two modalities: the first modality corresponds to the control samples by adding mineral K fertiliser to ensure 50 U K_2O (1 unit corresponds to 1 kg K_2O ha^{-1}). The second modality corresponds to the use of biochars to provide the same amount of K (1.77 g pot^{-1} for EGM 450, 1.46 g pot^{-1} for EGM500 and 1.50 g pot^{-1} for EGM 550, respectively). Biochar application amounts were calculated according to mineral K content contained in the biochar (Table 2). The two modality tests were replicated 4 times and received additional fertilisation during the ryegrass growing (10 mL of 109.86 g L^{-1} $NaH_2PO_4 \cdot 2H_2O$ solution, 10 mL of 73.75 g L^{-1} $Ca(NO_3)_2 \cdot 4H_2O$ solution and 10 mL of solution containing 0.450 g L^{-1} $FeSO_4 \cdot 7H_2O$, 0.452 g L^{-1} $MnSO_4 \cdot H_2O$, 0.202 g L^{-1} $ZnSO_4 \cdot 7H_2O$, 0.189 g L^{-1} $CuSO_4 \cdot 5H_2O$,

0.278 g L^{-1} $Na_2B_4O_7 \cdot 10H_2O$ and 0.013 g L^{-1} $(NH_4)_6Mo_7O_{24} \cdot 4H_2O$ and weekly nutritive solution (20 mL of 5.65 g L^{-1} NH_4NO_3, 5.33 g L^{-1} $Mg(NO_3)_2 \cdot 6H_2O$ and 2.06 g L^{-1} $(NH_4)_2SO_4$). Ryegrass was harvested at 4 weeks. The dry weight contents of the aerial part of the plant were measured to evaluate the impact of biochar amendment on plant growth.

Table 1. Plant growth experiments.

Modality	Experiment	K_2O Amendment (kg ha^{-1})
M1	Ryegrass seed (1.8) + Soil (1.2 kg) + Minerals	50 U K_2O
M2	Ryegrass seed (1.8) + Soil (1.2 kg) + Biochar	50 U K_2O

3. Results and Discussion

3.1. Thermo-Chemical Properties of EGM

Table 2 reports the results of the characterisation analyses of exhausted grape marc feedstock. The characterisation results are in line with the values found in the literature. The HHV and LHV values of EGM fall within the expected range for grape marc feedstocks. The HHV and LHV were assessed to about 23 and 21 MJ kg^{-1}, respectively. The macromolecular organic composition data show that EGM was richer in lignin (42 wt. %) than in cellulose (15 wt. %) and hemicellulose (5 wt. %). The high lignin content measured in the EGM sample may lead to a higher biochar yield [35].

The grape marc feedstock had a high carbon content, with lower amounts of oxygen, hydrogen and nitrogen. Mineral analysis indicates that the Ca, P and K contents in EGM are relatively high, with a value of a few thousands of ppm, which is in the classical range found for grape marc feedstocks. In the case of Al, Fe, Mg and Si, the values measured were higher than those found in the literature [8,21]. Depending on the grape marc sources, the soil composition of the Alsatian vineyard, the harvest and collection practices, the grape harvesting age, the handling operations and the wine-making process, a variability of mineral concentrations in exhausted grape marc can be expected. The high amounts of silicon, aluminium, iron and magnesium derive from the composition of the soil. Silicon, aluminium and iron tend to accumulate in the seeds fraction of the exhausted grape marcs and that constituted a part of the biomass feedstock.

The TG and DTG (differential thermogravimetric analysis) curves, relative to the thermal decomposition of grape marc under nitrogen, are shown in Figure 3 (the isothermal oxidation step has been omitted). The TG curve shows that the pyrolysis of grape marc is a three-step thermal decomposition process; the first weight loss at a temperature below 180 °C relates to residual moisture (dehydration stage). The second decomposition step (devolatilization or primary pyrolysis), in the temperature range of 180 to approximately 483 °C, corresponds to the loss of organic volatiles. At this temperature range, exhausted grape marc decomposes into char (primary char), condensable gases (vapours and precursors to bio-oil) and non-condensable gases. Finally, the third step (secondary pyrolysis), at a higher temperature (>500 °C), which is characterised by low and continuous mass loss, corresponds to the slow charring process of the solid residue [18,22,35,37].

The proximate analysis of the EGM sample (Table 2) gave a fixed carbon content of 22 wt. %, a volatile matter content of 72 wt. % and an ash content of 6 wt. %. It is clearly seen that exhausted grape marc residue contains more ash content and produces more fixed carbon than woody feedstocks [38].

The DTG curve of the EGM sample (Figure 3) shows a single peak with three shoulders (small plateau): the lower temperature shoulders at around 220 and 270 °C, represents the decomposition of low molecular weight sugars [39] and hemicellulose, respectively. The peak at 323 °C, where the maximum rate of decomposition occurs, was assigned to the degradation of cellulose [19,22]. The higher temperature shoulder at around 400 °C was attributed to the temperature of maximum decomposition of lignin [35].

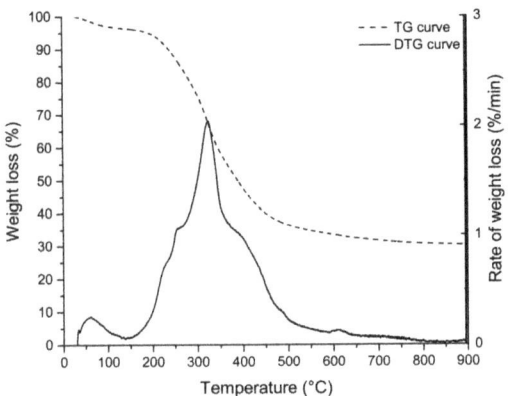

Figure 3. TG and DTG curves of EGM.

Table 2. Analysis and comparison of main properties of grape marc feedstocks.

Parameter	Present Study	Literature [1]	Reference
Energy properties (MJ kg^{-1}, dry basis)			
HHV	22.74	17.20–22.06	[3,8,10,12,14,19]
LHV	21.39	16.40–20.50	[3,8,10,12,14,19]
CHONS analysis (wt. %, dry basis)			
C	49	43–54.90	[3,8,10,12,14,21]
H	6.39	5.56–9.28	[3,8,10,12,14,21]
O	37	30.40–45.50	[3,8,10,12,14,21]
N	2	0.65–2.59	[3,8,10,12,14,21]
S	0.12	0.11–1.24	[3,8,10,12,14,21]
Mineral composition (mg kg^{-1}, dry basis)			
Al	503	50	[21]
Ca	6409	6220	[21]
Fe	478	110	[21]
Mg	1326	890	[21]
P	3669	2570	[21]
K	8038	6770–37,900	[8,21]
Si	2631	330	[21]
Na	249	102–1809	[8,21]
Cu	53	6–279	[8,21]
Organic composition (wt. %, dry basis)			
Cellulose	15.38	28.64	[13]
Hemicellulose	4.72	–	–
Lignin	42.46	41.98	[13]
Devolatilization			
T Start point [2] (°C)	180	–	–
T End point [3] (°C)	483	–	–
T max [4] (°C)	323	–	–
Rate max (wt. % min^{-1})	~2	–	–
Proximate analysis (wt. %, dry basis)			
Total volatile matter	71.9	63.60–72	[3,8,10,12,14,19]
Fixed carbon	22.4	20.68–28.20	[3,8,10,12,14,19]
Ash	5.7	3.46–8.20	[3,8,10,12,14,19]

[1] Grape marc originates from various sources and wine-making process. [2] Temperature at which devolatilization starts. [3] Temperature at which devolatilization ends. [4] Temperature at which the maximum rate of devolatilization occurs.

3.2. Pyrolysis Characteristics of EGM

A comparison of the thermal behaviour of the EGM sample at different pyrolysis temperature is presented in Figure 4. The mean objective here is to identify suitable low pyrolysis temperatures at which a high and effective conversion is possible. Most of the thermal decomposition of EGM occurs in the same temperature range under the different pyrolysis temperatures. The characteristic temperatures of the devolatilization stage involved in the thermal decomposition of EGM were approximately the same for all the runs. The temperature for which the maximum rate of reaction occurs (T_{max}) is almost identical for all the runs with a deviation of ± 1 °C from a mean temperature of 323 °C. For all the pyrolysis temperatures, there is a flat section at the higher temperatures (>500 °C) corresponding to the charring process.

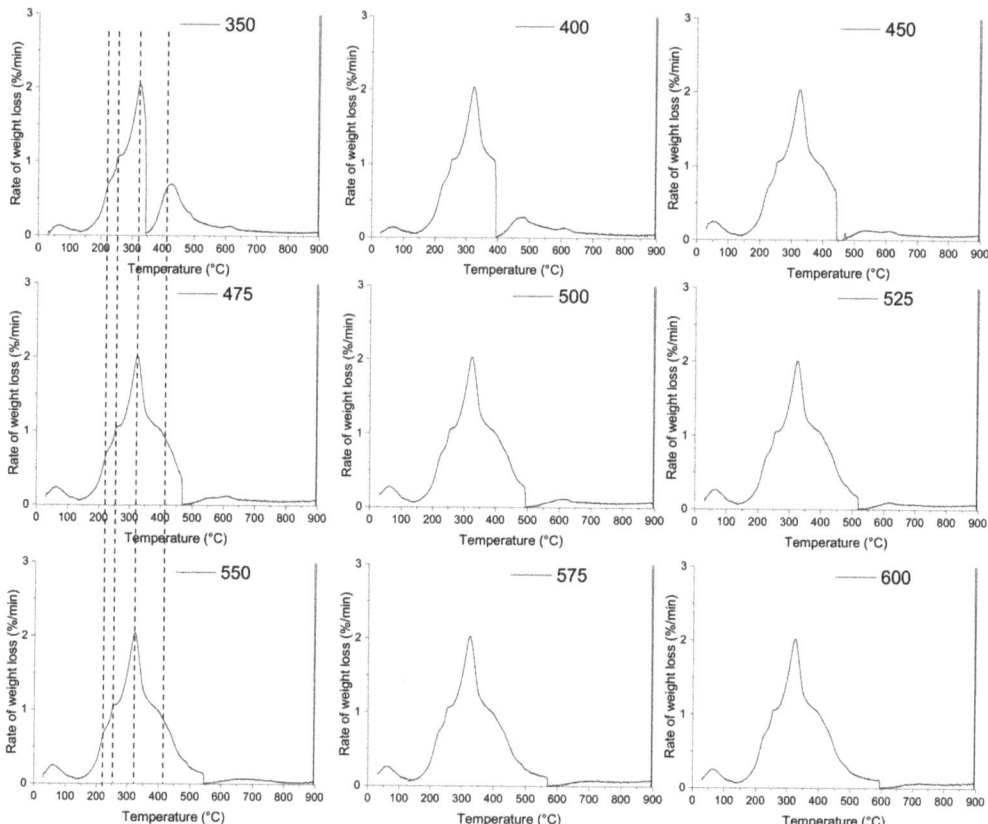

Figure 4. Effects of pyrolysis temperature on the pyrolysis behaviour of EGM.

However, some differences may be observed with respect to the thermal decomposition of lignin over the 350–450 °C pyrolysis temperature range. At a pyrolysis temperature $\geq$ 450 °C, the DTG curves of the EGM sample contain one broad peak with three shoulders. At a lower pyrolysis temperature $\leq$ 400 °C, the overlapping effects between cellulose, hemicellulose and lignin decompositions are limited, and two peaks can be distinguished.

Since cellulose and hemicellulose start to decompose at lower temperatures, their thermal decomposition was already complete before the pyrolysis temperature plateau. By increasing the pyrolysis temperature from 350 to 450 °C, lignin is decomposed at a higher extent and the DTG peak becomes broader. This means that a minimum temperature of 450 °C is needed to ensure the completion of the overall decomposition reactions for the used residence time. When increasing the pyrolysis temperature from 450 to 600 °C, the width of the DTG peaks is of an approximately equal size, suggesting that a higher pyrolysis temperature has little effect on the devolatilization process.

The pyrolysis yields for the volatile matter and char fraction were determined for all the investigated pyrolysis temperatures and their values are listed in Table 3. The solid char yield drastically decreases by increasing the pyrolysis temperature from 350 to 450 °C, while the volatile yield increases. At elevated temperatures, the pyrolysis yields remained stable, with a weight drop of ~1 wt. %. Another interesting observation can be made by plotting the mass loss curve vs. time, to estimate the time for char mass stabilisation (see Figure S1). The period for mass stabilisation at 450 °C was about 60 min, while for higher temperatures, the thermal decomposition lasted for 20–30 min. Thus, higher operating temperatures result in a shorter processing time.

Table 3. Product yields (wt. %, dry basis) in TGA [1].

Temperature	350 °C	400 °C	450 °C	475 °C	500 °C	525 °C	550 °C	575 °C	600 °C
Char Yield	50.6	41.9	37.3	35.9	35.2	34.5	33.9	33.2	32.6
Volatiles Yield	49.4	58.1	62.7	64.0	64.9	65.6	66.0	66.8	67.4

[1] Heating rate 5 °C min^{-1}.

In view of these findings, a temperature range from 450 to 550 °C and a one-hour residence time were selected for biochar production at a large scale. Indeed, temperatures higher than 450 °C will allow for a more effective decomposition of EGM organic components, while a longer products residence time (>30 min) permits a secondary reaction to occur between the char and volatiles, leading to a secondary char formation (maximising char aromaticity and stability) [11].

As mentioned earlier, pyrolysis experiments in a pilot plant were carried out at three different temperatures ranging from 450 to 550 °C, under a one-hour residence time. Depending on the pyrolysis temperature, different product yields were obtained, which must be attributed to their different pyrolysis behaviour. The average product yields are presented in Table 4. As the pyrolysis temperature is increased from 450 to 550 °C, the yields of char and liquid decreased, while the yield of gases increased. Obviously, higher temperatures promote secondary reactions of the tar, thus, increasing the yield of gases [40]. The char yields ranged from 41 to around 38 wt. %, which are slightly higher than the yields obtained in the TGA experiments. Similar char yields have been previously reported [14,21].

Table 4. Product yields (wt. %) in pilot plant.

Temperature	450 °C	500 °C	550 °C
Char yield	41.1	39.5	37.7
Liquid yield	18.7	24.4	20.2
Gas yield	40.2	36.1	42.1

Heating rate 10 °C min^{-1}.

3.3. Properties of EGM-Based Biochars

The elemental composition of EGM-based biochars is reported in Table 5. The carbon contents were high, ranging from 77.5 to 80%. The hydrogen contents were low, ranging from 2.5 to 3.32%. The nitrogen content ranged very tightly from 2.4 to 2.6%. The oxygen content ranged from 7.3 to 4.8%. The contents of C in the biochars tended to increase, while the contents of H and O tended to decrease with increased pyrolysis temperatures. These trends can be attributed to the enhanced thermal rearrangement process (aromatic growth) of the carbon structure with increased temperatures, leading to a highly carbonaceous and aromatic biochar.

Table 5. Physicochemical properties of EGM-based biochars.

	EGM450	EGM500	EGM550
CHONS analysis (wt. %, dry basis)			
Total C	77.50	80.00	79.90
H	3.32	2.90	2.50
O	7.29	5.36	4.76
N	2.57	2.42	2.54
S	<0.30	<0.30	<0.30
O/C molar ratio [1]	0.07	0.05	0.04
H/C molar ratio [1]	0.51	0.44	0.38
C/N molar ratio [1]	35.17	38.55	36.69
Mineral composition (mg kg^{-1}, dry basis)			
As	<0.80	<0.80	<0.80
Cr	2	5	12
Al	423	237	709
Ca	13,900	13,600	16,500
Fe	480	426	712
K	24,500	24,900	28,200
Mg	2590	2430	2950
Na	366	354	461
P	6460	6710	7560
Si	2060	3070	5060
Proximate analysis (wt. %, dry basis)			
Total volatile matter	23.06	19.10	17.20
Fixed carbon	67.92	71.88	72.80
Ash	9.02	9.02	10
Specific surface area (m^2 g^{-1})	170	193	219
Micropore volume (cm^3 g^{-1})	0.091	0.098	0.116
Mean pore size (nm)	0.56	0.56	0.61
pH	10.8	10.4	10.4

[1] Calculated by using total carbon.

Atomic element ratios were determined to predict some characteristics of the produced biochars such as aromaticity (H/C), stability (O/C) and potential N immobilisation (C/N) [41]. Both the H/C and O/C ratios decreased with increased pyrolysis temperatures, indicating a more condensed and stable biochar with fewer oxygen functional groups [41]. The C/N ratio varied slightly with increased pyrolysis temperatures, suggesting that those biochars will lead to increased N immobilisation in soils.

Analysis of the inorganic content in biochars compared with that of the EGM feedstock gave insight into the enrichment of biochars with inorganic elements, particularly Ca (13,600–16,500 ppm), Mg (2430–2950 ppm), P (6460–7560 ppm) and K (24,500–28,200 ppm). As shown in Table 5, increasing the pyrolysis temperature from 450 to 550 °C resulted in higher inorganic contents. This includes Al, Ca, Fe, K, Mg, Na, P and Si.

The TG and DTG thermograms of the three biochars are presented in Figure 5. For the three samples, the following two main regions of weight loss were observed: the lower temperature loss (25–150 °C) corresponds to the desorption of water, while the higher temperature loss, between 400 and 900 °C, could be attributed to the release of volatile organic compounds or/and loss of mineral compounds/salts not released during the pyrolysis process at a low temperature. As expected, the greater mass loss occurs for the biochars pyrolyzed at lower temperatures. By examining the DTG thermogram, EGM500 and EGM550 biochars gave a single DTG peak located at around 650 °C with a very small shoulder at 700 °C. EGM450 has an additional DTG peak located at a lower temperature, around 550 °C. This peak can be attributed to the presence of a small fraction of lignin that might not be totally decomposed during the pyrolysis process.

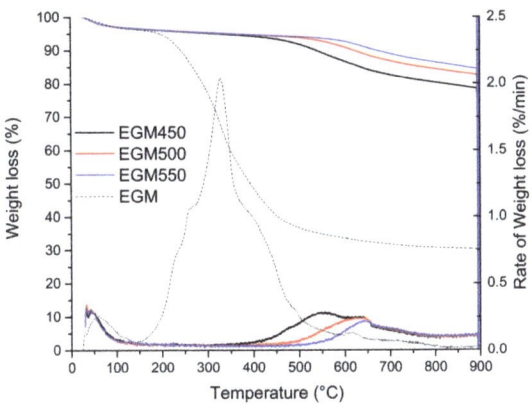

Figure 5. TG and DTG curves of EGM-based biochars.

Proximate analysis of the three biochar (Table 5) gave a high fixed carbon content ranging from 68 to 73 wt. %, a volatile matter ranging from 17 to 23 wt. % and a high ash content ranging from 9 to 10 wt. %. Fixed carbon and ash contents increased with the temperature while the volatile matter decreased. The high ash content (when compared with wood-derived biochars [42,43]) in the three biochars can be attributed to the high ash content found in the EGM feedstock, which is about 6 wt. %. Considering the above results, the EGM550 biochar seemed to undergo more complete carbonisation, resulting in a higher fixed carbon content, lower volatile matter and lower H/C ratio, and, therefore, higher stability.

The specific surface area, the micropore volume and the mean pore size of the different biochars were calculated starting for the CO_2 adsorption isotherms and presented in Table 5. The micropore volume increased with the increasing pyrolysis temperature from 0.091 to 0.116 $cm^3\ g^{-1}$. The BET surface area followed a similar trend, increasing from 170 to 219 $m^2\ g^{-1}$. Figure 6 shows the presence of micropores with a pore size between 0.4 and 1 nm. The micropore size distribution curve of the biochar samples showed a tri-modal pore size distribution. The ultra-microporous structure found in the biochars may arise from the release of gases during the pyrolysis process.

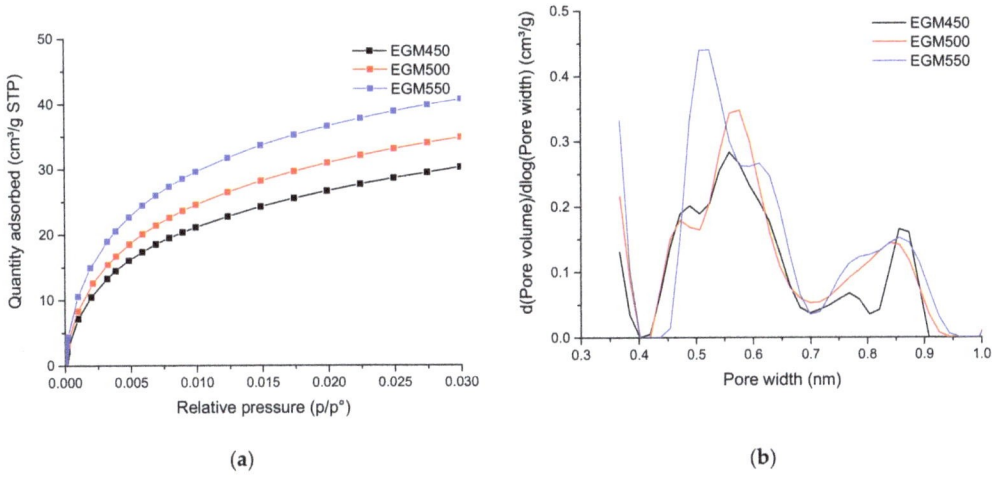

Figure 6. (**a**) Adsorption isotherms of CO_2 at 0° and (**b**) pore size distribution for EGM-based biochars.

The XRD patterns of the different biochar samples are shown in Figure 7. They show a broad band centred at around 23° 2θ (3.8 Å) and corresponding to amorphous carbon. Calcite ($CaCO_3$) and quartz (SiO_2) were the two major crystalline phases identified in the biochar samples. The presence of carbonate and quartz crystalline phases is consistent with their elemental composition. Calcite could have formed during the pyrolysis process via the hydration and carbonation of Ca oxide [44]. Some minor peaks of crystalline whewellite (calcium oxalate, $Ca(C_2O_4) H_2O$) were detected in the XRD pattern of the biochar produced at 450 °C. Calcium oxalate in biochar has been suggested to derive from the biomass feedstock. Whewellite peaks were absent in EGM500 and EGM550 due to its decomposition (at 479 °C) into calcite at a high pyrolysis temperature [45]. Most of these salts are highly soluble and could increase the biochars' alkalinity [44].

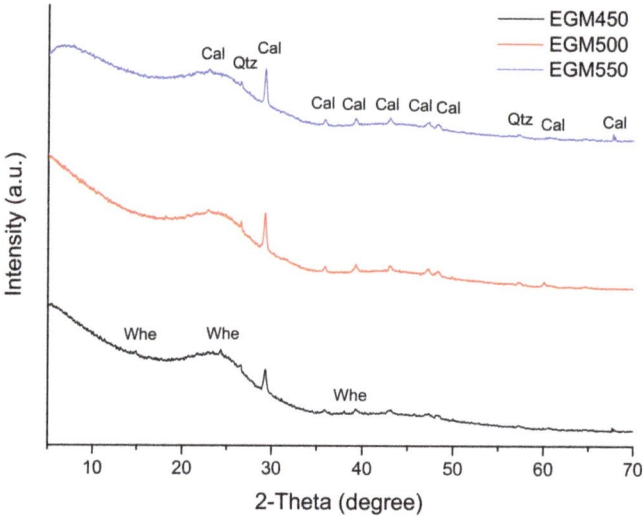

Figure 7. XRD patterns of EGM-based biochars (Cal: Calcite, Qtz: Quartz, Whe: Whewellite).

The SEM images (Figure 8) revealed a heterogeneous and porous structure of EGM-based biochar particles. The heterogeneity feature is derived from the complex composition of grape marc (seeds, stalks and skins). On the other hand, the porous nature of these materials originates from the precursor plant cellular structure of EGM and from the gas evolved during the pyrolysis process [41]. The EDX spectra (not shown) indicate that the chemical compositions of the different structures are similar. The presence of carbon, oxygen, calcium and potassium can be readily detected.

The pH values of the EGM-based biochars are given in Table 5. All the produced biochars had an alkaline pH (>10). This basic characteristic arises from both their chemical composition and their surface chemistry. Indeed, during pyrolysis, acidic functional groups (oxygen-containing groups) are removed and alkali and alkaline earth elements concentrate, resulting in a higher pH value [41].

3.4. Qualification of EGM-Based Biochar as a Potential Candidate for Soil Amendment

Biochar is recognised as a potential soil amendment. Its properties, such as high organic C, high mineral content of macronutrients (Ca, K and P) and micronutrients (B, Mn, Cu, Zn, Fe, Mo and Mg), porous structure, high and active surface area and high cation exchange capacity have been highlighted as soil improvers [41,46]. Moreover, the effect of biochar on soil pH is beneficial for ameliorating acid soils. In this study, biochars derived from EGM were found to meet the requirements relating to agricultural applications. The required characteristics include high organic carbon content, low ash and volatile matter contents, porous structure, high surface area and high pH value. In addition, the biochar material must meet the soil toxicity assessment thresholds. Thermal and proximate analysis revealed that the pyrolyzed biochars have a high fixed carbon (>68 wt. %), low volatile matter (<23 wt. %) and high ash content (9–10 wt.). The high fixed carbon content is a good indicator of the biochar's stability and ability in sequestering carbon. The volatile matter is generally considered as an organic contaminant that may be released from biochar in soil. Diminishing the volatile matter content in biochar is always essential to minimise the risk of soil contamination and its related effects on microbial populations and plant growth [41,46]. In terms of chemical composition, the produced biochars have shown a high carbon content ($\geq$77.5 wt. %), and very low values for both O/C and H/C ratios. Such information serves as a key indicator for the carbon sequestration potential of the produced biochars [28,41]. Mineral analysis results have revealed a high mineral content of macronutrients (Ca, K and P), in accordance with SEM-EDX analysis. These results are a good indicator of the nutrient value and the substantial fertiliser value of these biochars. CO_2 adsorption and SEM analysis provided valuable insight into the textural and structural properties of the biochars. The microporous nature of these materials and their relatively high surface areas would enhance their sorption capacity (water and added nutrients), while their microporosity will provide suitable habitats for micro-organisms colonisation. The biochars produced from EGM were all alkaline with pH, which made them more suitable for acid soils.

Figure 8. SEM images of (a) EGM450, (b) EGM500 and (c) EGM550 biochars.

The biochars properties show a qualitative dependence to the temperature of production. A clear correlation cannot be established due to the complex matrix of biochars and the influence of other parameters (inorganics catalytic impact, granulometry, etc.) on the pyrolysis process and consequently on the biochars characteristics. Nevertheless, certain behaviours can be underlined; concerning the elemental composition, higher pyrolysis temperatures led to a higher carbon content and lower H, O and N contents. For the proximate analysis, higher pyrolysis temperatures generated biochars with a higher fixed carbon content and a lower volatile matter content. For the textural properties, higher pyrolysis temperature resulted in accentuated microporous structures. Additionally, as the pyrolysis temperature increases, the content of minerals is increased due to concentration. In addition, the effect of the pyrolysis temperature on biochar properties, particularly the chemical and textural properties, was more pronounced at the highest temperature. The lowest pyrolysis temperature (450 °C) produces a less condensed biochar structure, which is expected to be more biodegradable. The highest pyrolysis temperature (550 °C) produces a more stable biochar with a high microporous structure.

3.5. Plant Growth Study

The impact of the biochars on ryegrass growth has been firstly evaluated by measuring the plant growth dry mass after 4 weeks (see Figure 9). The plant growth was clearly affected by the type of biochar added to the soil. When compared to the control samples (modality M1), the ryegrass dry mass obtained in the pots containing EGM550 biochar was lower (1.14 g) than that obtained with the control samples (1.61 g). No effect on the ryegrass dry mass could be evaluated in the presence of the EGM500 biochar, with an average plant growth dry mass of 1.62 g. On the other hand, a clear benefit related to the addition of EGM450 biochar has been observed; the dry mass was improved by 16% with respect to the control sample and reached 1.87 g. The reduction in the plant growth dry mass when EGM550 was added can be explained at first by the effects of biochar on soil N dynamics [47]. The plant-available N pool (by comparing the C/N ratio, which is often used to predict mineralisation and N release in soils) was lower; the application of biochar produced at a high temperature (550 °C) induced a higher net N mineralisation and lower N immobilisation than those produced at low temperatures ($\leq$550 °C); as a result, the biochar produced at a low temperature (450 °C) induced less mineral N leaching loss and greater soil N retention, resulting in a higher plant growth dry mass. Secondly, the carbon matrix reorganises at higher temperature and can incorporate certain minerals in the structure. They cannot be easily released as in the biochars prepared at lower temperature and then they cannot be assimilated by the plant.

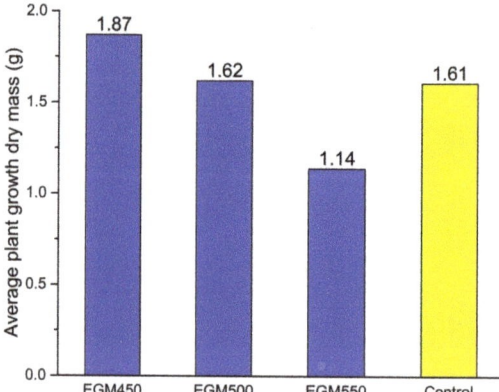

Figure 9. Average plant growth (on dry mass basis) for the pots containing EGM biochars prepared at different temperatures (blue) and the control sample (yellow).

The pyrolysis temperature has already been identified as a key parameter for optimising the biochar features adapted to agro applications [21]. In a previous study [48], an EGM pyrolysis temperature between 400 and 500 °C was identified as the best temperature range for producing biochars that are able to release the minerals at an optimum rate, as determined by leaching experiments at a laboratory scale. The present results consolidate the previous observations in terms of plant growth. Even if the present plant growth experiments are targeted on K loading, synergetic effects, such as the water retention, the realising of other minerals and the biological and structural modification of the soil, can affect the growth performance.

4. Conclusions

Biochar production and characterisation provided in-depth knowledge of the chemical and physical characterisation of the produced biochars. Through the results of this study, it was found that the pyrolysis temperature impacts the pyrolysis yields, the biochar thermal stability, the development of surface area, the mineral composition and, to a lesser extent, the ash content. The pyrolysis of exhausted grape marc at a medium temperature, between 450 and 550 °C, could be considered as a sustainable pathway to manage winery industry wastes, while achieving the economic efficiency of the production process once implemented in different applications. Biochars prepared at a pyrolysis temperature below 500 °C have been demonstrated to be promising as an agricultural amendment. The use of biochar in this field is crucial in decreasing the environmental pollution caused by synthetic amendment, reducing the global-carbon emission and improving the soil quality (i.e., water retention).

Supplementary Materials: The following are available online at https://www.mdpi.com/article/10.3390/su132011187/s1, Figure S1. Effects of pyrolysis temperature on the pyrolysis behaviour of EGM (weight loss vs time).

Author Contributions: Conceptualisation, L.L. and S.B.; methodology, L.L., S.B. and N.T.; validation, S.B. and L.L.; formal analysis, K.F. and N.T.; investigation, M.B.A., L.R., K.F. and L.L.; resources, S.B. and L.L.; writing—original draft preparation, K.F.; writing—review and editing, S.B.; visualisation, K.F.; supervision, L.L., S.B., K.F. and M.Z.; project administration, S.B. and L.L.; funding acquisition, S.B. All authors have read and agreed to the published version of the manuscript.

Funding: This research was partially funded by the Region Grand Est, 3BR "biomolecules et biomatériaux pour la bioéconomie Régionale vers une valorization zero déchets".

Institutional Review Board Statement: Not applicable.

Informed Consent Statement: Not applicable.

Data Availability Statement: Data is contained within the article or Supplementary Materials.

Acknowledgments: The authors acknowledge the Carnot MICA institute for funding the pilot scale pyrolysis tests and IS2M technical platforms for the physicochemical characterisations. The authors are very grateful to L. Michelin (XRD), L. Josien (SEM/EDX), and C. Vaulot (CO_2 adsorption experiments) for their contributions.

Conflicts of Interest: The authors declare no conflict of interest.

References

1. Gómez-Brandón, M.; Lores, M.; Insam, H.; Domínguez, J. Strategies for Recycling and Valorization of Grape Marc. *Crit. Rev. Biotechnol.* **2019**, *39*, 437–450. [CrossRef]
2. Nakai, J. Food and Agriculture Organization of the United Nations and the Sustainable Development Goals Natural Re-Sources Management Officer, FAO From the Millennium Development Goals to the Sustainable Development Goals. *Sustain. Dev.* **2018**, *22*, 3–11.
3. Muhlack, R.A.; Potumarthi, R.; Jeffery, D.W. Sustainable Wineries through Waste Valorisation: A Review of Grape Marc Utilisation for Value-Added Products. *Waste Manag.* **2018**, *72*, 99–118. [CrossRef]
4. Ahmad, B.; Yadav, V.; Yadav, A.; Rahman, M.U.; Yuan, W.Z.; Li, Z.; Wang, X. Integrated Biorefinery Approach to Valorize Winery Waste: A Review from Waste to Energy Perspectives. *Sci. Total. Environ.* **2020**, *719*, 137315. [CrossRef]
5. Panorama Des Industries Agroalimentaires. Available online: https://agriculture.gouv.fr/Le-Panorama-Des-Industries-Agroalimentaires (accessed on 20 April 2021).
6. State of the World Vitivinicultural Sector in 2020. Available online: https://www.oiv.int/public/medias/7909/oiv-State-of-the-World-Vitivinicultural-Sector-in-2020.Pdf (accessed on 20 April 2021).
7. Ruiz-Moreno, M.J.; Raposo, R.; Cayuela, J.M.; Zafrilla, P.; Piñeiro, Z.; Rojas, J.M.M.; Mulero, J.; Puertas, B.; Girón, F.; Guerrero, R.F.; et al. Valorization of Grape Stems. *Ind. Crop. Prod.* **2015**, *63*, 152–157. [CrossRef]
8. Toscano, G.; Riva, G.; Duca, D.; Pedretti, E.F.; Corinaldesi, F.; Rossini, G. Analysis of the Characteristics of the Residues of the Wine Production Chain Finalized to Their Industrial and Energy Recovery. *Biomass-Bioenergy* **2013**, *55*, 260–267. [CrossRef]
9. Bordiga, M.; Travaglia, F.; Locatelli, M. Valorisation of Grape Pomace: An Approach That Is Increasingly Reaching Its Maturity—A Review. *Int. J. Food Sci. Technol.* **2019**, *54*, 933–942. [CrossRef]
10. Zhang, N.; Hoadley, A.; Patel, J.; Lim, S.; Li, C. Sustainable Options for the Utilization of Solid Residues from Wine Production. *Waste Manag.* **2017**, *60*, 173–183. [CrossRef]
11. Basu, P. Chapter 3—Pyrolysis and Torrefaction. In *Biomass Gasification and Pyrolysis*; Academic Press: Boston, FL, USA, 2010; pp. 65–96. ISBN 978-0-12-374988-8.
12. Encinar, J.; Beltran, F.J.; Bernalte, A.; Ramiro, A.; González, J.F.G. Pyrolysis of Two Agricultural Residues: Olive and Grape Bagasse. Influence of Particle Size and Temperature. *Biomass- Bioenergy* **1996**, *11*, 397–409. [CrossRef]
13. Demiral, I.; Ayan, E.A. Pyrolysis of Grape Bagasse: Effect of Pyrolysis Conditions on the Product Yields and Characterization of the Liquid Product. *Bioresour. Technol.* **2011**, *102*, 3946–3951. [CrossRef]

14. Marculescu, C.; Ciuta, S. Wine Industry Waste Thermal Processing for Derived Fuel Properties Improvement. *Renew. Energy* **2013**, *57*, 645–652. [CrossRef]
15. Spigno, G.; Marinoni, L.; Garrido, G.D. State of the Art in Grape Processing By-Products. In *Handbook of Grape Processing By-Products*; Elsevier BV: London, UK, 2017; pp. 1–27. [CrossRef]
16. Basso, D.; Patuzzi, F.; Castello, D.; Baratieri, M.; Rada, E.C.; Weiss-Hortala, E.; Fiori, L. Agro-Industrial Waste to Solid Biofuel through Hydrothermal Carbonization. *Waste Manag.* **2016**, *47*, 114–121. [CrossRef] [PubMed]
17. Xu, R.; Ferrante, L.; Briens, C.; Berruti, F. Flash Pyrolysis of Grape Residues into Biofuel in a Bubbling Fluid Bed. *J. Anal. Appl. Pyrolysis* **2009**, *86*, 58–65. [CrossRef]
18. Valente, M.; Brillard, A.; Schönnenbeck, C.; Brilhac, J.-F. Investigation of Grape Marc Combustion Using Thermogravimetric Analysis. Kinetic Modeling Using an Extended Independent Parallel Reaction (EIPR). *Fuel Process. Technol.* **2015**, *131*, 297–303. [CrossRef]
19. Casazza, A.A.; Aliakbarian, B.; Lagazzo, A.; Garbarino, G.; Carnasciali, M.M.; Perego, P.; Busca, G. Pyrolysis of Grape Marc before and After the Recovery of Polyphenol Fraction. *Fuel Process. Technol.* **2016**, *153*, 121–128. [CrossRef]
20. Tounsi, M.S.; Ouerghemmi, I.; Wannes, W.A.; Ksouri, R.; Zemni, H.; Marzouk, B.; Kchouk, M.E. Valorization of Three Varieties of Grape. *Ind. Crop. Prod.* **2009**, *30*, 292–296. [CrossRef]
21. Ibn Ferjani, A.; Jeguirim, M.; Jellali, S.; Limousy, L.; Courson, C.; Akrout, H.; Thevenin, N.; Ruidavets, L.; Muller, A.; Bennici, S. The Use of Exhausted Grape Marc to Produce Biofuels and Biofertilizers: Effect of Pyrolysis Temperatures on Biochars Properties. *Renew. Sustain. Energy Rev.* **2019**, *107*, 425–433. [CrossRef]
22. Torres-Garcia, E.; Brachi, P. Non-Isothermal Pyrolysis of Grape Marc. *J. Therm. Anal. Calorim.* **2019**, *139*, 1463–1478. [CrossRef]
23. Zacharof, M.-P. Grape Winery Waste as Feedstock for Bioconversions: Applying the Biorefinery Concept. *Waste Biomass Valorization* **2016**, *8*, 1011–1025. [CrossRef]
24. Rivera, O.M.P.; Leos, M.D.S.; Solis, V.E.; Domínguez, J.M. Recent Trends on the Valorization of Winemaking Industry Wastes. *Curr. Opin. Green Sustain. Chem.* **2021**, *27*, 100415. [CrossRef]
25. Balat, M.; Balat, M.; Kırtay, E.; Balat, H. Main Routes for the Thermo-Conversion of Biomass into Fuels and Chemicals. Part 1: Pyrolysis Systems. *Energy Convers. Manag.* **2009**, *50*, 3147–3157. [CrossRef]
26. Fahmy, T.Y.A.; Fahmy, Y.; Mobarak, F.; El-Sakhawy, M.; Abou-Zeid, R.E. Biomass Pyrolysis: Past, Present, and Future. *Environ. Dev. Sustain.* **2018**, *22*, 17–32. [CrossRef]
27. Xie, T.; Reddy, K.; Wang, C.; Yargicoglu, E.; Spokas, K. Characteristics and Applications of Biochar for Environmental Remediation: A Review. *Crit. Rev. Environ. Sci. Technol.* **2015**, *45*, 939–969. [CrossRef]
28. Schmidth, H.P.; Bucheli, T.; Kammann, C.; Glaser, B.; Abiven, S. *EBC (2012) "EuropeanEuropean Biochar Certificate—Guidelines for a Sustainable Production of Biochar"*; EBC: Arbaz, Switzerland, 2021.
29. Ding, Y.; Liu, Y.; Liu, S.; Li, Z.; Tan, X.; Huang, X.; Zeng, G.; Zhou, L.; Zheng, B. Biochar to Improve Soil Fertility. A Review. *Agron. Sustain. Dev.* **2016**, *36*, 36. [CrossRef]
30. Verheijen, F.; Jeffery, S.; Bastos, A.C.; Van Der Velde, M.; Diafas, I. *Biochar Application to Soils—A Critical Scientific Review of Effects on Soil Properties, Processes and Functions*. EUR 24099 EN; European Commission: Luxembourg, 2010; Volume 8, ISBN 9789279142932.
31. Kookana, R.S.; Sarmah, A.K.; Van Zwieten, L.; Krull, E.; Singh, B. Biochar Application to Soil. Agronomic and Environmental Benefits and Unintended Consequences. In *Advances in Agronomy*; Elsevier Inc.: Amsterdam, The Netherlands, 2011; Volume 112, pp. 103–143. ISBN 9780123855381.
32. Kavitha, B.; Reddy, P.V.L.; Kim, B.; Lee, S.S.; Pandey, S.K.; Kim, K.-H. Benefits and Limitations of Biochar Amendment in Agricultural Soils: A Review. *J. Environ. Manag.* **2018**, *227*, 146–154. [CrossRef]
33. Guo, X.-X.; Liu, H.-T.; Zhang, J. The Role of Biochar in Organic Waste Composting and Soil Improvement: A Review. *Waste Manag.* **2020**, *102*, 884–899. [CrossRef]
34. Van Soest, P.J. Use of Detergents in the Analysis of Fibrous Feeds. II. A Rapid Method for the Determination of Fiber and Lignin. *J.-Assoc. Off. Anal. Chem.* **1990**, *73*, 491–497. [CrossRef]
35. Groenli, M.G. A Theoretical and Experimental Study of the Thermal Degradation of Biomass. Master's Thesis, Norges teknisknaturvitenskapelige Universitet, Trondheim, Norway, 1996.
36. Lombaert, V. Micro-cultures, Stanford-De Ment Methods [determination of the short term availability of fertilizers, diagnosis of soil fertility]. *Dossiers Agronomiques d'Aspach-le-Bas* **1987**, *2*, 59–86.
37. Werner, K.; Pommer, L.; Broström, M. Thermal Decomposition of Hemicelluloses. *J. Anal. Appl. Pyrolysis* **2014**, *110*, 130–137. [CrossRef]
38. Mohan, D.; Pittman, C.U.; Steele, P.H. Pyrolysis of Wood/Biomass for Bio-Oil: A Critical Review. *Energy Fuels* **2006**, *20*, 848–889. [CrossRef]
39. Örsi, F. Kinetic Studies on the Thermal Decomposition of Glucose and Fructose. *J. Therm. Anal. Calorim.* **1973**, *5*, 329–335. [CrossRef]
40. Grammelis, P.; Margaritis, N.; Kourkoumpas, D.-S. 4.27 Pyrolysis Energy Conversion Systems. *Compr. Energy Syst.* **2018**, 1065–1106. [CrossRef]
41. Rajapaksha, A.U.; Mohan, D.; Igalavithana, A.D.; Lee, S.S.; Ok, Y.S. Definitions and Fundamentals of Biochar. In *Biochar: Production, Characterization, and Applications*; Ok, Y.S., Uchimiya, S.M., Chang, S.X., Bolan, N., Eds.; Urbanization, Industrialization, and the Environment; CRC Press: Boca Raton, FL, USA, 2016; pp. 4–16. ISBN 978-1-4822-4230-0.

42. Wang, S.; Dai, G.; Yang, H.; Luo, Z. Lignocellulosic Biomass Pyrolysis Mechanism: A State-of-the-Art Review. *Prog. Energy Combust. Sci.* **2017**, *62*, 33–86. [CrossRef]
43. Gholizadeh, M.; Hu, X.; Liu, Q. A Mini Review of the Specialties of the Bio-Oils Produced from Pyrolysis of 20 Different Bio-Masses. *Renew. Sustain. Energy Rev.* **2019**, *114*, 109313. [CrossRef]
44. Singh, B.; Camps-Arbestain, M.; Lehmann, J. *Biochar: A Guide to Analytical Methods*; Csiro Publishing: Clayton Southy, Australia, 2017; ISBN 9781486305094.
45. Frost, R.L.; Weier, M.L. Thermal Treatment of whewellite—A Thermal Analysis and Raman Spectroscopic Study. *Thermochim. Acta* **2004**, *409*, 79–85. [CrossRef]
46. Bruckman, V.J.; Varol, E.A.; Uzun, B.B.; Liu, J. *Biochar: A Regional Supply Chain Approach in View of Climate Change Mitigation*; Cambridge University Press: Cambridge, UK, 2016; ISBN 9780333227794.
47. Chan, K.Y.; Xu, Z. Biochar: Nutrient Properties and Their Enhancement. In *Biochar for Environmental Management*; Routledge: Oxfordshire, UK, 2012; pp. 99–116.
48. Ibn Ferjani, A.; Jellali, S.; Akrout, H.; Limousy, L.; Hamdi, H.; Thevenin, N.; Jeguirim, M. Nutrient Retention and Release from Raw Exhausted Grape Marc Biochars and an Amended Agricultural Soil: Static and Dynamic Investigation. *Environ. Technol. Innov.* **2020**, *19*, 100885. [CrossRef]

Article

Two-Stage Continuous Process for the Extraction of Silica from Rice Husk Using Attrition Ball Milling and Alkaline Leaching Methods

Ji Yeon Park [1,2], Yang Mo Gu [1,2], Seon Young Park [1,2], Ee Taek Hwang [3], Byoung-In Sang [2], Jinyoung Chun [1,*] and Jin Hyung Lee [1,*]

1. Korea Institute of Ceramic Engineering and Technology (KICET), Heungdeok-gu, Cheongju-si 28160, Korea; qkrwl660@kicet.re.kr (J.Y.P.); rndidah123@kicet.re.kr (Y.M.G.); young005@kicet.re.kr (S.Y.P.)
2. Division of Chemical Engineering & Bio Engineering, Hanyang University, Seoul 04763, Korea; biosang@hanyang.ac.kr
3. Department of Food Biotechnology, Dong-A University, Busan 49315, Korea; ethwang@dau.ac.kr
* Correspondence: jchun@kicet.re.kr (J.C.); leejinh1@kicet.re.kr (J.H.L.); Tel.: +82-55-792-2626 (J.C.); +82-43-913-1502 (J.H.L.)

Abstract: A two-stage continuous process was developed for improved silica extraction from rice husk. The two-stage continuous process consists of attrition ball milling and alkaline leaching methods. To find the optimum conditions for the continuous process, the effects of alkaline leaching parameters, such as the alkaline solution type and reaction conditions, on the silica extraction yield were investigated in a batch process. The use of NaOH showed a slightly higher silica yield than KOH. The optimum reaction conditions were found to be 0.2 M, 80 °C, 3 h, and 6% (w/v) for the reaction concentration, temperature, duration time, and solid content, respectively. Attrition ball milling was used to make micron-sized rice husk particles and to improve the fluidity of the rice husk slurry. The two-stage continuous process was performed using optimum conditions as determined based on the results of the batch experiment. The two-stage continuous extraction was stably operated for 80 h with an 89% silica yield. During the operation, the solid content remained consistent at 6% (w/v). The obtained silica was characterized using inductively coupled plasma–optical emission spectrometry (ICP–OES), X-ray diffraction (XRD), and the Brunauer–Emmett–Teller (BET) method.

Keywords: silica; rice husk; alkaline leaching; continuous process; biomass; bio-based material

Citation: Park, J.Y.; Gu, Y.M.; Park, S.Y.; Hwang, E.T.; Sang, B.-I.; Chun, J.; Lee, J.H. Two-Stage Continuous Process for the Extraction of Silica from Rice Husk Using Attrition Ball Milling and Alkaline Leaching Methods. *Sustainability* 2021, *13*, 7350. https://doi.org/10.3390/su13137350

Academic Editors: Dirk Enke, Hossein Beidaghy Dizaji, Volker Lenz and Thomas Zeng

Received: 6 June 2021
Accepted: 29 June 2021
Published: 30 June 2021

Publisher's Note: MDPI stays neutral with regard to jurisdictional claims in published maps and institutional affiliations.

Copyright: © 2021 by the authors. Licensee MDPI, Basel, Switzerland. This article is an open access article distributed under the terms and conditions of the Creative Commons Attribution (CC BY) license (https://creativecommons.org/licenses/by/4.0/).

1. Introduction

Rice is a major agricultural product across the world, and its annual production was approximately 996 million tons in 2018 [1]. Rice husk accounts for 20% of rice byproducts [2] and has various applications in different industries, e.g., (a) as an industrial fuel for paddy processing and in the generation of process steam in power plants; (b) as a fertilizer and substrate or pet food fiber; (c) as an ingredient for the preparation of activated carbon or substrate for silica and silicon compound production, and (d) as raw material for brick production [3,4]. Rice husk is composed of approximately 70–80% organic substances such as cellulose, hemicellulose, and lignin, and the remaining 20–30% comprises inorganic compounds [5,6]. A major inorganic component is silica, which accounts for approximately 95% of the inorganic compounds. The silica in rice husk is amorphous and has a colloidal state in water. Silica is an industrial material that is highly utilized as an additive for catalysts, insulation, toothpaste [7], coating solutions [8,9], and cosmetics [10]. The use of "biosilica" (rice husk-derived silica) as an alternative for silica in various industrial applications would mitigate high energy consumption, natural resource depletion, and greenhouse gas emissions [3].

Two approaches are used to extract silica from rice husk: combustion and chemical treatment. Direct combustion is the most popular method and is conducted in open fire stoves or boilers. During burning, rice husk is oxidized, resulting in ash products. This is the simplest method to obtain inorganic compounds from rice husk. The inorganic compounds, so-called rice husk ash, can be converted into soluble sodium silicate by reacting with aqueous sodium hydroxide [11]. Lee et al. used sulfuric acid to remove organic compounds from rice husk before combustion [5]. Sulfuric acid dissolves most celluloses and hemicelluloses, which are discarded by separating liquids from solids. The acid treatment improved the purity of silica finally obtained. Hincapié-Rojas et al. obtained submicron silica particles from rice husk by subsequent treatments of combustion, acid leaching, and mechanical ball milling [12]. Souza et al. compared the use of hot organic acid and boiling water before combustion to obtain high-quality silica [13]. Chemical extraction is adopted for environmentally friendly extraction. The chemical extraction method consists of acid treatment and an alkaline leaching step [14]. In the chemical extraction method, several chemical routes are used to achieve highly efficient silica extraction [14–17]. Zulkifli et al. combusted rice husk to obtain rice husk ash, followed by acid and alkaline leaching [17]. Chun et al. directly treated rice husk with sulfuric acid to leach metallic impurities, followed by combustion to remove organic residual compounds; afterward, high-purity silica was dissolved in sodium hydroxide to control the size and pores in silica particles [14]. In their study, 99.8% silica purity was obtained. Costa and Paranhos converted rice husk to rice husk ash (RHA) by combustion. The rice husk ash was dissolved in concentrated sulfuric acid, followed by treatment with alkaline solution to obtain sodium silicate. Nanosilica particles were synthesized by precipitation using phosphoric acid [15]. Song et al. employed the Taguchi method to obtain surfactant-free synthesis of high surface area silica nanoparticles from rice husk [16]. The Taguchi method was efficient for designing factorial experiments with a minimum number of experiments. Regardless of the specific chemical route used, the alkaline leaching step is critical for obtaining high-purity silica from rice husk [6].

This study developed a two-stage continuous silica extraction process from rice husk using attrition ball milling and alkaline leaching methods. A continuous process has several advantages over a batch process, namely production of a narrow specification product, reduced production cost, and increased productivity. Rice husk has a very low density, within the range 90–150 kg/m^3 [18], and conveying is usually conducted by a pneumatic conveying system [19]. In this study, rice husk was ground into micron-sized particles and mixed with a sodium hydroxide solution to make a rice husk slurry. The rice husk slurry can be easily conveyed by a fluid pump and continuously reacted to leach silica from rice husk. In addition, alkaline leaching was performed under an atmospheric pressure, which is safe to apply in a rice mill where rice husk is generated. Therefore, this study is an initial step toward the field application of a silica extraction process using rice husk. The circular bioeconomy has gained attention as a key concept for sustainable technical cycles. The circular bioeconomy focuses on the valorization of biomass in integrated production chains and making use of residues [20]. Currently, biomass valorization focuses on valorizing the organic fraction of biomass [21]. However, the valorization of ash content is also important and has the potential to extract more value from biomass. In this respect, this study is worthwhile to extend the area of biomass valorization and, ultimately, promote the facilitation of circular bioeconomy.

2. Materials and Methods
2.1. Materials

Rice husk was kindly supplied by a rice processing facility in the Chungbuk region, Rep. Korea, which was harvested in 2019. Sodium hydroxide powder (97%), acetic acid (99.5%), and potassium hydroxide (93.0%) were purchased from Daejung Chemicals & Metals Inc. (Goryeong, Korea). Sodium hydroxide powders were dissolved in distilled water and used in the experiments; the others were used as received without further purification.

2.2. Alkaline Leaching Process

Before using the rice husk, it was washed with deionized water three times and dried at 80 °C overnight. After drying, the rice husk was immersed in an alkaline solution (sodium hydroxide or potassium hydroxide) and thoroughly mixed to allow sufficient soaking in the solution. The sample was moved to a heating oven (ThermoStable™ "OF-105", Daihan Scientific, Wonju, Korea) set at a specific temperature for reaction over a given reaction time. After reaction, the solids were separated from the solution using vacuum filtration (Circulating Aspirator (WJ-15, SIBATA, Saitama, Japan)) and filter paper (Whatman No. 41, 20~25 μm, Maidstone, UK). To measure the leached ash, acetic acid was added to the solution to adjust the pH to 7.0, which was stirred at 300 rpm overnight. The precipitation was washed three times with deionized water at 4000 rpm for 10 min. The washed precipitation was dried at 80 °C overnight. The organics such as hemicellulose and lignin were leached during the alkaline leaching process and contained in the precipitation. Therefore, the washed precipitation was calcined at 900 °C for 6 h to remove the organics in the precipitation. The silica yield was calculated using Equation (1) below:

$$Silica\ extraction\ yield\ (\%) = \frac{(Weight\ of\ ash\ precipitated\ \times\ silica\ purity)}{(Weight\ of\ Rice\ husk\ ash\ \times\ silica\ purity)} \times 100 \quad (1)$$

To find the optimum alkaline leaching conditions, four experimental parameters—the solid content, alkaline reaction concentration, temperature, and duration—were optimized.

2.3. Attrition Ball Mill

An attrition pulverizer (Korea powder system Co., Ltd., Incheon, Korea) previously developed for lignocellulosic pretreatment [22] was used to prepare micron-sized rice husk particles. One-third of the inner space was filled with rice husk, while another third was filled with grinding steel balls (10 mm in diameter). Alkaline solvent was added to the grinding jar. The rice husk was pulverized under wet-grinding conditions at 300 rpm for 20 or 30 min. After milling, the rice husk was transferred to 1 mm shaking sieve (Aanlysette3, Fritsch GmbH, Idar-Oberstein, Germany) and shaken for 1 min to separate the pulverized rice husk particles from the grinding balls.

2.4. Two-Stage Continuous Silica Extraction Process

A schematic diagram of the continuous silica extraction process is shown in Figure 1. The continuous extraction process consists of two steps: pulverization and alkaline reaction. At the pulverization step, rice husk was pulverized to make fine rice husk particles and to increase the fluidity. In the continuous process, rice husk was pulverized in an alkaline solvent at 300 rpm for 20 min and stored in a reservoir after separating from grinding balls. The reservoir was stirred at 600 rpm using an electronic overhead stirrer (MS 3060D, MTOPS, Yangju, Korea) to prevent the rice husk particles from settling down. The rice husk slurry in the reservoir was continuously fed into a reactor using a peristaltic pump (BT100S, Lead Fluid Technology, Co., Ltd., Baoding, China). The reactor was stirred at 400 rpm and 80 °C. The outlet sample was collected and separated using a vacuum filter and filter paper (Whatman No. 41, 20–25 μm, Maidstone, UK) for calculating the silica yield, which was calculated as described in Section 2.2. For measuring the solid content, 10 g of the rice husk slurry was sampled at the outlet of the reservoir every 8 h. The sample was kept in a heating oven (ThermoStable™ "OF-105", Daihan Scientific, Wonju, Korea) set to 105 °C for 24 h. The solid content was calculated by using the weight difference before and after drying.

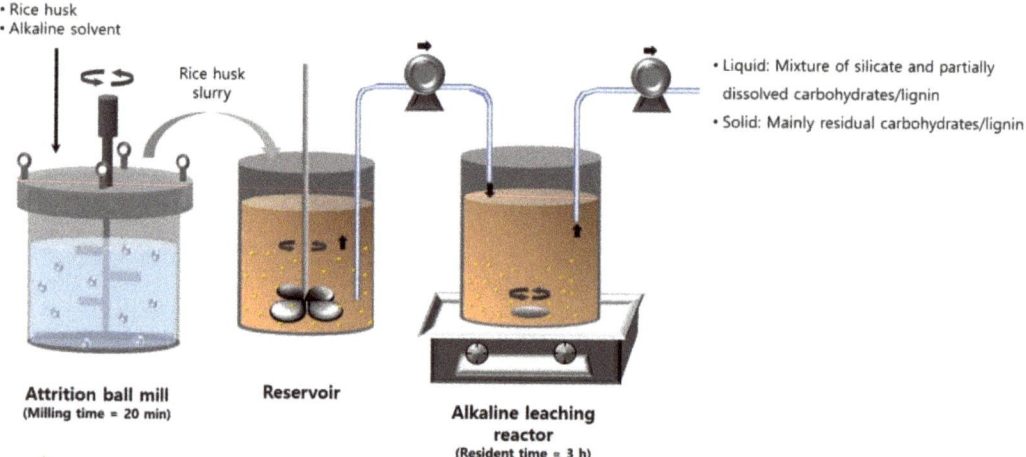

Figure 1. Scheme of the continuous silica extraction process.

2.5. Analytical Methods

The compositions of rice husk, carbohydrates and lignin, were measured according to the standard procedure provided by National Renewable Energy Laboratory (NREL) [23]. The compositions of rice husk were compared between before and after extracting to calculate the quantities of extracted carbohydrates and lignin. The inorganic chemical composition was determined by using inductively coupled plasma–optical emission spectrometry (ICP–OES; Optima 5300DV, PerkinElmer, MA, USA). X-ray diffraction patterns were obtained using X-ray diffraction (XRD; D/Max 2500/PC, Rigaku, Tokyo, Japan). The surface areas of the obtained silica were calculated from the measured isotherms according to the Brunauer–Emmett–Teller (BET) method, and the pore volumes were taken at the $P/P_0 = 0.995$ single point using a Micromeritics Tristar 3200 system (Micromeritics Inc., Norcross, GA, USA). The pore size distributions of the silica were calculated using the Barrett–Joyner–Halenda (BJH) method from the adsorption branches of the isotherms. The rheological properties of the pulverized rice husk slurry were analyzed using a stress-controlled rotational rheometer (MCR 702, Anton Paar, Graz, Austria) with a C-PTD200 (Cup-Peltier Temperature Device).

3. Results and Discussion

3.1. Optimization of Alkaline Leaching Conditions

For extracting silica from rice husk, two alkaline solutions, NaOH and KOH, were used and their performance compared. Both are popularly used alkaline solutions and have similar properties. The performances of various concentrations of NaOH and KOH solutions were compared to assess which would be optimal for leaching. Figure 2a presents a comparison of the performances of the two alkaline solutions regarding silica leaching from rice husk depending on their concentrations. Both showed similar extracting yields, but there was a slight difference. At 0.1 M concentration, the silica extraction yields were 1% and 30% for NaOH and KOH, respectively. The use of KOH showed a higher extraction yield at 0.1 M concentration. As the concentration of the alkaline solution increased, the extracting yields increased up to a certain point. Over this point, the extracting yields were saturated and did not increase further, even when the alkaline concentration increased. When NaOH was used, the silica yield became saturated starting from 0.2 M, with an approximately 79% yield. When KOH was used, saturation of the extracting yield was approximately 77%, starting from 0.5 M KOH. Therefore, NaOH can be used at a lower concentration than KOH while obtaining a slightly higher yield.

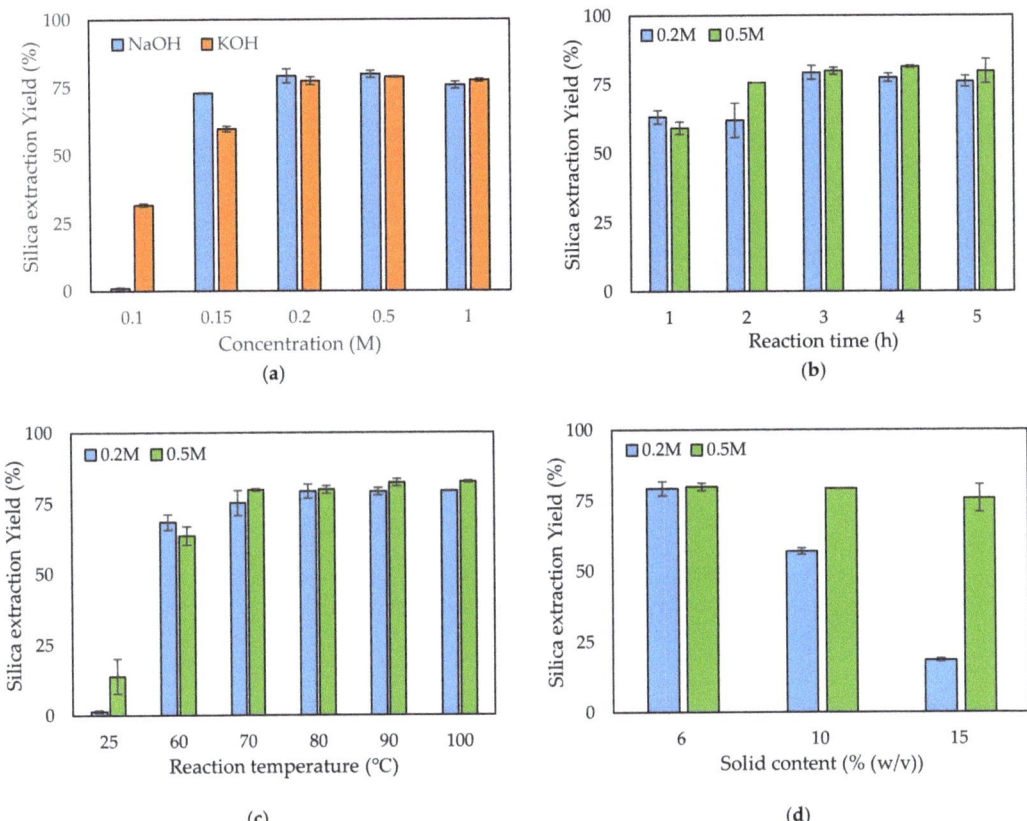

Figure 2. Comparison of the silica extraction yield depending on (**a**) the type of alkaline solution, (**b**) the alkaline leaching reaction time, (**c**) the temperature, and (**d**) the solid content.

When using NaOH, similar silica extraction yields were found for 0.2 and 0.5 M—79.3% and 79.9%, respectively. Therefore, both concentrations were tested for optimizing the reaction time, temperature, and solid content. Under the conditions of 80 °C reaction temperature and 6% (w/v) solid content, five reaction times—1, 2, 3, 4, and 5 h—were investigated to determine the optimum reaction time. As the reaction time increased, the silica yield increased up to 3 h (Figure 2b). At reaction times over 3 h, the yield did not increase further (Figure 2b). At 3 h reaction time, no significant difference in yield was observed between use of 0.2 and 0.5 M concentrations. The optimum reaction temperature was determined among six chosen temperatures. At the temperature of 25 °C, only a small quantity of silica leached from the rice husk into the NaOH solution: 1.4% and 14% for 0.2 and 0.5 M, respectively (Figure 2c). At 60 °C, the silica yields increased to 68.4% and 63.6% for 0.2 and 0.5 M, respectively. Over 70 °C, the silica yield did not significantly increase, even when the reaction temperature was increased. The yield over 70 °C was approximately 80%. In the reaction temperature tests, there was no significant difference between 0.2 and 0.5 M. The solid content is related to reaction volume, which determines the reactor size. As the solid content increased, the silica yield decreased because of a lack of NaOH compared to Si (Figure 2d). Typically, there was tendency for a higher decrease in silica yield for 0.2 M compared with 0.5 M. The highest silica yield was found at 6% (w/v): 79.3% and 79.9% for 0.2 and 0.5 M, respectively.

3.2. Preparation of Rice Husk Slurry for Continuous Process

For application in a continuous process, rice husk should be continuously supplied to a reactor. Rice husk has a very low density and is usually conveyed by a pneumatic conveying system. In this study, rice husk slurry was prepared to easily convey the sample by a fluid pump. To prepare the rice husk slurry, rice husk was pulverized in NaOH solution by attrition ball milling. The rice husk had a diameter of approximately 6–7 mm before milling. After ball milling, the size of rice husk was drastically reduced, the extent of which was mainly related to milling time. However, the alkaline concentration also slightly affected the size reduction. The mean diameter of the rice husk particles was 228.9 and 139.0 μm for 20 and 30 min of milling, respectively, when it was treated with 0.2 M NaOH. When 0.5 M NaOH was used, the mean diameter was 218 and 78.1 μm for 20 and 30 min of milling, respectively. The size distribution of the rice husk after 20 min of milling showed a bimodal curve, with peaks being observed at approximately 77 and 777 μm (Figure 3). However, the peak around 777 μm reduced and shifted to 77 μm as the milling time was increased to 30 min. This result indicates that the size of rice husk particles became more homogenous as the milling time increased. The increase of milling time effectively reduced the portion of larger particle sizes. The concentration of NaOH also affected particle size distribution, but not as much as milling time. When comparing 0.2 and 0.5 M NaOH with 20 min milling time, 0.5 M NaOH showed a lower peak on 777 μm and a higher shoulder on 23 μm when compared to 0.2 M NaOH. In the 30 min milling condition, both samples of 0.2 and 0.5 M NaOH showed a monomodal curve. However, the graph of 0.5 M NaOH treatment showed a higher peak than 0.2 M NaOH. This result indicates that the size distribution of the rice husk particles was affected mainly by milling time and partially by NaOH concentration.

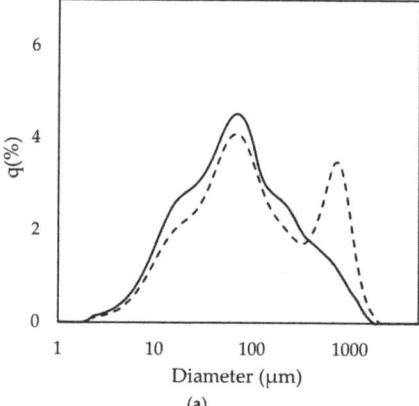
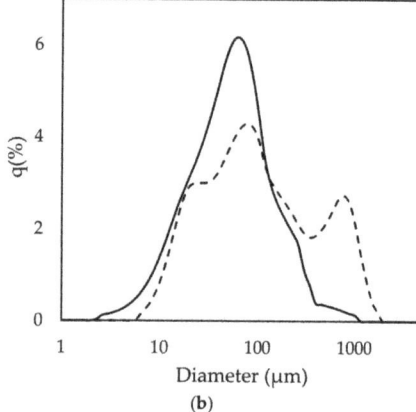

Figure 3. Size distribution of the pretreated rice husk under the conditions of (a) 0.2 M NaOH and (b) 0.5 M NaOH. Dash and line indicate 20 and 30 min milling, respectively.

Figure 4 shows the rheological properties of the pulverized rice husk slurry with a 6% (w/v) solid content. The viscosity of the slurry solutions decreased as the shear rates increased. The viscosities of all samples decreased as the shear rate increased, indicating shear thinning. The untreated sample showed higher shear stress and viscosity than the treated samples (Figure 4). The shear stress and viscosity of a slurry are closely related to the particle size [24–26]. In general, the shear stress and viscosity of the slurry increased as the particle size increased, especially at low shear rates. In this study, the untreated slurry contained 6–7 mm rice husk particles. The rice husk particles were soaked with alkaline solution, meaning the larger particles could have been heavier than the smaller particles because of soaking up more of the alkaline solution and, thus, needed stronger force to be moved. Therefore, the untreated rice husk slurry showed higher shear stress and viscosity through the range of shear rate (Figure 4). As the particle size decreased, the shear stress and viscosity reduced. In Figure 4, the ball-milled samples show drastically reduced shear stress and viscosity. This indicates that the ball-milled rice husk slurries needed less force to be moved than the untreated slurry. The ball mill used in this study improved the fluidity of the rice husk slurry and made conveying rice husk easy.

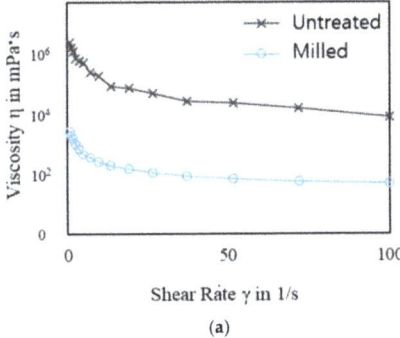 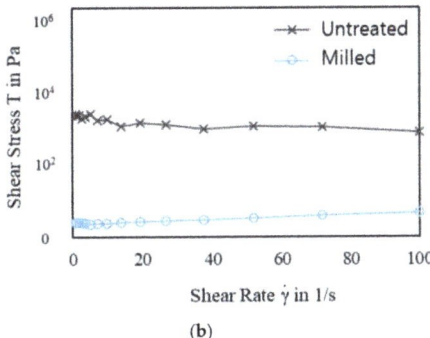

Figure 4. Rheological properties of the rice husk slurry before/after ball milling and shown as a function of shear rate: (a) viscosity and (b) shear stress. The ball-milled rice husk slurry was prepared by ball milling for 20 min with 0.2 M NaOH.

3.3. Continuous Silica Extraction Process

The continuous silica extraction process was performed using conditions based on the results of the batch experiments. The milled rice husk slurry with 0.2 M NaOH solution was stored in a reservoir and continuously supplied to a reactor. For stable operation in a continuous process, the solid content should be steadily supplied to the reactor because the quantity of raw material for silica should be constant during the process. Initially, a separate set of experiments was performed to measure the solid content. At every 8 h, samples were collected from the outlet of the reservoir. The first sample, obtained 8 h after starting the continuous process, showed a 6% (w/v) solid content. During the period of the continuous process, the solid content was steady, approximately 6% (w/v) (Figure 5). As previously mentioned, the ball mill used in this study improved the fluidity and enabled a steady supply of the rice husk slurry.

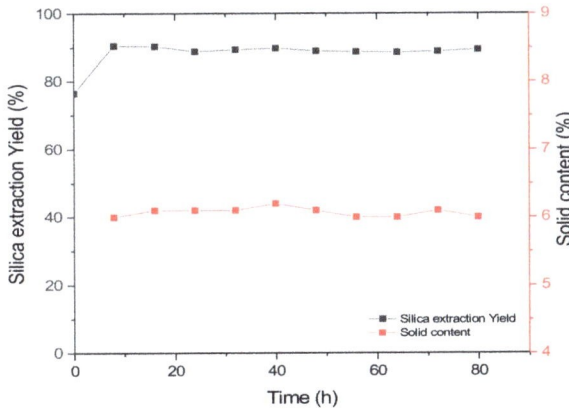

Figure 5. Silica extraction yield and solid content in the continuous process depending on process time.

Initially, the rice husk slurry was reacted for 3 h before starting the continuous process. After the reaction, the yield of the silica extraction was 77% (Figure 5). After starting the continuous supply of the rice husk slurry, the silica yield increased to 90%. This could be due to extended reaction time for the initially filled sample. After starting the continuous process, the silica extraction yield slightly decreased, but reached a steady state after 24 h. After reaching a steady state, the silica yield was constant, indicating that the process was stable. In this study, the continuous silica extraction was performed for 80 h and the process was stable during the operation, with 89% silica yield.

Usually, the alkaline pretreatment was performed for delignification. Therefore, the continuous process used in this study could remove carbohydrates and lignin. The compositions of rice husk before the NaOH leaching were 52.8 wt% carbohydrates, 29.6 wt% lignin, and 12.9 wt% ash (Figure 6). The continuous silica extraction reaction used in this study leached 21.3% carbohydrates, 30.6% lignin, and 84.2% ash into the NaOH solution. After precipitation and calcination, about 89% of silica in the rice husk was recovered. This two-stage alkaline leach strategy is capable to produce a de-ashed rice husk slurry which is more suitable for further biorefinery, a silica-rich by-product for high-value applications, and an extracted carbohydrates/lignin mixture for further valorization.

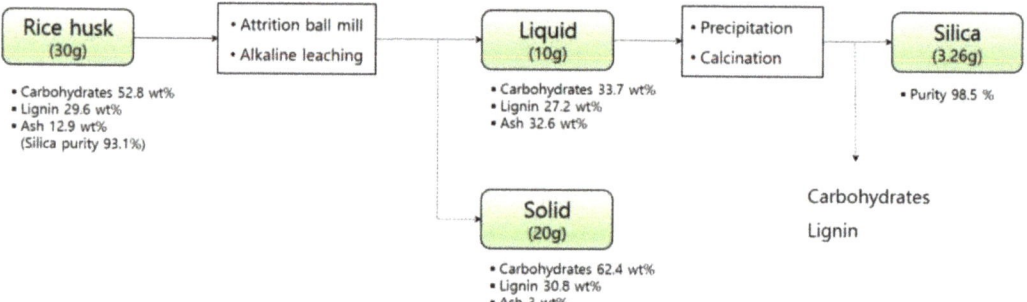

Figure 6. Mass balance of the continuous silica extraction process used in this study.

3.4. Characterization of the Silica Obtained from the Continuous Extraction Process

The extracted silica from the continuous process was characterized after precipitation. Originally, the purity of silica in the rice husk ash was only 93.1% (Table 1). The rice husk ash contained high impurities such as CaO, MaO, and K_2O. After alkaline leaching, the silica purity increased to 98.5%. The main impurity was Na_2O, which increased from 0.08% to 0.96% after the NaOH leaching. Considering the increase in Na_2O content after the NaOH leaching, it is possible that the sodium in the NaOH solution was precipitated and that it could be reduced by applying stringent washing steps.

Table 1. Inorganic composition of the raw material ash and the extracted ash.

Sample	SiO_2	Al_2O_3	CaO	MgO	Na_2O	K_2O	SO_3
Rice husk ash (wt%)	93.1	0.05	1.52	0.65	0.08	0.84	0.09
NaOH-treated (wt%)	98.5	0.07	0.01	0.01	0.96	0.04	0.01

The crystallinity of the obtained silica was investigated using XRD. The XRD patterns of the silica samples obtained from both the rice husk ash and the continuous processes showed a broad diffraction near to 20°, which is typical for amorphous silica (Figure 7). In the XRD pattern, no crystalline structure of silica was found, and continuous alkaline leaching did not cause any crystallinity changes. The phase of silica is determined by the combustion temperature [27]. Over certain temperature, the phase transformation starts but the crystallization temperature varied depending on the composition of rice husk ash.

In this study, the calcination was performed at 900 °C for 6 h to remove residual moisture and volatile compounds. The calcination did not cause any crystallinity changes of rice husk silica.

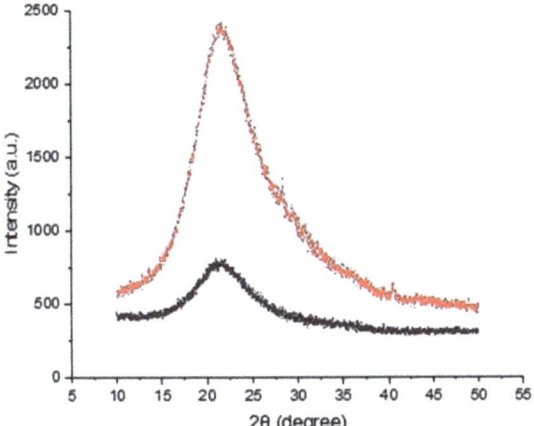

Figure 7. X-ray patterns of silica samples obtained from rice husk ash (black) and the continuous process (red).

The surface area and pore volume of the silica obtained from the continuous process were 1.973 m^2/g and 0.004 cm^3/g, respectively (Figure 8). A previous study reported that the presence of metal impurities, such as Na and K, causes surface melting and agglomeration in the particles during combustion [28]. The surface melting and agglomeration led to reduced surface area and pore volume. In this study, the content of Na$_2$O in the silica obtained increased due to the NaOH leaching and it could be the reason for low surface area and pore volume. This study did not control the structure of silica. Therefore, an additional process is required to obtain a highly specific surface area or well-defined nanostructures.

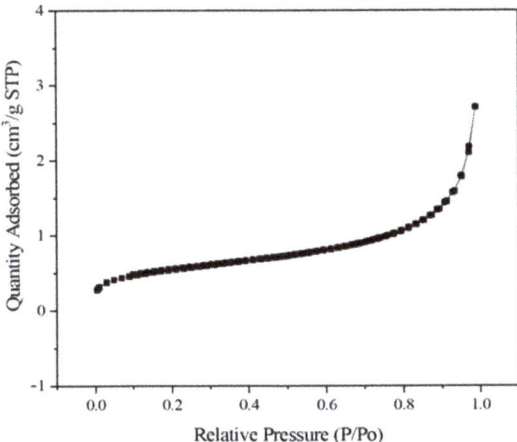

Figure 8. N$_2$ physisorption isotherms of silica obtained from the continuous process.

4. Conclusions

A process for two-stage continuous silica extraction from rice husk was successfully developed using alkaline leaching and attrition ball mill methods. The alkaline leaching conditions obtained in batch experiments were employed in the continuous process. The attrition ball mill treatment was used to obtain a rice husk slurry, which improved fluidity of the sample. By applying alkaline leaching conditions and the ball mill-treated rice husk slurry, continuous silica extraction from rice husk was stably operated, with 89% silica yield and 6% (w/v) solid content for 80 h. An improvement in silica purity was obtained from the continuous process, which increased to 98.5% when compared to rice husk ash. The continuous process did not change the crystallinity or surface properties of the silica.

The continuous extraction process developed in this study would be beneficial for product uniformity and process capacity. It is very easy to operate once the system has been set up. Therefore, we expect that this method can be used in the field for the mass production of rice husk silica.

Author Contributions: Conceptualization, J.H.L. and B.-I.S.; methodology, J.H.L. and J.C.; validation, E.T.H. and B.-I.S.; formal analysis, J.H.L. and J.C.; investigation, J.Y.P., S.Y.P. and Y.M.G.; data curation, J.Y.P.; writing—original draft preparation, J.Y.P. and J.H.L.; writing—review and editing, J.H.L. and J.C.; project administration, J.H.L.; funding acquisition, J.H.L. All authors have read and agreed to the published version of the manuscript.

Funding: This work was supported by the Korea Institute of Planning and Evaluation for Technology in Food, Agriculture, and Forestry (IPET) through the Agro and Livestock Products Safety Flow Management Technology Development Program, funded by the Ministry of Agriculture, Food, and Rural Affairs (MAFRA) (319109-02).

Institutional Review Board Statement: Not applicable.

Informed Consent Statement: Not applicable.

Data Availability Statement: Not applicable.

Conflicts of Interest: The authors declare no conflict of interest.

References

1. FAOSTAT. 2019. Available online: http://www.fao.org/faostat/en/#data/QC (accessed on 12 November 2020).
2. Esa, N.M.; Ling, T.B.; Peng, L.S. By-products of rice processing: An overview of health benefits and applications. *J. Rice Res.* **2013**, *1*, 107. [CrossRef]
3. Pode, R. Potential applications of rice husk ash waste from rice husk biomass power plant. *Renew. Sustain. Energy Rev.* **2016**, *53*, 1468–1485. [CrossRef]
4. Babaso, P.N.; Sharanagouda, H. Rice Husk and Its Applications: Review. *Int. J. Curr. Microbiol. Appl. Sci.* **2017**, *6*, 1144–1156. [CrossRef]
5. Lee, J.H.; Kwon, J.H.; Lee, J.-W.; Lee, H.-S.; Chang, J.H.; Sang, B.-I. Preparation of high purity silica originated from rice husks by chemically removing metallic impurities. *J. Ind. Eng. Chem.* **2017**, *50*, 79–85. [CrossRef]
6. Hossain, S.S.; Mathur, L.; Roy, P. Rice husk/rice husk ash as an alternative source of silica in ceramics: A review. *J. Asian Ceram. Soc.* **2018**, *6*, 299–313. [CrossRef]
7. Mason, S.; Young, S.; Araga, M.; Butler, A.; Lucas, R.; Milleman, J.L.; Milleman, K.R. Stain control with two experimental dentin hypersensitivity toothpastes containing spherical silica: A randomised, early-phase development study. *BDJ Open* **2019**, *5*, 8. [CrossRef]
8. Caldona, E.B.; Sibaen, J.W.; Tactay, C.B.; Mendiola, S.L.D.; Abance, C.B.; Añes, M.P.; Serrano, F.D.D.; De Guzman, M.M.S. Preparation of spray-coated surfaces from green-formulated superhydrophobic coatings. *SN Appl. Sci.* **2019**, *1*, 1657. [CrossRef]
9. Mozumder, M.S.; Mourad, A.-H.I.; Pervez, H.; Surkatti, R. Recent developments in multifunctional coatings for solar panel applications: A review. *Sol. Energy Mater. Sol. Cells* **2019**, *189*, 75–102. [CrossRef]
10. Fytianos, G.; Rahdar, A.; Kyzas, G.Z. Nanomaterials in Cosmetics: Recent Updates. *Nanomaterials* **2020**, *10*, 979. [CrossRef]
11. Foletto, E.L.; Gratieri, E.; De Oliveira, L.H.; Jahn, S.L. Conversion of rice hull ash into soluble sodium silicate. *Mater. Res.* **2006**, *9*, 335–338. [CrossRef]
12. Rojas, D.F.H.; Gomez, P.P.; Rivera, A.R. Synthesis and characterisation of submicron silica particles from rice husk. *Green Mater.* **2018**, *6*, 15–22. [CrossRef]
13. De Souza, M.; Magalhães, W.; Persegil, M. Silica Derived from Burned Rice Hulls. *Mater. Res.* **2002**, *5*, 467–474. [CrossRef]
14. Chun, J.; Gu, Y.M.; Hwang, J.; Oh, K.K.; Lee, J.H. Synthesis of ordered mesoporous silica with various pore structures using high-purity silica extracted from rice husk. *J. Ind. Eng. Chem.* **2020**, *81*, 135–143. [CrossRef]
15. Costa, J.A.; Paranhos, C.M. Systematic evaluation of amorphous silica production from rice husk ashes. *J. Clean. Prod.* **2018**, *192*, 688–697. [CrossRef]

16. Song, S.; Cho, H.-B.; Kim, H.T. Surfactant-free synthesis of high surface area silica nanoparticles derived from rice husks by employing the Taguchi approach. *J. Ind. Eng. Chem.* **2018**, *61*, 281–287. [CrossRef]
17. Zulkifli, N.S.C.; Ab Rahman, I.; Mohamad, D.; Husein, A. A green sol–gel route for the synthesis of structurally controlled silica particles from rice husk for dental composite filler. *Ceram. Int.* **2013**, *39*, 4559–4567. [CrossRef]
18. Singh, B. Rice husk ash. In *Waste and Supplementary Cementitious Materials in Concrete*; Woodhead Publishing: Cambridge, UK, 2018; pp. 417–460.
19. He, C.; Chen, X.; Wang, J.; Ni, H.; Xu, Y.; Zhou, H.; Xiong, Y.; Shen, X. Conveying characteristics and resistance characteris-tics in dense phase pneumatic conveying of rice husk and blendings of rice husk and coal at high pressure. *Powder Technol.* **2012**, *227*, 51–60. [CrossRef]
20. Stegmann, P.; Londo, M.; Junginger, M. The circular bioeconomy: Its elements and role in European bioeconomy clusters. *Resour. Conserv. Recycl. X* **2020**, *6*, 100029. [CrossRef]
21. Vea, E.B.; Romeo, D.; Thomsen, M. Biowaste valorization in a future circular bioeconomy. *Procedia CIRP* **2018**, *69*, 591–596. [CrossRef]
22. Gu, Y.M.; Byun, H.R.; Kim, Y.-H.; Park, D.-Y.; Lee, J.H. Assessing the potential of facile biofuel production from corn stover using attrition mill treatment. *Water-Energy Nexus* **2019**, *2*, 46–49. [CrossRef]
23. Sluiter, A.; Hames, B.; Ruiz, R.; Scarlata, C.; Sluiter, J.; Templeton, D. *Determination of Structural Carbohydrates and Lignin in Biomass*; National Renewable Energy Laboratory: Golden, CO, USA, 2008.
24. He, M.; Wang, Y.; Forssberg, E. Slurry rheology in wet ultrafine grinding of industrial minerals: A review. *Powder Technol.* **2004**, *147*, 94–112. [CrossRef]
25. Senapati, P.K.; Panda, D.; Parida, A. Predicting Viscosity of Limestone–Water Slurry. *J. Miner. Mater. Charact. Eng.* **2009**, *8*, 203–221. [CrossRef]
26. Tangsathitkulchai, C.; Austin, L. Rheology of concentrated slurries of particles of natural size distribution produced by grinding. *Powder Technol.* **1988**, *56*, 293–299. [CrossRef]
27. Chun, J.; Lee, J.H. Recent Progress on the Development of Engineered Silica Particles Derived from Rice Husk. *Sustainability* **2020**, *12*, 10683. [CrossRef]
28. Zareihassangheshlaghi, A.; Dizaji, H.B.; Zeng, T.; Huth, P.; Ruf, T.; Denecke, R.; Enke, D. Behavior of Metal Impurities on Surface and Bulk of Biogenic Silica from Rice Husk Combustion and the Impact on Ash-Melting Tendency. *ACS Sustain. Chem. Eng.* **2020**, *8*, 10369–10379. [CrossRef]

 *sustainability*

Article

Wood Ashes from Grate-Fired Heat and Power Plants: Evaluation of Nutrient and Heavy Metal Contents

Hans Bachmaier *, Daniel Kuptz and Hans Hartmann

Technology and Support Centre in the Centre of Excellence for Renewable Resources (TFZ), Schulgasse 18, 94315 Straubing, Germany; Daniel.Kuptz@tfz.bayern.de (D.K.); Hans.Hartmann@tfz.bayern.de (H.H.)
* Correspondence: Johannes.Bachmaier@tfz.bayern.de; Tel.: +49-9421-300-160

Abstract: Ashes from biomass heat (and power) plants that apply untreated woody biofuels may be suitable for use as fertilizers if certain requirements regarding pollutant and nutrient contents are met. The aim of this study was to examine if both bottom and cyclone ashes from 17 Bavarian heating plants and one ash collection depot are suitable as fertilizers ($n = 50$). The range and average values of relevant nutrients and pollutants in the ashes were analyzed and evaluated for conformity with the German Fertilizer Ordinance (DüMV). Approximately 30% of the bottom ashes directly complied with the heavy metal limits of the Fertilizer Ordinance. The limits were exceeded for chromium(VI) (62%), cadmium (12%) and lead (4%). If chromium(VI) could be reduced by suitable treatment, 85% of the bottom ashes would comply with the required limit values. Cyclone ashes were high in cadmium, lead, and zinc. The analysis of the main nutrients showed high values for potassium and calcium in bottom ashes, but also relevant amounts of phosphorus, making them suitable as fertilizers if pollutant limits are met. Quality assurance systems should be applied at biomass heating plants to improve ash quality if wood ashes are used as fertilizers in agriculture.

Keywords: wood ash; fertilizer; heat and power plants; heavy metals; nutrients; German fertilizer legislation

1. Introduction

Combustion of wood in heat (and power) plants generates solid residues in the form of ashes [1,2]. In the Federal State of Bavaria (i.e., Southeast Germany), a total of 30,000 to 60,000 t/a of wood ashes from untreated wood accumulates each year from plants with an installed capacity of more than 1 MW_{therm} (calculated from the 2018 Energy Wood Market Report of the Bavarian State Institute of Forestry (LWF)) [3]. Due to the physical and chemical properties of these combustion by-products, suitable utilization strategies might be recommended for their use as raw materials in the bioeconomy.

Depending on the point of origin of wood ashes in the heat (and power) plant, a distinction can be made between different ash fractions. The ash accumulating in the boiler is called "bottom ash" or "coarse ash". In most cases, the ash from the heat exchangers is also considered as part of the bottom ash. After the hot flue gas passes through the heat exchanger, the air is usually cleaned by a cyclone in which the "cyclone ash" (also called "coarse fly ash") is separated. If the plant has an electrostatic precipitator, a fabric filter or a flue gas condensation system, a third ash fraction, i.e., the so-called "filter ash" (also called "fine fly ash") or the "condensate sludge" is generated [1]. The following article focuses on bottom ash and cyclone ash from grate-fired boilers as these are the most common ash fractions in Bavarian heat (and power) plants.

The chemical composition of individual ash fractions depends on the fuel quality and the plant technology [1,2]. Chemical elements such as plant nutrients (e.g., Ca, Mg) or pollutants such as heavy metals vary in wood fuels depending on the species, but also on bark content, the share of green biomass (i.e., needles/leaves), growing conditions, the

Citation: Bachmaier, H.; Kuptz, D.; Hartmann, H. Wood Ashes from Grate-Fired Heat and Power Plants: Evaluation of Nutrient and Heavy Metal Contents. *Sustainability* **2021**, *13*, 5482. https://doi.org/10.3390/su13105482

Academic Editors: Dirk Enke, Hossein Beidaghy Dizaji, Volker Lenz and Thomas Zeng

Received: 12 April 2021
Accepted: 4 May 2021
Published: 13 May 2021

Publisher's Note: MDPI stays neutral with regard to jurisdictional claims in published maps and institutional affiliations.

Copyright: © 2021 by the authors. Licensee MDPI, Basel, Switzerland. This article is an open access article distributed under the terms and conditions of the Creative Commons Attribution (CC BY) license (https://creativecommons.org/licenses/by/4.0/).

degree of external contamination or dry or wet ash removal from the combustion unit [4–9]. Major and trace elements are volatile to varying degrees at temperatures that prevail within the combustion chamber [1,10–12]. For instance, heavy metals such as Cd, Pb, Zn, and Hg are highly volatile, while elements such as Cr or Cu have low volatility. Therefore, the elements accumulate differently in bottom, cyclone or filter ash [10,13–15].

The material use of wood ashes as raw materials for the bioeconomy poses certain challenges [16–18]. Complex legal frameworks, difficult ash logistics due to decentralized accumulation, fluctuating product qualities and aspects of storage and occupational safety are just some of the points that must be considered in this context [11,19–21]. In addition, many utilization pathways are still under development or at the pilot stage [14,18,22,23] and are not applied regularly. Consequently, ashes from biomass heat (and power) plants are usually not perceived as by-products of the energetic use of wood. Thus, it is often not regarded as a valuable intermediate for further processing but is classified as waste that must be disposed at significant costs [24]. A survey conducted under the AshUse-project showed that in the opinion of the heating plant operators the challenges for implementing recycling of a functioning material include legal uncertainties, fluctuating ash qualities and low economic revenues. In addition, operators often lack knowledge about quality management strategies, such as how a defined ash quality can be reliably maintained and verified [21]. The planned further work at TFZ will therefore focus on the area of quality management in the production of wood ash at biomass heating (power) plants.

Wood ashes are already used to some extent as raw materials for various purposes in Germany and in other European countries, and to some extent also in Northern America, depending on the technical, economic, and legal circumstances and considering environmental aspects. A relatively widespread application is the use of ashes as a fertilizer or as an additive for fertilizer production for agricultural and forestry applications [11,15,20,25–29]. In Germany and Austria, suitable ashes are also added to composts [30,31]. In Austria, this pathway is limited to a very low blending rate of 2% ash to the compost, which makes this process uneconomical and no relevant quantities of ash are recycled via this route [30]. Instead, approximately 40% of the annually produced ash in Austria is processed by the cement and building materials industry [30]. The use of ashes in road construction has successfully been tested in research projects in Austria and Finland [22,32,33]. Tejada et al. (2019) [34] investigated wood ashes as a source of raw materials in urban mining.

Heavy metal contamination, and in particular Cd in ashes, is seen as a major concern in the use of ashes as fertilizers [16,35,36]. Limit values in other European countries are considerably higher, especially for Cd [19]. There are regulations on specific parameters of ash in individual countries. For example, in Germany, there is a limit value for Cr(VI) for application on arable land [11,37], and in Denmark there is a conductivity limit value for the eluate from ashes [38].

Many authors document the chemical composition of ashes used as fertilizers or soil conditioners [10,11,19,20,24,26,28,29]. This is carried out by considering national conditions regarding prevailing plant technology and legal requirements for the application of the ashes. Due to the higher limits for heavy metals in ashes for fertilizer purposes in many countries [19], higher contaminated mixed bottom and fly ashes or ashes from fluidized bed combustion plants are also sometimes used as fertilizers [13,15,39,40]. In terms of composition, these are often not comparable to the predominantly pure bottom ashes from grate-fired furnaces, as they mainly occur in the study area [21] and are eligible for use as fertilizer and soil conditioner under the German fertilizer law. Ash qualities of German biomass heating plants are documented by Reichle et al. (2009) [16], Wilpert (2016) [11], and Schilling (2020) [10]. Wilpert (2016) [11] and Schilling (2020) [10] examined bottom ashes that were collected in the context of a quality assessment of bottom ashes for fertilizing purposes. Therefore, rather less contaminated ashes than average may have been used. The values of Reichle et al. (2009) [16] originate from before 2003, without giving any further details on the sampled plants. Data on the ordinary ash quality of combustion plants according to the current state of the art are therefore missing. However, these data are necessary to estimate the bioeconomic potential of an increased ash utilization.

Due to the high solubility of calcium oxide (CaO) in the ash, a pH shock is feared when spreading in the forest, which negatively affects the soil flora and soil fauna. Therefore, the ash often is pretreated before application. This process, the so-called "ash stabilisation" or "ash hardening", includes the addition of water followed by a storage period of several months. Moistening and contact with atmospheric carbon dioxide causes a variety of chemical transformations. Most importantly, the easily soluble calcium oxide (CaO) transforms into the poorly soluble calcium carbonate ($CaCO_3$) [11,12,24,38]. In large piles, this reaction occurs only on the surface if there is no mixing [12,28].

In Germany, wood ash is mixed with lime dolomite and is then used for soil improvement on arable and forestry land [11,41,42]. Wilpert (2020) [11] points out that the use of wood ash-lime mixtures is particularly recommended where improved potassium supply is desired and the alkaline effect is present. Since the solubility of the alkali salts remains high even after ash hardening, the hardened ash should be protected from rain during storage in order to prevent nutrient leaching. Another positive effect of humidifying the ash and storing it for several months is the conversion of any toxic chromium(VI) into the harmless chromium(III). Schilling (2020) [10] and Polandt-Schwandt (1999) [9] observed this effect in the case of ash from combustion plants with a wet ash discharge system where the hot ash was placed in a water bath and then discharged moist. Pelletizing and granulation of wood ash also serve to reduce the reactivity of the wood ash. Auxiliary materials such as cement or organic binders can be used in this process [28,43]. Pelletized or granulated ash can be applied with conventional fertilizer spreaders [40]. Moistening of the ash is not recommended if the ash is to be used as a substitute for quicklime— e.g., in road construction. In this case, the ash must be stored dry [33].

Many authors emphasize the liming effect of ashes and mixtures with ashes on agricultural and forest soils [11,24,28,44]. Katzensteiner et al. (2011) [45] describe the plant availability of calcium and potassium from wood ashes as "high", magnesium availability as "medium" and phosphate availability as "low". "Low" in this context means that less than 10% of the total phosphate from wood ashes is available to the plant in the year of application. In pot experiments, Kebli et al. (2017) [46] and Maltas et al. (2014) [47] demonstrated the uptake of potassium from wood ash in ryegrass and sunflowers. In the case of sunflowers, P uptake from the ashes was also observed.

An important prerequisite for approval as a fertilizer in Germany is compliance with the heavy metal limits in the German Fertilizer Ordinance (DüMV) [42] and, if applicable, the minimum nutrient contents required, depending on the fertilizer. If the ash is mixed with biowaste or compost, the limit values of the German Biowaste Ordinance (BioAbfV) [48] must also be complied with in certain cases [31]. Table 1 summarizes these limit values. The DüMV also contains limit values for organic compounds (perfluorinated tensides, dioxins and dioxin-like substances). These compounds are usually absent in ash from biomass heating (and power) plants [10] and were not investigated in this study. Currently, both bottom ashes and cyclone ashes (if the cyclone is not the last precipitation unit in the plant) may be used for fertilizer production according to DüMV. Compared to the application of wood ashes on farmland, 50% higher heavy metal limit values apply to the application on forestry land. The Cr(VI) limit only applies to ash fertilization on arable land.

Much uncertainty remains around the variability of wood ashes among plants or within the same plant and which of these ashes might be suitable for application as fertilizers in agriculture or for liming of forest soils. The aim of this study was to assess the range and average values of nutrients and pollutants in ashes from individual Bavarian biomass heat (and power) plants. This is an important prerequisite for an increase in ash utilization as it is in line with the Bavarian bioeconomy strategy [49].

Table 1. Limit values or maximum contents (in brackets) for wood ashes according to the current German Fertilizer Ordinance (DüMV) and Biowaste Ordinance (BioAbfV) (d.b. = dry basis).

Element	Unit	Current National Limits		
		DüMV	DüMV (Forest)	BioAbfV
Arsenic	mg/kg d.b.	40	60	-
Lead	mg/kg d.b.	150	225	150
Cadmium	mg/kg d.b.	1.5	2.25	1.5
Chromium, total	mg/kg d.b.	-	-	100
Chromium(VI)	mg/kg d.b.	2	-	-
Copper	mg/kg d.b.	(900)	(2000)	100
Nickel	mg/kg d.b.	80	120	50
Mercury	mg/kg d.b.	1	1.5	1
Thallium	mg/kg d.b.	1	1.5	0
Zinc	mg/kg d.b.	(5000)	(5000)	400
PFT	mg/kg d.b.	0.1	0.15	-
I-TE Dioxines and dl-PCB	ng WHO-TEQ/kg d.b.	30	45	-

For this purpose, mainly bottom ashes but also mixtures of bottom and cyclone ashes (due to individual ash handling at certain plants) and pure cyclone ashes from heat (and power) plants with a thermal output of more than 1 MW were sampled and analyzed. The cyclone ashes are used for comparison with the bottom ashes and for an estimate of how the distribution of ash constituents could be influenced by plant operation.

2. Materials and Methods

A total of 17 biomass heat (and power) plants with an installed thermal capacity between 0.8 and 31.6 MW$_{therm}$, as well as one centralized ash collection depot of several small heating plants (plant ID 18), were selected for sampling. Table 2 gives an overview of the sampled heat (and power) plants with thermal outputs, fuels used, ash samples and plant IDs. The quality of the ashes varied due to different fuels, plant types or operating parameters of the furnace. At 17 sites, pure bottom ash could be sampled. At one plant (plant ID 1, TFZ), pure cyclone ash was sampled too. At five points, mixtures of bottom ash and cyclone ash could be sampled, leading to a total of 50 ash samples (Table 2). Depending on the ash management procedure, the storage duration of bottom ashes at the heating plants varied considerably and ranged from a few days to several weeks. For ten plants, sampling took place in two different heating seasons (winter 2018/2019 and winter 2019/2020). Seven plants and the central ash collection depot were sampled only once (n = 1). At the heating plant of TFZ (plant ID 1), a series of a total of 20 ash samples was obtained over an entire heating period (12 × bottom ashes, 8 × cyclone ashes). For the general analysis of variability between plants, mean values on ash quality per plant were calculated, while individual samples were used to assess heterogeneity within one plant.

Table 2. Overview of sampled heating (power) plants with thermal output, fuels (a = wood chips, b = wood pellets both from untreated wood) and composition of bottom ashes and mixtures of bottom and cyclone ashes.

Parameter	Unit	Plant ID																		
		1	2	3	4	5	6	7	8	9	10	11	12	13	14	15	16	17	18	4
Fuel		a	a	a	a	a	a	a	b	a	a	a	a	a	a	a	a	a	a	a
Thermal power per combustion unit	kW_{therm}	650	750	750	1050	1250	1500	1900	1950	2000	3000	3000	3500	3690	12,100	18,350	31,600	800	<1 MW	1050
Total thermal power	kW_{therm}	1300	1500	1500	2100	2500	1500	1900	1950	2000	3000	3000	3500	3690	12,100	27,850	31,600	800		2100
Bottom ash (n)		12	1	1	2	2	2	1	1	1	2	2	1	2	2	2	2	-	-	-
Mixed bottom and cyclone ash (n)		-	-	-	-	-	-	-	-	-	-	-	-	-	-	-	-	2	1	1
Cyclone ash (n)		8	-	-	-	-	-	-	-	-	-	-	-	-	-	-	-	-	-	-
		Composition of bottom ashes																Composition of mixed bottom and cyclone ashes		
Arsenic	mg/kg d.b.	5.0	5.0	5.0	5.0	5.0	5.5	7.5	5.0	27.0	5.0	6.3	5.0	5.0	5.0	5.0	10.7	10.0	5.0	11.0
Lead	mg/kg d.b.	4.7	5.0	2.0	5.4	3.7	16.0	3.0	12.0	2.0	5.4	11.0	2.3	4.5	45.5	3.9	150.5	38.0	180.0	15.0
Cadmium	mg/kg d.b.	0.4	0.4	0.2	0.7	0.3	2.5	0.7	1.3	0.3	1.0	4.1	1.0	0.4	0.4	0.3	1.0	15.4	15.0	7.6
Chromium, total	mg/kg d.b.	40	37	81	38	39	46	37	200	120	40	58	39	76	37	60	49	47	130	47
Chromium (VI)	mg/kg d.b.	2.0	1.8	32.8	2.0	6.3	3.8	2.1	42.3	16.0	2.0	3.7	1.6	6.7	1.0	1.8	1.0	1.8	21.8	4.0
Copper	mg/kg d.b.	51	53	100	69	51	68	66	150	120	45	47	36	64	29	54	44	55	110	94
Nickel	mg/kg d.b.	45	18	41	28	50	31	38	39	43	32	41	36	37	28	62	21	42	44	36
Mercury	mg/kg d.b.	0.05	0.05	0.05	0.06	0.05	0.09	0.05	0.05	0.05	0.05	0.05	0.10	0.05	0.05	0.05	0.06	0.05	0.06	0.09
Thallium	mg/kg d.b.	0.37	0.40	0.20	0.30	0.30	0.50	0.20	0.20	0.20	0.30	0.30	0.20	0.30	0.30	0.30	0.30	1.45	0.89	0.55
Zinc	mg/kg d.b.	110	46	120	173	37	500	490	95	64	170	420	260	142	161	145	205	760	1,500	990
Lime (CaO)	wt% d.b.	23.9	38.0	21.0	17.5	50.5	31.5	16.0	43.0	18.0	46.5	24.5	32.0	22.0	31.0	19.0	19.5	44.0	31.0	20.0
Alkaline active substances (CaO)	wt% d.b.	33.7	49.0	30.6	30.2	69.1	34.3	19.9	64.0	19.2	37.4	26.4	31.8	27.4	17.7	23.8	21.3	13.5	21.0	11.0
Phosphate, total (P$_2$O$_5$)	wt% d.b.	3.2	4.1	2.9	2.3	3.7	2.9	2.1	5.2	2.5	2.2	2.9	6.5	2.9	0.9	2.4	1.3	2.5	3.2	2.5
Potassium oxide, total (K$_2$O)	wt% d.b.	10.2	11.0	8.1	5.5	5.1	6.4	4.5	12.0	4.8	5.5	6.0	3.5	8.1	2.5	7.0	3.4	5.6	11.0	6.1
Magnesium oxide, total (MgO)	wt% d.b.	4.2	5.0	4.1	2.9	6.3	4.2	2.2	5.8	2.3	7.9	3.1	3.5	3.8	1.7	3.6	2.0	3.2	4.3	3.3
Total sulfur (S)	wt% d.b.	0.11	0.14	0.04	0.07	0.04	0.23	0.15	0.16	0.09	0.15	0.34	0.10	0.11	0.07	0.05	0.25	0.54	1.40	0.41
Boron (B)	mg/kg d.b.	0.0	0.0	0.0	2.6	0.0	82.5	0.0	1.6	1.4	0.0	1.0	0.0	0.0	0.3	0.6	1.8	240.0	280.0	220.0
Iron (Fe)	mg/kg d.b.	26,083	9,500	18,000	14,000	5,750	15,500	16,000	5,200	9,600	13,500	14,000	12,000	17,000	11,300	21,000	14,500	47,000	94,000	51,000
Cobalt (Co)	mg/kg d.b.	22	4	13	11	12	10	11	28	14	8	17	13	15	10	19	9	14,000	17,000	16,000
Manganese (Mn)	mg/kg d.b.	17,567	2,600	12,000	8,250	25,000	11,550	5,500	40,000	17,000	6,250	15,500	13,000	11,500	1,450	13,000	1,700	11,350	23,000	13,000
Molybdenum (Mo)	mg/kg d.b.	2.6	7.0	5.0	3.5	3.9	4.1	5.0	5.0	5.0	4.6	3.5	5.0	3.5	3.5	3.5	3.6	3.5	3.1	5.0
Sodium (Na)	mg/kg d.b.	4,442	2,700	5,500	4,050	1,135	3,250	2,800	1,900	2,200	2,600	2,700	3,200	6,000	2,550	6,550	2,300	2,250	3,500	4,000
Moisture content	wt%.	0.1	0.0	0.5	13.9	0.2	0.1	0.4	1.0	1.8	0.3	0.3	0.7	0.4	24.8	7.6	30.4	0.5	0.2	29.3
Loss of ignition	wt% d.b.	0.0	0.0	0.0	2.6	0.0	0.0	0.0	1.6	1.4	0.0	1.0	0.0	0.0	0.3	0.6	1.8	0.0	2.2	7.2
pH		12.7	13.1	12.8	12.9	12.8	12.8	12.7	13.3	12.4	12.8	12.9	12.9	12.8	12.7	12.6	12.7	12.8	13.1	12.5

Sampling was carried out directly at the heating plants in accordance with LAGA PN 98 [50]. Thereby, it was necessary to prepare a representative sample for laboratory analysis from several individual samples of the ashes stored at their respective locations. The minimum volume of an individual on-site sample and of the laboratory sample prepared by sample combination, homogenization and sample division depends on the maximum grain size of the ash and was between 0.5 and 10 L. Fine-grained ashes have a lower required minimum volume than coarse-grained ashes. The minimum number of incremental samples results from the basic quantity of stored bottom ash or cyclone ash. For example, up to a volume of 30 m^3, at least eight individual samples should be taken according to LAGA PN 98. During sampling, the individual samples were recorded photographically (Figure 1).

Figure 1. Sampling of bottom ash according to LAGA PN 98 [50].

To obtain the laboratory sample from the individual samples, the samples were combined and thoroughly mixed with a shovel. After that, the mixed sample was divided into four even parts. Two of the four parts were discarded. The two remaining quarters were combined again, carefully mixed and the laboratory sample of approx. 8 L was taken from the mixture. Each sampling was documented on a sampling protocol.

The TFZ heating plant was an exception regarding sampling. Here, twelve individual samples of bottom ash and eight individual samples of cyclone ash were collected to assess variability of this plant over a complete heating season. To compare variation among heating plants, results from the bottom ash analyses were combined mathematically by calculating a theoretical mixed sample for the entire heating season.

The chemical analyses were performed by Wessling GmbH, Neuried, Germany. The analysis included the following ash components with a fertilizing effect—the macronutrients calcium (Ca), phosphorus (P), potassium (K) and sulfur (S), as well as the micronutrients cobalt (Co), iron (Fe), manganese (Mn), molybdenum (Mo), sodium (Na) and selenium (Se). The following heavy metals were analyzed: arsenic (As), lead (Pb), cadmium (Cd), chromium (Cr), both as total content and as chromium(VI), copper (Cu), nickel (Ni), mercury (Hg), thallium (Th), and zinc (Zn). In addition, pH, moisture content and loss of ignition of the ashes were measured. Elemental concentrations of the ash were determined mostly according to ISO standards. The dry residue was determined according to DIN EN

12879 [51]. The ash samples were dissolved with aqua regia (DIN ISO 11466 1997-06) [52] and analyzed by plasma mass spectrometry (ICP-MS) (DIN EN ISO 17294-2 (2005-02) [53]. Cr(VI) was determined according to DIN 19734 (1999-01) [54] The pH value in the solid was analyzed according to DIN ISO 10390 (2005-12) [55] and the alkaline active components according to VDLUFA Method Book Volume II.2, Method 4.5.1 [56].

3. Results and Discussion

Table 2 summarizes the results for the bottom ashes and the mixtures of bottom and cyclone ashes. The results are given per heating plant and ordered from left to right with ascending boiler output. In this order, the different combustion plants were also provided with IDs. All plants except one used wood chips from natural wood as fuel, and one plant used wood pellets. The analysis of the ashes included heavy metals, nutrients, pH, moisture content and loss of ignition. The results refer to the dry mass and are given either as concentrations (mg/kg d.b.) or as mass fractions (wt% d.b.).

3.1. Quality of Bottom Ash

First, the results of the heavy metals in bottom ashes were evaluated in more detail (Figure 2), followed by analysis of the nutrient contents (Figure 3). The mean values for the relevant chemical elements and the physical ash properties per plant are given in Table 2.

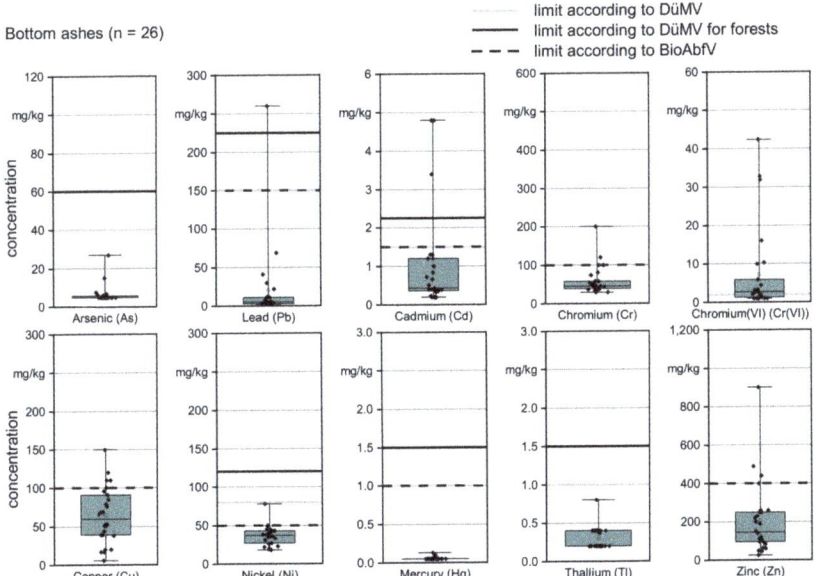

Figure 2. Heavy metal contents in mg/kg (on dry basis) of the 26 bottom ash samples as point clouds and as boxplots with 25% and 75% quantiles (box) and minimum to maximum values (whisker). Horizontal lines show the respective limit values according to the German DüMV and BioAbfV.

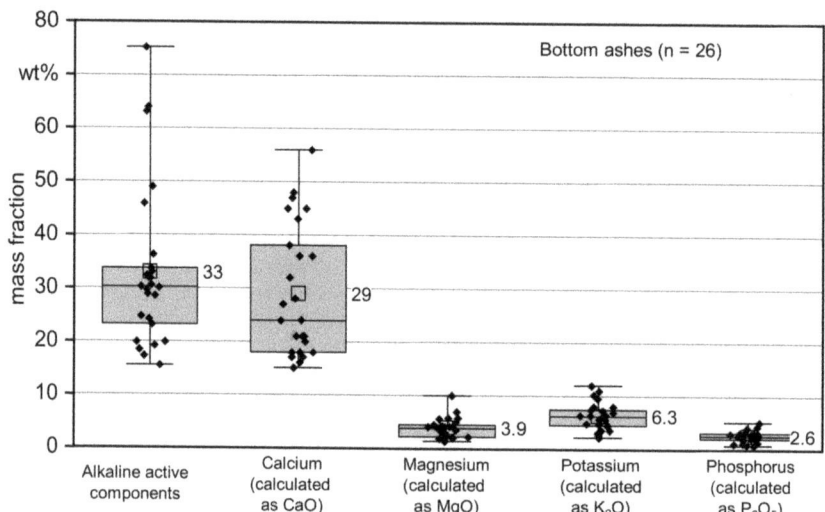

Figure 3. Main nutrients of the 26 bottom ashes (based on dry matter) as point clouds as well as boxplots with 25% and 75% quantiles (box) and minimum to maximum values (whisker). The numbers next to the boxplots are the mean values.

The results refer to dry basis (d.b.) and, for each element, the individual results are shown as a cloud of points and as a boxplot with a minimum and maximum. The twelve ash samples from the TFZ heating plant are included in the evaluation as one mean value to avoid weighting effects. This results in a total number of $n = 26$ for the evaluation of the variation of bottom ashes among plants. In addition, the limit values for agricultural and forestry use according to the German DüMV and the limit values of the German BioAbfV are indicated in Figure 2. Table 3 gives the results in numbers.

Table 3. Analytical results of heavy metal contents of 26 bottom ashes from Bavarian biomass heat (and power) plants with an installed capacity of >1 MW (d.b. = dry basis).

Parameter	Unit	Min	1st Quantile	Median	Mean	3rd Quantile	Max
Arsenic (As)	mg/kg d.b.	5	5	5	6.5	5.6	27
Lead (Pb)	mg/kg d.b.	2	2.8	5	20.2	11	260
Cadmium (Cd)	mg/kg d.b.	0.2	0.4	0.4	1	1.2	4.8
Chromium, total (Cr)	mg/kg d.b.	29	39	45	59	58	200
Chromium(VI) (Cr(VI))	mg/kg d.b.	1	1.3	2.7	7.2	5.5	42.3
Copper (Cu)	mg/kg d.b.	5.8	39	60	64	90	150
Nickel (Ni)	mg/kg d.b.	18	28	37	37	43	78
Mercury (Hg)	mg/kg d.b.	0.1	0.1	0.1	0.1	0.1	0.1
Thallium (Th)	mg/kg d.b.	0.2	0.2	0.2	0.3	0.4	0.8
Zinc (Zn)	mg/kg d.b.	26	95	145	205	250	900

The limit values of the DüMV were exceeded in one sample for Pb (3%) and in three samples for Cd (8%). These exceedances apply to the DüMV limit values of both agricultural and forestry applications, although a 50% higher heavy metal content is permissible for forestry applications (Table 1). Schilling (2020) [10] examined 334 ash samples from 12 plants. He found exceedances for Pb in 1.9% of cases and for Cd in 1.6% of cases. The author documented exceedances for Cr(VI) in just 6% of cases. This deviates strongly from the values in the present study. For an application on agricultural land, a limit value for Cr(VI) of 2.0 mg/kg applies, according to DüMV. This limit is exceeded by 62% of the examined bottom ashes. Additionally, Reichle et al. (2009) [16] point out that the Cr(VI) limit is frequently exceeded in bottom ash from wood combustion. The authors recommend paying particular attention to Cr(VI) during the recycling of wood ash.

Ten heating plants were sampled twice in the current study, i.e., during winter 2018/2019 and during winter 2019/2020, and only two heating plants complied with the limit value for Cr(VI) in both samples. These were plants with a wet ash removal system (plant IDs 4, 14, 16), whereas all other plants used dry ash removal systems. For the plants that used dry ash removal, at least one sample per heating plant exceeded the limit value for Cr(VI). In three plants, the limit value was exceeded both times. Moistening of bottom ashes provides the conditions for a chemical reduction of Cr(VI) into Cr(III) [9]. Pohlandt-Schwandt (1999) [9] and Schilling (2020) [10] state that wet bottom ashes are low in Cr(VI). Therefore, moistening of bottom ashes is already often applied as a quality management tool to improve bottom ash quality [9,57].

In contrast to DüMV, there is no limit value for Cr(VI) in the BioAbfV. However, some of the other limit values in the BioAbfV are lower compared to DüMV and some bottom ashes exceeded the values for copper (19%, $n = 5$), nickel (8%, $n = 2$) and zinc (15%, $n = 4$).

The DüMV limit value for Cd was exceeded by two plants (plant IDs 6 and 11). Nickel and copper limits of BioAbfV were exceeded in each of the heating plants that were sampled twice in one of the samples, while the BioAbfV limit value for zinc was exceeded by both samples at one heating plant. Thereby, Zn was exceeded in all three ash samples with exceeded Cd. Kovacs et al. (2018) [8] and Schilling (2020) [10] show that there is a negative correlation between the concentration of volatile metals such as Cd, Pb or Zn and the temperature in the combustion chamber. Therefore, higher temperature combustion could probably solve the problem of Cd in bottom ash. Schilling (2020) [10] observed a complete volatilization of Cd at an average temperature of above 750 °C in the combustion chamber. The boiling temperature of Cd is 767 °C.

In total, only eight of the bottom ashes sampled complied with all heavy metal limit values according to the DüMV and the BioAbfV (Table 1), directly. Assuming that Cr(VI) can be sufficiently reduced by suitable treatments, e.g., by moistening the ashes [9,10], 85% of the ashes ($n = 22$) complied with the limit values of the DüMV. A total of 54% of the ashes ($n = 14$) also complied with the requirements of the BioAbfV regarding the maximum permissible heavy metal concentrations.

Bottom ashes contain many nutrients that are relevant for plant growth [11,24,26,28]. The sum of the basic components (metal oxides and carbonates [24]) and the individual values for calcium (calculated as CaO), potassium (calculated as potassium oxide K_2O), magnesium (calculated as magnesium oxide MgO) and phosphorus (calculated as phosphate P_2O_5) are shown in Figure 3 as point clouds and box plots. Table 4 shows the results in figures together with the contents of the additional trace nutrients and other parameters.

First, a comparison is made with publications on ash quality from Germany and Austria. Since here the wood qualities and the technology of the CHP plants are quite similar to the plants investigated. Reichle et al. (2009) [16] reported average nutrient contents for bottom ash of 25 to 45 wt% for calcium oxide (CaO), 3 to 6 wt% for magnesium oxide (MgO) and potassium oxide (K_2O), each, and of 2 to 3 wt% for phosphate (P_2O_5). In the current study, higher values were measured, especially for potassium oxide. Here, the mean value is 6.3 wt% (d.b.) and 50% of the analytical results were between 4.5 and 7.5 wt% (d.b.). Obernberger (1997) [58] also gives a higher value for K_2O than Reichle et al. (2009) [16] with 6.7 wt% d.b. as the average value for the content of potassium oxide in 12 bottom ashes from the combustion of wood chips. The mean phosphate content in Obernberger (1997) [58] is 3.6 wt% (d.b.) and thus about one percentage point higher than results in this study.

Table 4. Analytical results of trace nutrients and other parameters of 26 bottom ashes from Bavarian biomass heat (and power) plants with an installed capacity of >1 MW (d.b. = dry basis).

Parameter	Unit	Min	1st Quantile	Median	Mean	3rd Quantile	Max
Alkaline active components (CaO)	wt% d.b.	15.5	23.5	30.1	32.9	33.5	75.2
Calcium (calculated as CaO)	wt% d.b.	15	18.5	24	28.9	37.5	56
Magnesium (calculated as MgO)	wt% d.b.	1.4	2.4	3.9	3.9	4.5	10
Potassium (calculated as K_2O)	wt% d.b.	2.3	4.6	6.3	6.3	7.5	12
Phosphorus (calculated as P_2O_5)	wt% d.b.	0.9	2.1	2.6	2.6	3.1	5.2
Total sulfur (S)	wt% d.b.	0.0	0.1	0.1	0.1	0.2	0.4
Boron (B)	mg/kg d.b.	0	92	145	148	195	330
Iron (Fe)	mg/kg d.b.	5000	12,000	14,500	14,057	16,750	26,083
Cobalt (Co)	mg/kg d.b.	4.1	9.1	12.5	13.0	15.8	28.0
Manganese (Mn)	mg/kg DM	1000	4075	12,500	11,772	15,750	40,000
Molybdenum (Mo)	mg/kg d.b.	2.0	2.7	5.0	4.1	5.0	7.0
Sodium (Na)	mg/kg d.b.	970	2425	2750	3362	4181	8000
Selenium (Se)	mg/kg d.b.	5.0	5.0	5.0	5.0	5.0	5.0
pH value	-	12.3	12.7	12.8	12.8	12.8	13.3
Moisture content	wt%	0.0	0.0	0.5	6.2	9.9	32.7
Loss on ignition	wt% d.b.	0.0	0.0	0.0	0.6	1.3	3.6

The results indicate that the nutrient contents in bottom ash from wood combustion can fluctuate over a wide range of values. The pH values of the ashes examined vary between pH 12.3 (minimum) and pH 13.3 (maximum) (Table 2). They thus fluctuate quite closely around the mean value of pH 12.8 and lie within the range of pH 11 to pH 13 given by Reichle et al. (2009) [16] for wood ashes.

Most of the ashes were very dry, the median value of the moisture content is 0.5 wt%. Nurmesniemi et al. (2012) [15] also notes this value for bottom ashes. Only the two plants with wet ash removal raised the mean moisture content to 6.2 wt%. For the plants with a wet ash removal system, the moisture content varied between 21 and 33 wt%.

Most of the ashes were completely combusted and showed only a low loss of ignition, which amounted to 0.6 wt% on average and reached a maximum of 3.6 wt%. Thus, all ashes remained below the value of 5 wt%. Therefore, it can be assumed that there are no organic pollutants in the ash [16].

Looking at ash qualities that have been published beyond Germany and Austria, similar contents for CaO, MgO, P_2O_5 and K_2O have been reported by Okmanis et al. (2015) [40] and Ingerslev et al. (2011) [20]. Considerably lower nutrient levels have been published by Nurmesniemi et al. (2012) [13] and Hannam et al. (2018) [16] for bottom ashes. Except for Cr (partly originating from the steels in the combustion chamber [20], the ash constituents originate from the fuels [7,20]. These differences can therefore be partly due to different fuel compositions. However, the main causes are differences in combustion technology and different temperatures in the combustion chamber.

Table 5 correlates the bottom ash contents of the present study with the thermal power of the combustion unit classified in <1 MW, 1 to 10 MW and >10 MW. The nutrient levels of alkaline active substances (CaO), MgO, P_2O_5 and K_2O decrease with increasing furnace power due to higher temperatures in the combustion chamber. This is consistent with the research of Okmanis et al. (2015) [40] who examined the ash from heating plants in Lithuania. Additionally, Wilpert et al. (2016) [11] shares this observation and suggests a mixture of ashes from large and smaller heating plants to increase the nutrient content in fertilizers from wood ash.

Table 5. Chemical composition (mean value and range) of the bottom ash according to thermal power of the combustion unit.

Parameter	Unit	Thermal Power of Combustion Unit in MW$_{therm}$, n = Number of Units—Each Represented with One or Two Samples		
		<1 (n = 3)	1–10 (n = 9)	>10 (n = 3)
Arsenic	mg/kg d.b.	5.0 (5.0–5.0)	7.6 (5.0–27.0)	6.9 (5.0–10.7)
Lead	mg/kg d.b.	3.9 (2.0–5.0)	6.5 (2.0–16.0)	66.6 (3.9–150.5)
Cadmium	mg/kg d.b.	0.3 (0.2–0.4)	1.2 (0.3–4.1)	0.6 (0.3–1.0)
Chromium, total	mg/kg d.b.	53 (37–81)	69 (37–200)	48 (37–60)
Chromium(VI)	mg/kg d.b.	12.2 (1.8–32.8)	8.7 (1.6–42.3)	1.3 (1.0–1.8)
Copper	mg/kg d.b.	68 (51–100)	76 (45–150)	42 (29–54)
Nickel	mg/kg d.b.	35 (18–45)	37 (28–50)	37 (21–62)
Mercury	mg/kg d.b.	0.05 (0.05–0.05)	0.06 (0.05–0.10)	0.05 (0.05–0.06)
Thallium	mg/kg d.b.	0.3 (0.2–0.4)	0.3 (0.2–0.5)	0.3 (0.3–0.3)
Zinc	mg/kg d.b.	92 (46–120)	235 (37–500)	170 (145–205)
Lime (CaO)	wt% d.b.	28 (21–38)	30 (16–51)	23 (19–31)
Alkaline active substances (CaO)	wt% d.b.	37.7 (30.6–49.0)	36.0 (19.2–69.1)	20.9 (17.7–23.8)
Phosphate, total (P$_2$O$_5$)	wt% d.b.	3.4 (2.9–4.1)	3.0 (2.1–5.2)	1.5 (0.9–2.4)
Potassium oxide, total (K$_2$O)	wt% d.b.	9.8 (8.1–11.0)	6.4 (4.5–12.0)	4.3 (2.5–7.0)
Magnesium oxide, total (MgO)	wt% d.b.	4.4 (4.1–5.0)	4.2 (2.2–7.9)	2.4 (1.7–3.6)
Total sulfur (S)	wt% d.b.	0.10 (0.04–0.14)	0.14 (0.04–0.34)	0.12 (0.05–0.25)
Boron (B)	mg/kg d.b.	0.02 (0.00–0.05)	8.91 (0.00–82.50)	0.90 (0.30–1.76)
Iron (Fe)	mg/kg d.b.	17,861 (9500–26,083)	12,255 (5200–17,000)	15,600 (11,300–21,000)
Cobalt (Co)	mg/kg d.b.	13 (4–22)	14 (8–28)	13 (9–19)
Manganese (Mn)	mg/kg d.b.	10,722 (2600–17,567)	15,355 (5500–40,000)	5383 (1450–13,000)
Molybdenum (Mo)	mg/kg d.b.	4.9 (2.6–7.0)	4.3 (3.5–5.0)	3.5 (3.5–3.6)
Sodium (Na)	mg/kg d.b.	4213.9 (2700.0–5500.0)	2983.5 (1135.0–6000.0)	3800.0 (2300.0–6550.0)

The bottom ashes which, apart from Cr(VI), do not exceed any other limit values of the DüMV, all contain more than 15 wt% (d.b.) CaO and thus meet the requirement for a "lime fertilizer made from ash from the combustion of vegetable matter". A recycling path established in Bavaria and Baden-Württemberg consists of mixing ashes of this quality with lime or lime dolomite to form "carbonic acid lime". The ash content may not exceed 30 wt%. Theoretically, it would also be possible to mix this lime fertilizer from ash with biowaste. However, minimum nutrient content limits in the finished product of 3 wt% N, 3 wt% P$_2$O$_5$ or 3 wt% K$_2$O in the dry matter would then have to be met. According to Kehres (2016) [31], these contents are generally not achieved by mixtures of bottom ash and biowaste.

For a large part of the bottom ash (69%), the classification as "PK fertilizer from ash from the incineration of vegetable matter" would be possible, since at least 2 wt% P$_2$O$_5$ and 3 wt% K$_2$O are contained in their dry matter.

Four of the ashes (corresponding to approximately 15%) contain at least 10 wt% (d.b.) K$_2$O and would thus fulfil the requirement for a "potassium fertilizer from ashes of the combustion of vegetable matter".

Wood ash can also be used in composting. If the resulting "organic-mineral fertilizers" are to be spread on agricultural land in accordance with DüMV, the limit values of the BioAbfV must also be met. Taking into account the exceedances of Cr(VI) according to the DüMV, a total of 54% of the bottom ash examined also complies with the limit values of the BioAbfV. However, the limit values of the BioAbfV do not have to be met if the application takes place on land, for which the BioAbfV does not apply, such as in gardening and landscaping or if substrates or topsoil materials are produced from the mixture of ash and compost [31]. This latter recovery path would thus be possible for 85% of the bottom ash investigated, as long as a reduction in the Cr(VI) content can be assumed.

3.2. Distribution of Element Loads between Bottom Ash to Cyclone Ash (TFZ Heating Plant)

At the TFZ heating plant, the distribution of the element loads between bottom ash and cyclone ash was investigated. For this purpose, individual samples of bottom ash and cyclone ash were sampled simultaneously at eight points during the same heating period.

Volatile ash components, such as Cd, Pb, Zn and Hg, evaporate at the high temperatures in the combustion chamber [8,10,13,15,16]. For this reason, volatile components can be discharged from the hot ash bed and accumulate in the cyclone ash through condensation. This results in increased concentrations of these elements in the cyclone ash compared to the bottom ash. By using the data set of samples obtained at the TFZ heating plant, this correlation should be directly verifiable.

Table 6 shows the heavy metal and nutrient concentrations in the bottom ash in direct comparison with the corresponding cyclone ash. The mean value and the standard deviation of the eight samples taken in pairs are given in each case. Pairs of mean values that differ significantly are printed in bold. Means were compared using the Wilcoxon signed-rank test. At the points where there is no standard deviation, all samples had fallen below the detection or determination limit with respect to this element. The specified detection or quantification limit was then used as the concentration. For the elements As and Hg, which also occur at very low concentrations in the cyclone ash, this can lead to a distortion in the calculation of the element loads, since this procedure means that a similarly high value must be assumed in both the bottom ash and the cyclone ash. In fact, it can be assumed that the proportion of the two volatile elements As and Hg is higher in the cyclone ash than in the bottom ash. However, the detection limit of the analysis via the external laboratory does not allow this conclusion to be drawn.

Table 6. Mean heavy metal and nutrient concentrations (including standard deviation) in bottom ashes and in the associated cyclone ashes from eight paired samplings at the TFZ heating plant.

Parameter	Unit	Bottom Ash		Cyclone Ash	
		Mean *	Standard Deviation	Mean *	Standard Deviation
Arsenic	mg/kg d.b.	<5.0		8.7	2.5
Lead	mg/kg d.b.	<5.0		123.3	65.6
Cadmium	mg/kg d.b.	0.5	0.1	56.3	14.7
Chromium, total	mg/kg d.b.	41.1	7.3	54.4	10.4
Chromium(VI)	mg/kg d.b.	1.9	1.3	2.1	0.7
Copper	mg/kg d.b.	32.9	27.8	77.1	20.5
Nickel	mg/kg d.b.	46.0	9.1	51.9	8.7
Mercury	mg/kg d.b.	<0.1		0.4	0.2
Thallium	mg/kg d.b.	<0.4		5.5	2.1
Zinc	mg/kg d.b.	110	25.4	3787	788.1
Lime (CaO)	wt% d.b.	20.9	5.9	25.3	8.8
Alkaline active substances (CaO)	wt% d.b.	30.7	10.4	40.5	6.1
Phosphate, total (P_2O_5)	wt% d.b.	3.3	1.8	3.9	1.4
Potassium oxide, total (K_2O)	wt% d.b.	10.1	3.3	8.5	3.0
Magnesium oxide, total (MgO)	wt% d.b.	4.0	0.9	5.3	1.1
Total sulfur (S)	wt% d.b.	<0.1		1.8	0.3

Table 6. *Cont.*

Parameter	Unit	Bottom Ash		Cyclone Ash	
		Mean *	Standard Deviation	Mean *	Standard Deviation
Boron (B)	mg/kg d.b.	191.3	40.8	435.0	72.6
Iron (Fe)	mg/kg d.b.	**16,125**	3370	**35,125**	11,263
Cobalt (Co)	mg/kg d.b.	20.6	6.5	29.9	13.4
Manganese (Mn)	mg/kg d.b.	15,850	7284	24,213	11,097
Molybdenum (Mo)	mg/kg d.b.	**2.2**	0.2	**4.7**	0.9
Sodium (Na)	mg/kg d.b.	**4525**	1253	**2788**	491

* Significantly different pairs of mean values are printed in bold.

The interpretation of the results in Table 4 is based on the calculated absolute element loads related to the total mass of the respective element in the ash (Figure 4). In order to make quantitative statements about how the actual loads of the individual elements are distributed between the bottom ash and the cyclone ash, it is first necessary to make reasonable assumptions about the mass ratio between bottom ash and the associated cyclone ash. For fixed-bed furnaces, a proportion of 10 to 30 wt% of cyclone ash is usually reported [45,58–60]. Fine fly ash is not considered in the following analysis.

The actual proportion of cyclone ash depends on various factors, such as the turbulence of the primary air in the combustion bed or the fineness of the fuel, for comparison of the ash fractions from wood chips or sawdust shows [1]. With these assumptions, it is possible to derive from the eight pairwise analyses of the bottom ash and the cyclone ash at the TFZ heating plant how the fractions of heavy metals and nutrients are distributed between the ash fractions. In addition to the 1:1 mixing ratio (bottom bar chart), Figure 4 shows the distribution of the loads at 10, 20 and 30 wt% cyclone ashes of the total ash.

Heavy metal compounds containing Pb, Cd, Tl, Hg and Zn, are highly volatile [8,13] and are predominantly found in the cyclone ash in all calculations. Consequently, even at the lowest assumed cyclone ash content of 10 wt% of total ash, Cd accumulates in the cyclone ash of up to 93 wt%. Should high concentrations of highly volatile elements be observed in bottom ashes that are considered for utilization as fertilizer, an increase in the temperature in the combustion bed could result in a reduction in these elements in bottom ashes and an increase in cyclone ashes.

For As, no clear effect could be seen in the data presented here. As the concentrations of As in the investigated bottom and cyclone ashes were overall very low, the limit of determination often had to be used as the concentration in the ash fractions. Cu, Cr, Ni and the main nutrients Mg, P, Ca and K are less volatile and, depending on the calculation performed here, are found in only 11 to 50 wt% in the cyclone ash. Therefore, they predominantly remain in the bottom ash.

Obernberger (1997) [58] shows basically similar element ratios between bottom ash and cyclone ash for wood chips. However, the reported concentrations in the cyclone ash were consistently lower compared to the results presented here (with exception of K and P), which may be due to different combustion chamber and cyclone temperatures of the heating plants investigated. The combustion chamber temperatures near the combustion bed are not known for the TFZ heating plant. Lanzerstorfer (2017) [13] observed that at combustion chamber temperatures between 830 and 920 °C, Cd, Pb and Zn accumulate in the fly ashes, while most nutrients (Ca, Mg, P_2O_5) remain in the bottom ash. Both Lanzerstorfer (2017) [13] and Schilling (2020) [10] note a higher volatility for potassium, which leads to K losses from the bottom ash. These increased K losses could not be observed at the TFZ heating plant, suggesting that the combustion temperatures are sufficiently high to remove the volatile heavy metals and at the same time low enough to avoid high potassium losses.

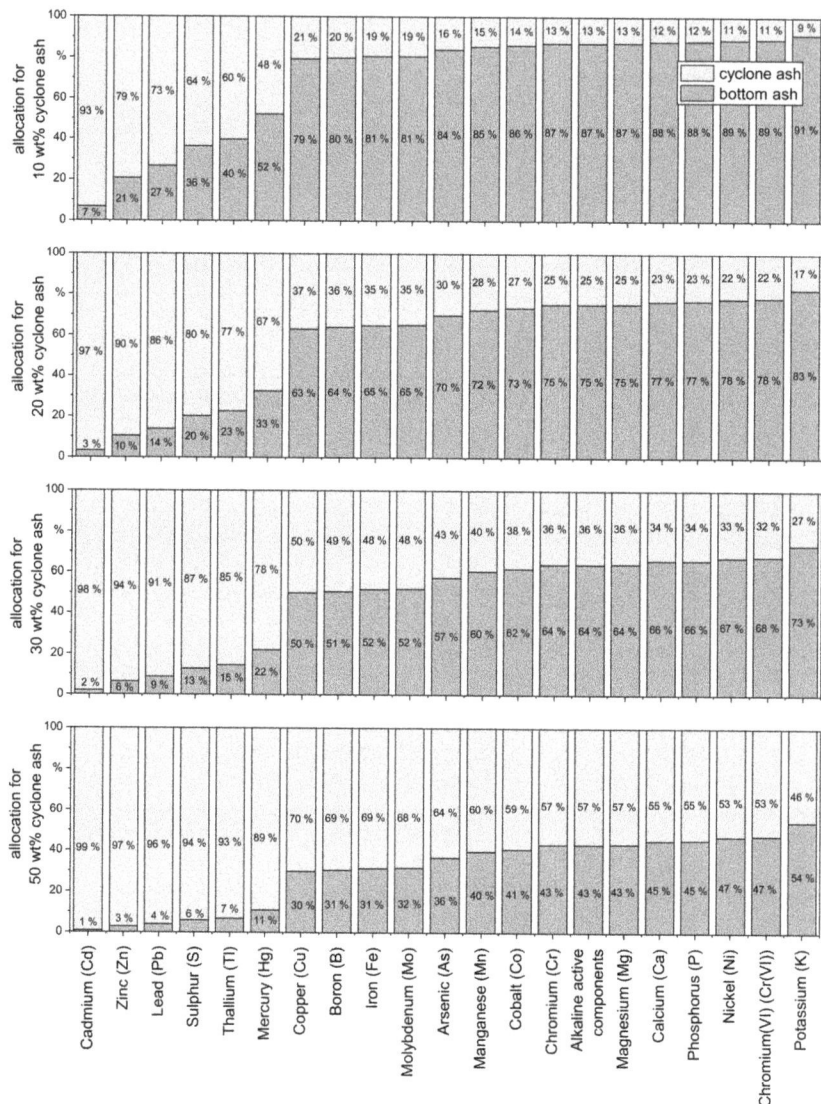

Figure 4. Ratio of element loads in bottom ash and cyclone ash of the TFZ heating plant at hypothetical mixing ratios with cyclone ash contents of 10, 20, 30 and 50 wt%.

3.3. Quality of Mixtures of Bottom Ash with Cyclone Ash

In some heating plants, the bottom ash and the cyclone ash are collected in the same container due to the plant design. The composition of these ashes is shown in Table 2 (right columns). All five samples of these mixed ashes exceed the DüMV limit value for Cd. Further exceedances occurred for Cr(VI) ($n = 4$), thallium ($n = 1$) and lead ($n = 1$). None of the bottom ashes mixed with cyclone ash can meet the requirements regarding the heavy metal limit values of the DüMV or the BioAbfV. They are therefore not eligible as a source material for fertilizers. These ashes are excluded from being spread on agricultural and forestry land in Germany. If the aim is to recycle bottom ashes, it is recommended that these ash fractions are collected and reused separately. When using other fuels, e.g., when firing agricultural fuels such as straw, a mixture of bottom ash and cyclone ash can often comply with the limit values of the DüMV [42]. This is due to the generally lower heavy metal content of agricultural fuels compared to wood fuels.

4. Conclusions

The energetic use of untreated wood in biomass heat (and power) plants produces combustion residues in the form of ash. The increased use of by-products and residues contributes to the conservation of natural resources. It has been shown that the bottom ashes produced are basically suitable for use as fertilizers or as a raw materials for fertilizers despite the low pollutant limits in the German DüMV.

On average, the bottom ashes examined contained 33 wt% alkaline active components, 29 wt% Calcium (calculated as CaO), 3.9 wt% Magnesium (calculated as MgO), 6.3 wt% Potassium (calculated as K_2O) and 2.6 wt% of phosphorus (calculated as P_2O_5).

However, quality assurance of the ashes and compliance with the relevant legal requirements due to possible exceedances of the heavy metal limits prescribed by the German Fertilizer Ordinance are crucial. The limits were exceeded in the bottom ashes for chromium(VI) (62%), cadmium (12%) and lead (4%). Mixing of the bottom ashes with cyclone ashes led, in all cases, to the heavy metal limit values being exceeded, especially for cadmium. The following measures contribute to the quality assurance of ashes for fertilization purposes:

- As prescribed by the German Fertilizer Ordinance, only untreated wood should be used in biomass heat (and power) plants, since waste wood can contain elevated concentrations of heavy metals.
- Fluctuations in fuel quality or the combustion conditions can change the heavy metal contents in bottom ash. Regular sampling and chemical analysis of the ash is therefore necessary.
- Since some of the heavy metals are volatile under the usual combustion conditions in biomass heating (power) plants, care should be taken to ensure that the combustion temperatures in the boiler are constantly high enough. An average temperature of over 750 °C, for example, reliably leads to sufficiently low cadmium contents in the bottom ash.
- As has been shown, mixing of bottom ash and cyclone ash leads to an increase in heavy metals. Separate collection of these ash fractions is essential.

The frequently exceeded limit value for chromium(VI) in the German Fertilizer Ordinance can be reduced by moistening and storing the bottom ashes. In this process, chromium(VI) converts into the harmless chromium(III).

The present study maps the ash quality of typical biomass heating plants according to the state of the art in Germany. The evaluation of the results is carried out according to the regulations applicable in Germany for the use of biomass ash for fertilizer purposes. Other combustion techniques, other fuels and other legal regulations may lead to different assessments.

Author Contributions: H.B. designed and performed the experiments, derived the models and analyzed the data. Both D.K. and H.H. contributed to the final version of the manuscript. D.K. supervised the project. All authors have read and agreed to the published version of the manuscript.

Funding: This research was funded by the Bavarian State Ministry of Food, Agriculture and Forestry, grant number G2/KS/17/02.

Institutional Review Board Statement: Not applicable.

Informed Consent Statement: Not applicable.

Data Availability Statement: Not applicable.

Conflicts of Interest: The authors declare no conflict of interest.

Abbreviations

BioAbfV = German Biowaste Ordinance; d.b. = on dry basis; DIN = German Institute for Standardization; DüMV = German Fertilizer Ordinance; ISO = International Organization for Standardization; LAGA = German Federal/State Working Group on Waste; TFZ = Technology and Support Centre in the Centre of Excellence for Renewable Resources; VDLUFA = Association of German Agricultural Analytic and Research Institutes.

References

1. Kaltschmitt, M.; Hartmann, H.; Hofbauer, H. *Energie aus Biomasse. Grundlagen, Techniken und Verfahren*; Springer: Berlin/Heidelberg, Germany, 2016; p. 1755.
2. Vassilev, S.V.; Baxter, D.; Vassileva, C.G. An overview of the behaviour of biomass during combustion: Part II. Ash fusion and ash formation mechanisms of biomass types. *Fuel* **2014**, *117*, 152–183. [CrossRef]
3. Gößwein, S.; Hiendlmeier, S. *Energieholzmarkt Bayern 2016. Untersuchung des Energieholzmarktes in Bayern hinsichtlich Aufkommen und Verbrauch*; LWF: Freising, Germany, 2018; p. 131.
4. Dietz, E.; Kuptz, D.; Blum, U.; Schulmeyer, F.; Borchert, H.; Hartmann, H. *Qualität von Holzhackschnitzeln in Bayern. Gehalte ausgewählter Elemente, Heizwert und Aschegehalt, Berichte aus dem TFZ, Straubing, Nr. 46*; Technologie- und Förderzentrum im Kompetenzzentrum für Nachwachsende Rohstoffe (TFZ); Bayerische Landesanstalt für Wald und Forstwirtschaft (LWF): Freising-Weihenstephan, Germany, 2016; p. 141.
5. Kuchler, C.; Zimmermann, D.; Kuptz, D.; Dietz, E.; Rist, E.; Riebler, M.; Schön, C.; Mack, R.; Blum, U.; Borchert, H.; et al. Contamination of wood chips with mineral soils–Fuel quality and combustion behaviour. In Proceedings of the 52nd International Symposium on Forestry Mechanization (FORMEC), Sopron, Hungary/Forchtenstein, Austria, 6–9 October 2019; University of Sopron Press: Sopron, Hungary, 2019; pp. 320–329.
6. FRED Feste Regenerative Energieträger Datenbank. Straubing: Technologie- und Förderzentrum im Kompetenzzentrum für Nachwachsende Rohstoffe (TFZ). Available online: https//:www.fred.bayern.de/ (accessed on 10 June 2020).
7. Werkelin, J.; Skrifvars, B.-J.; Hupa, M. Ash-forming elements in four Scandinavian wood species: Part 1: Summer harvest. *Biomass Bioenergy* **2005**, *29*, 451–466. [CrossRef]
8. Kovacs, H.; Dobo, Z.; Koos, T.; Gyimesi, A.; Nagy, G. Influence of the flue gas temperature on the behavior of metals during biomass combustion. *Energy Fuels* **2018**, *32*, 7851–7856. [CrossRef]
9. Pohlandt-Schwandt, K. Treatment of wood ash containing soluble chromate. *Biomass Bioenergy* **1999**, *16*, 447–462. [CrossRef]
10. Schilling, S. Steuerungsmöglichkeiten der Qualität und Eignung von Holzaschen für deren Einsatz bei der Waldkalkung. Master's Thesis, Albert-Ludwigs-Universität Freiburg, Freiburg, Germany, 2020.
11. Wilpert, K.V.; Hartmann, P.; Schäffer, J. Quality control in a wood ash re-cycling concept for forests. *VGB Powertech* **2016**, *96*, 67–72.
12. Steenari, B.-M.; Karlsson, L.G.; Lindqvist, O. Evaluation of the leaching characteristics of wood ash and the influence of ash agglomeration. *Biomass Bioenergy* **1999**, *16*, 119–136. [CrossRef]
13. Lanzerstorfer, C. Grate-fired biomass combustion plants using forest residues as fuel. Enrichment factors for components in the fly ash. *Waste Biomass Valorization* **2017**, *8*, 235–240. [CrossRef]
14. Lienemann, P.; Vock, W. *Elementgehalte in Holzaschen und Validierung der Holzaschenkontrolle*; ZHAW: Wädenswil, Switzerland, 2013; p. 96.
15. Nurmesniemi, H.; Manskinen, K.; Pöykiö, R.; Dahl, O. Forest fertilizer properties of the bottom ash and fly ash from a large-sized (115 MW) industrial power plant incinerating wood-based biomass residues. *J. Univ. Chem. Technol. Metall.* **2012**, *47*, 43–52.
16. Reichle, E.; Müller, R.; Schmoeckel, G.; Müller, C.; Wendland, M.; Geiger, H.; Stetter, U.; Zormaier, F. Verwertung und Beseitigung von Holzaschen. Merkblatt. Stand: 01.08.2009; Bayerisches Landesamt für Umwelt (LfU); Bayerische Landesanstalt für Wald und Forstwirtschaft (LWF); Bayerische Landesanstalt für Landwirtschaft (LfL): Augsburg, Germany, 2009; p. 19.
17. Schrägle, R. Innovative Ansätze und Perspektiven der Verwertung von Holzaschen. In Proceedings of the 12th Internationaler BBE-Fachkongress Holzenergie, Augsburg, Germany, 27–28 September 2012; pp. 1–41.
18. Stetter, U.; Zormaier, F. *Verwertung und Beseitigung von Holzaschen. Neues LfU-Merkblatt Greift Altes Thema auf*; LWF aktuell: Geneva, Switzerland, 2010; Volume 74, pp. 28–30.
19. Hannam, K.D.; Venier, L.; Allen, D.; Deschamps, C.; Hope, E.; Jull, M.; Kwiaton, M.; McKenney, D.; Rutherford, P.M.; Hazlett, P.W. Wood ash as a soil amendment in Canadian forests: What are the barriers to utilization? *Can. J. For. Res.* **2018**, *48*, 442–450. [CrossRef]
20. Lama, I.; Sain, D. A case study review of wood ash land application programs in North America, TAPPI Journal February 2021. *TAPPI J.* **2021**, *20*, 111–120.

21. Bachmaier, H.; Kuptz, D.; Hartmann, H. Ash management at biomass heating plants in Southern Germany. In *Setting the Course for a Biobased Economy–Papers of the 27th European Biomass Conference, Proceedings of the 27th European Biomass Conference, Lisbon, Portugal, 27–30 May 2019*; Carvalho, M.D.G., Scarlat, N., Grassi, A., Helm, P., Eds.; ETA-Florence Renewable Energies: Florence, Italy, 2019; pp. 1814–1817.
22. Bohrn, G.; Stampfer, K. Untreated wood ash as a structural stabilizing material in forest roads. *Croat. J. For. Eng.* **2014**, *35*, 81–89.
23. Lahtinen, P. Utilization of biomass ashes in infrastructure construction in Finland. In Proceedings of the 4th Central European Biomass Conference, Graz, Austria, 15–18 January 2014; pp. 1–36.
24. Ingerslev, M.; Hansen, M.; Pedersen, L.B.; Skov, S. Effects of wood chip ash fertilization on soil chemistry in a Norway spruce plantation on a nutrient-poor soil. *For. Ecol. Manag.* **2014**, *334*, 10–17. [CrossRef]
25. Hansen, M.T. Options for increased use of ash from biomass combustion. In Proceedings of the Scandinavian Bio Mass Ash Workshop 2019. BMA Workshop, Copenhagen, Denmark, 25 March 2019; pp. 1–17.
26. Saarsalmi, A.; Smolander, A.; Kukkola, M.; Moilanen, M.; Saramäki, J. 30-Year effects of wood ash and nitrogen fertilization on soil chemical properties, soil microbial processes and stand growth in a Scots pine stand. *For. Ecol. Manag.* **2012**, *278*, 63–70. [CrossRef]
27. Brais, S.; Bélanger, N.; Guillemette, T. Wood ash and N fertilization in the Canadian boreal forest: Soil properties and response of jack pine and black spruce. *For. Ecol. Manag.* **2015**, *348*, 1–14. [CrossRef]
28. Karltun, E.; Saarsalmi, A.; Ingerslev, M.; Mandre, M.; Andersson, S.; Gaitnieks, T.; Ozolincius, R.; Varnagiryte-Kabasinskiene, I. Wood ash recycling–possibilities and risks: Chapter 4. In *Sustainable Use of Forest Biomass for Energy: A Synthesis with Focus on the Baltic and Nordic Region*; Röser, D., Asikainen, A., Raulund-Rasmussen, K., Stupak, I., Eds.; Springer: Berlin/Heidelberg, Germany, 2008; pp. 79–108, ISBN 978-1-4020-5054-1.
29. Ernfors, M.; Sikström, U.; Nilsson, M.; Klemedtsson, L. Effects of wood ash fertilization on forest floor greenhouse gas emissions and tree growth in nutrient poor drained peatland forests. *Sci. Total Environ.* **2010**, *408*, 4580–4590. [CrossRef]
30. Walter, B.; Mostbauer, P.; Karigl, B. *Biomasse–Aschenströme in Österreich*; Umweltbundesamt–Report, Nr. REP-0561; Umweltbundesamt GMBH: Wien, Austria, 2016; p. 56.
31. Kehres, B. Zumischung von Holzasche bei der Kompostierung. *H&K Aktuell* **2016**, *5*, 1–3.
32. Korpjiarvi, K. The development in the use of ashes in Finland. In Proceedings of the Scandinavian Biomass Ash Workshop 2019. BMA Workshop, Copenhagen, Denmark, 25 March 2019; pp. 1–22.
33. Obernberger, I.; Supancic, K. *Fact-Sheet: Einsatz von Holzasche als Bindemittel zur Bodenstabilisierung z.B. im Straßenbau*; BIOS BIOENERGIESYSTEME GmbH: Wien, Austria, 2015; p. 8.
34. Tejada, J.; Grammer, P.; Kappler, A.; Thorwarth, H. Trace element concentrations in firewood and corresponding stove ashes. *Energy Fuels* **2019**, *33*, 2236–2247. [CrossRef]
35. Mortensen, L.H.; Rønn, R.; Vestergård, M. Bioaccumulation of cadmium in soil organisms—With focus on wood ash application. *Ecotoxicol. Environ. Saf.* **2018**, *156*, 452–462. [CrossRef]
36. Saarsalmi, A.; Mälkönen, E.; Piirainen, S. Efffects of Wood Ash Fertilization on Forest Soil Chemical Properties. *Filva Fenn.* **2001**, *35*, 355–368.
37. Kehres, B. *Verwertung von Holzaschen auf Flächen*; 08.03.2013. 2., überarb. Fassung; Bundesgütegemeinschaft Kompost e. V., Ed.; BGK Information: Köln-Gremberghoven, Gernamy, 2013; p. 14.
38. Maresca, A.; Krüger, O.; Herzel, H.; Adam, C.; Kalbe, U.; Astrup, T.F. Influence of wood ash pre-treatment on leaching behaviour, liming and fertilising potential. *Waste Manag.* **2019**, *83*, 113–122. [CrossRef] [PubMed]
39. Ingerslev, M.; Skov, S.; Sevel, L.; Pedersen, L.B. Element budgets of forest biomass combustion and ash fertilization–A Danish case-study. *Biomass Bioenergy* **2011**, *35*, 2697–2704. [CrossRef]
40. Okmanis, M.; Lazdina, D.; Lazdiņš, A. The composition and use value of tree biomass ash. *Rural Sustain. Res.* **2015**, *34*, 32–37. [CrossRef]
41. Bundesverband Mineralischer Rohstoffe e. V. (MIRO). *Deutscher Nachhaltigkeitspreis 2019–Preisträger und Projekte. Die deutsche Gesteinsindustrie–modern, Effizient, Nachhaltig*; Bundesverband Mineralischer Rohstoffe e. V. (MIRO): Duisburg, Germany, 2019; p. 39.
42. Bundesministerium für Ernährung, Landwirtschaft und Verbraucherschutz. Verordnung über das Inverkehrbringen von Düngemitteln, Bodenhilfsstoffen, Kultursubstraten und Pflanzenhilfsmitteln; BGBI, 2017; Part I; Volume 68, pp. 1305–1349. Available online: https://www.gesetze-im-internet.de/d_mv_2012/BJNR248200012.html (accessed on 16 June 2020).
43. Ettl, R.; Weis, W.; Göttlein, A. Laborversuch zur Bewertung von Organo-Asche-Presslingen und einem Kalk-Asche-Gemisch als mögliche Produkte für eine nährstoffliche Kreislaufwirtschaft in Wäldern. *Forstarchiv* **2010**, *81*, 12–20.
44. Saarsalmi, A.; Smolander, A.; Moilanen, M.; Kukkola, M. Wood ash in boreal, low-productive pine stands on upland and peatland sites: Long-term effects on stand growth and soil properties. *For. Ecol. Manag.* **2014**, *327*, 86–95. [CrossRef]
45. Katzensteiner, K.; Holzner, H.; Obernberger, I. *Richtlinien für den sachgerechten Einsatz von Pflanzenaschen zur Verwertung auf land- und forstwirtschaftlich genutzten Flächen*; Bundesministerium für Land- und Forstwirtschaft, Umwelt und Wasserwirtschaft, Eds.; Fachbeirat für Bodenfruchtbarkeit und Bodenschutz: Wien, Austria, 2011; 74p.
46. Kebli, H.; Maltas, A.; Sinaj, S. Landwirtschaftlisches Potenzial von Asche aus rezykliertem Holz. *Agrar. Schweiz* **2017**, *8*, 30–37.
47. Maltas, A.; Sinaj, S. Holzasche: Ein neuer Dünger für die Landwirtschaft. *Agrar. Schweiz* **2014**, *5*, 232–239.

48. Bundesministerium für Umwelt, Naturschutz und Reaktorsicherheit (BMU); Bundesministerium für Ernährung, Landwirtschaft und Forsten (BMELF); Bundesministerium für Gesundheit (BMG). *Verordnung über die Verwertung von Bioabfällen auf landwirtschaftlich, forstwirtschaftlich und gärtnerisch genutzten Böden). Bioabfallverordnung—BioAbfV, in der Fassung vom 1. August, Vorschriftensammlung Version 02*; Stuttgart: Gewerbeaufsicht Baden-Württemberg, Germany, 2012; pp. 1–58.
49. *Sachverständigenrat Bioökonomie Bayern*; Empfehlungen zur Förderung der Bioökonomie in Bayern: Straubing, Germany, 2017.
50. Länderarbeitsgemeinschaft Abfall. *LAGA PN 98—Grundregeln für die Entnahme von Proben aus festen und stichfesten Abfällen sowie abgelagerten Materialien Richtlinie für das Vorgehen bei physikalischen, chemischen und biologischen Untersuchungen im Zusammenhang mit der Verwertung/Beseitigung von Abfällen*; LAGA: Potsdam, Germany, 2019; p. 69.
51. Deutsches Institut für Normung (DIN). *Characterization of Sludges–Determination of the Loss on Ignition of Dry Mass*, German version (EN 12879:2000); Beuth-Verlag: Berlin, Germany, 2001.
52. Deutsches Institut für Normung(DIN). *Soil Quality–Extraction of Trace Elements Soluble in Aqua Regia (ISO 11466:1995)*; Beuth-Verlag: Berlin, Germany, 1997.
53. Deutsches Institut für Normung (DIN). *Water Quality–Application of Inductively Coupled Plasma Mass Spectrometry (ICP-MS)–Part 2: Determination of 62 Elements (ISO 17294-2:2003)*, German version (EN ISO 17294-2:2004); Beuth-Verlag: Berlin, Germany, 2005.
54. Deutsches Institut für Normung (DIN). *Soil Quality–Determination of Chromium(VI) in Phosphate Extract (DIN 19734:1999)*; Beuth-Verlag: Berlin, Germany, 1999.
55. Deutsches Institut für Normung (DIN). *Soil Quality–Determination of pH (ISO 10390:2005)*; Beuth-Verlag: Berlin, Germany, 2005.
56. Verband deutscher landwirtschaftlicher Untersuchungs- und Forschungsanstalten (VDLUFA). *Methodenbuch II. 2 Die Untersuchung von Sekundärrohstoffdüngern, Kultursubstraten und Bodenhilfsstoffen*; VDLUFA: Darmstadt, Germany, 2000.
57. Eberl, G. Veränderung von Chrom(VI)-Gehalten in Holzaschen durch Bewässerung. Vortrag. In Proceedings of the 21th Österreichischer Biomassetag, Kufstein, Austria, 6–7 November 2018; pp. 1–14.
58. Obernberger, I. Aschen aus Biomassefeuerungen–Zusammensetzung und Verwertung. In Proceedings of the Thermische Biomassenutzung–Technik und Realisierung. Internationale Tagung, Salzburg, Austria, 23–24 April 1997; pp. 199–222.
59. Ministerium für Umwelt und Verkehr Baden-Württemberg. Schadstoffströme bei der Entsorgung von Holzasche. Schadstoffströme bei der Verbrennung naturbelassener Hölzer und holzartiger Biomassen im Hinblick auf die Ascheentsorgung. *Reihe Abfall* **2003**, *76*, 79.
60. Zimmermann, S.; Hässig, J.; Landolt, W. *Literaturreview Holzasche–Wald. Nährstoffentzug durch Holzernte, abiotische und biotische Wirkungen*; Eidgenössische Forschungsanstalt für Wald, Schnee und Landschaft (WSL): Birmensdorf, Switzerland, 2010; p. 80.

Article

Nitrogen Migration during Pyrolysis of Raw and Acid Leached Maize Straw

Huan Li [1,2,3], Huawei Mou [1,2,3], Nan Zhao [1,2,3], Yaohong Yu [1,2,3], Quan Hong [1,2,3], Mperejekumana Philbert [1,2,3], Yuguang Zhou [1,2,3,4,*], Hossein Beidaghy Dizaji [5,*] and Renjie Dong [1,2,3,6]

1. Bioenergy and Environment Science & Technology Laboratory, College of Engineering, China Agricultural University, Beijing 100083, China; huanli828@cau.edu.cn (H.L.); mouhuawei@163.com (H.M.); nan.zhaoca@outlook.com (N.Z.); syyh0613@163.com (Y.Y.); hong_quan97@163.com (Q.H.); philbertson2@yahoo.com (M.P.); rjdong@cau.edu.cn (R.D.)
2. Key Laboratory of Clean Production and Utilization of Renewable Energy, Ministry of Agriculture and Rural Affairs, Beijing 100083, China
3. National Center for International Research of BioEnergy Science and Technology, Ministry of Science and Technology, Beijing 100083, China
4. Prataculture Machinery and Equipment Research Center, College of Engineering, China Agricultural University, Beijing 100083, China
5. Thermo-chemical Conversion Department, DBFZ Deutsches Biomasseforschungszentrum Gemeinnützige GmbH, Torgauer Straße 116, 04347 Leipzig, Germany
6. Yantai Institute, China Agricultural University, No. 2006 Binhai Zhonglu, Laishan District, Yantai 264670, China
* Correspondence: zhouyg@cau.edu.cn (Y.Z.); hossein.beidaghy@dbfz.de (H.B.D.); Tel.: +86-10-62737885 (Y.Z.)

Abstract: Solid biofuel is considered as a possible substitute for coal in household heat production because of the available and sustainable raw materials, while NO_x emissions from its combustion have become a serious problem. Nitrogen-containing compounds in pyrolysis products have important effects on the conversion of fuel-N into NO_x-N. Understanding these converting pathways is important for the environmentally friendly use of biomass fuels. The nitrogen migration during pyrolysis of raw and acid leached maize straw at various temperatures was investigated in this study. Thermal gravimetric analysis and X-ray photoelectron spectroscopy were used to investigate the performances of thermal decomposition and pyrolysis products from samples. The main nitrogen functional groups in biomass and biochar products were N-A (amine-N/amide-N/protein-N), pyridine-N, and pyrrole-N, according to the findings. The most common gaseous NO_x precursor was NH_3, which was produced primarily during the conversion of N-A to pyridine-N and pyrrole-N. The formation of HCN mainly came from the secondary decomposition of heterocyclic-N at high temperatures. Before the pyrolysis temperature increased to 650 °C, more than half of the fuel-N was stored in the biochar. At the same pyrolysis temperature, acid-leached maize straw yielded more gas-N and char-N than the raw biomass. The highest char-N yield of 76.39 wt% was obtained from acid-leached maize straw (AMS) pyrolysis at 350 °C. Low pyrolysis temperature and acid-leaching treatment can help to decrease nitrogen release from stable char structure, providing support for reducing nitrogenous pollutant emissions from straw fuel.

Keywords: maize straw; acid leaching; ash; pyrolysis; nitrogen conversion

1. Introduction

Biomass energy is a form of renewable energy derived directly or indirectly from plant photosynthesis. Solid biomass fuel has been gaining popularity as a way to reduce reliance on fossil fuels while also dealing with climate change [1]. However, biomass fuel combustion is reported to result in high NO_x (NO, NO_2, N_2O, etc.) emissions [2]. The NO_x produced from fuel combustion can be divided into thermal-NO_x, fuel-NO_x, and prompt-NO_x according to their origins. The formation mechanism of prompt-NO_x is complicated and its amount is relatively small. Thermal-NO_x is primarily formed by the reaction of nitrogen and oxygen at high temperatures (>1300 °C) [3]. While the furnace temperature of household biomass stove is mostly lower than 700 °C (refer to previous studies [4,5]), fuel-NO_x [6,7] accounted for most of all NO_x [6,7]. Fuel-N can be converted into various nitrogen-containing functional groups during the pyrolysis process, and these compounds can react with oxygen to produce various NO_x species [8]. For NO_x regulation, it is critical to have a thorough understanding of the transformation and migration behavior of fuel-N during biomass pyrolysis.

Both biomass characteristics (species, contents of nitrogen, volatile matter and ash content, etc.) and pyrolysis operation conditions (temperature, heating rate, air supply, etc.) influence the pyrolysis performance of the biomass. Researchers investigated the migration course of nitrogen elements in raw biomass, proteins, and N-containing compounds (formed by protein mixed with hemicellulose, cellulose, and lignin) under various operating conditions [9–11]. With the increase of temperature, NO_x precursors are primarily formed by the following three ways: the pyrolysis of unstable fuel-N, secondary decomposition of char, and tar products [12]. The main NO_x gaseous precursors are NH_3, HCN, and HNCO, with HNCO accounting for little and being easily transferred to the other two [13,14]. The dominant N-species are determined by the biomass type and origin. In some biomass, such as sewage sludge, nitrogen mainly exists in the form of heterocyclic-N, HCN can be found to be the main NO_x precursor during pyrolysis [8]. However, the majority of nitrogen in plant residues is found in proteins and free amino acids, with only a small amount in the form of nucleic acid, chlorophyll, enzymes, vitamins, alkaloids, and inorganic nitrogen [15], NH_3 will become the dominant nitrogen-containing precursor [16].

Agricultural and woody residues are the two main materials of solid biofuels. According to the literature, agricultural residues (straw fuels) may have 5–20 times higher ash content than those of woody biomass, which could contribute to inefficient combustion [17]. Besides, alkali and alkaline earth metallic species (AAEMs) in ash (K, Ca, Na, Mg, etc.) are thought to influence the devolatilization and combustion process [18,19]. Acid-leaching treatment can reduce the content of ash and AAEMs, effectively improve the combustion performance and reduce ash melting, slagging, and corrosion [20–23], but the impact of AAEMs on the NO_x precursor generations of biomass pyrolysis remains unclear. Deep research on the effects of de-ashing and demineralization by acid-leaching on nitrogen transfer and conversion would help the clean utilization of straw biomass fuel.

Therefore, nitrogen migration behaviors during pyrolysis of raw and acid-leached maize straw were investigated in this report. To reduce the ash and AAEMs contents in maize straw, pre-treatment of acid-leaching with CH_3COOH was introduced. The thermal gravimetric analysis (TGA) and pyrolysis experiments at various temperatures were performed to evaluate the decomposition characteristics of the two biomass samples. X-ray photoelectron spectroscopy (XPS) was used to investigate the distribution and migration of nitrogen functional groups in biomass and corresponding biochar, and the effect of acid-leaching on NO_x precursors generation was further analyzed.

2. Materials and Methods

2.1. Materials and Pretreatment Process

In this study, raw maize straw (RMS) was obtained from Shangzhuang experimental station of China Agricultural University, Beijing, China. The samples were crushed and sieved through a 100 mesh sieve, then dried in an oven at 105 °C for 12 h.

Various organic and inorganic acids have been used for acid treatment in previous researches. Strong acids, such as sulfuric acid and nitric acid were reported to enhance dehydration reactions in biomass. Hydrochloric acid is also a commonly leaching solution, but chloride ions will affect the subsequent HCN concentration measurement. To effectively remove AAEMs, while reducing the damage to the biomass structure and not introducing other ions [24], CH_3COOH solution was adopted in this study. The solution was prepared by adding 30 mL of glacial acetic acid (1.05 g/mL, 99.5%) into 970 mL of distilled water and mix thoroughly, which was converted to a mass concentration of 3.13%. Acid-leached maize straw (AMS) was pretreated by soaking 10 g of raw maize straw samples in 200 mL acid solution for 2 h at room temperature while stirring. After removing the acid liquor, the acid-leached biomass was washed with deionized water to neutral pH and then dried at 105 °C for 12 h.

Ultimate analysis of the samples was performed by Vario EL cube (Elementar, Langenselbold, Germany), and oxygen content was detected by difference (Table 1). Proximate analysis of the biomass fuels was carried following the existing Chinese standards [25,26]. The ash content for the biomass sample was reduced from 12.69 wt% to 8.67 wt% after acid treatment. The main residual material in the ash should be silica, since more than half of the ash of the maize straw consists of SiO_2 [27], and it is hardly removed by acid leaching. The contents of cellulose, hemicellulose, and lignin in biomass were determined using the methods mentioned by Guo et al. [28]. To better understand the effect of the pre-treatment, inductively coupled plasma with optical emission spectroscopy (ICP-OES) (ICPOES730, Agilent, Santa Clara, CA, USA) was performed on RMS and AMS to provide selected elemental compositions of the fuel ashes (Table 1). It should be noted that the acid leaching significantly decreased the content of K, Na, and Mg, which were abundantly found in biomass ash.

Table 1. Basic characteristics of the two biomass samples.

Characteristic		Biomass Sample		Removal Ratio %
		RMS	AMS	
Ultimate analysis wt%, dry basis	C	41.83	44.69	
	H	5.08	6.18	
	S	0.58	0.68	
	N	1.39	1.13	
	O [1]	51.13	47.33	
Proximate analysis wt%, dry basis	Volatile matter	68.68	75.70	
	Fixed carbon	16.94	13.01	
	Ash	12.69	8.67	
Composition%	Hemicellulose	31.11	36.49	
	Cellulose	29.03	34.47	
	Lignin	3.72	3.62	
Inorganic elements [2] mg/kg	K	21,245.4	92.1	99.57
	Na	1474.9	61.5	95.83
	Mg	2171.9	91.7	95.78
	P	2254.1	364.4	83.84
	Fe	379.1	203.3	46.38
	Ca	3979.7	2444.4	38.58

[1] By difference. [2] By ICP-OES analysis.

2.2. Pyrolysis System and Experimental Procedure

The slow pyrolysis system is shown in Figure 1. A gas supply, gas flowmeter, tubular resistance furnace (OTF-1200X, Hefei, China), condensation unit, filter unit, and N-containing gas collection unit were all included in the system. At the start of each experiment, a 5.00 ± 0.02 g sample was mounted in the center of the furnace tube (constant temperature zone length: 150 mm) in a quartz boat (length: 80 mm). The system was purged for 40 min (400 mL/min) with high purity Ar (99.999%). Subsequently, the flow rate was adjusted to 200 mL/min, and the reactor was heated to the desired temperature (350 °C, 450 °C, 550 °C, or 650 °C) with a heating rate of 10 °C/min and then kept steady for 30 min to ensure the completion of pyrolysis. The outlet of the tubular quartz reactor was connected with the condensing device placed in the ice-water mixture. After that, the pyrolysis gas was filtered by absorbent cotton and entered the nitrogen-containing gas collection bottle containing 200 mL absorbent solution. The two kinds of NO_x precursors, NH_3 and

HCN, were absorbed by 5 g/L HBO$_3$ and 8 g/L NaOH solutions, respectively [29], and converted to the corresponding ions (NH$_3$→NH$_4^+$, HCN →CN$^-$). To prevent interference, NH$_3$ and HCN collection experiments were carried out separately. Each experiment was duplicated under the same condition and the average value was adopted.

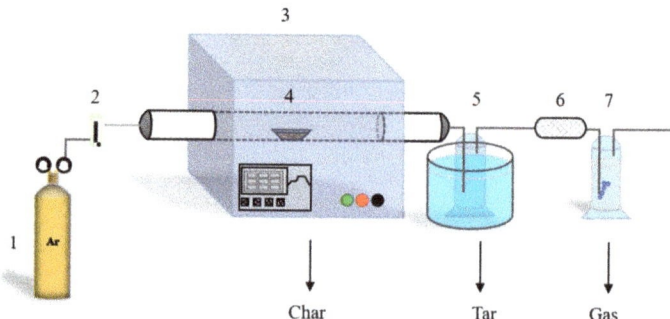

Figure 1. Schematic diagram of pyrolysis system: 1-gas supply, 2-gas flowmeter, 3-horizontal tubular resistance furnace, 4-quartz boat with samples, 5-condensation unit, 6-filter unit, 7-N-containing gas collection unit.

2.3. TG Analysis

The TG analysis of raw and acid-leached maize straw was performed by a thermal gravimetric analyzer (TGA 2, Mettler-Toledo, Greifensee, Switzerland) in the Ar atmosphere (50 mL/min). In each experiment, 5.00 ± 0.05 mg of the straw sample was heated from 30 °C to 900 °C at a constant heating rate of 10 °C/min. The experiments were performed twice to test the repeatability.

2.4. XPS Analysis

The nitrogen functional groups in the biomass and corresponding biochar were characterized using XPS analysis. The sample was uniformly glued to the conductive adhesive tape and degassed for 12 h. An X-ray photoelectron spectrometer (ESCALAB 250Xi, Thermo Fisher Scientific, Waltham, MA, USA) with a monochromatic Al K$_\alpha$ X-ray source (150 W, hv = 1486.6 eV) was used, the following were the operating conditions: a spot size of 650 µm, a voltage of 14.8 kV, a current of 1.6 A, a constant analyzer pass energy mode (100 eV for survey scans and 20 eV for narrow scans) and a pressure less than 10^{-10} mbar. The XPS results were evaluated by Thermo Avantage software. The Shirley-type background was subtracted and all spectra were calibrated based on the principal C1s peak at 285.0 eV [30–33]. Thereafter, the N1s signal was curve-resolved using peaks with a 70% Gaussian and 30% Lorentzian line shape and the FWHM of 1.4 eV [32]. Peaks at 398.8 ± 0.2 eV, 399.9 ± 0.2 eV, 400.4 ± 0.2 eV, 401.4 ± 0.2 eV and 402–405 eV corresponding to the energy positions were considered as N functional groups of pyridine-N (N-6), amine-N/amide-N/protein-N (N-A), pyrrole-N (N-5), quaternary-N/inorganic-N (N-Q/N-IN) and N-oxide (N-X), respectively [12,34–36].

2.5. Scanning Electron Microscopy (SEM)

The sample powers were fixed on the conductive adhesive tape and coated with a gold film. The morphology images of two biomass and corresponding char products were performed using an SEM device (SU3500, HITACHI, Tokyo, Japan, an accelerating voltage of 10 kV).

2.6. Calculations

The difference between the initial and final weight of containers was used to calculate char and tar production. According to the forms of pyrolysis product, nitrogen compounds were classified into char-N, gas-N (NH$_3$-N, HCN-N) and rest-N (tar-N and other N-containing products). The following equations were used to measure the yields of char-N and nitrogen functional groups (group-N) in char products [36]:

$$Y_{char-N} = \frac{m_{char} \times w_{char-N}}{m_{biomass} \times w_N} \times 100\% \tag{1}$$

$$Y_{group-N} = \frac{A_{group-N}}{\sum A_{group-N}} \times Y_{char-N} \tag{2}$$

where Y_{char-N} and $Y_{group-N}$ mean the yields of char-N and group-N in char products, respectively (wt%), $m_{biomass}$ and m_{char} mean the mass of biomass and corresponded char products, respectively (g), w_N and w_{char-N} mean the N contents from biomass and char products, respectively (wt%), and $A_{group-N}$ means the peak area of each nitrogen functional group in char according to the XPS analysis.

Based on relevant Chinese standards, the absorption solutions of NO$_x$ gaseous precursors were analyzed using a UV2800 ultraviolet spectrophotometer (UNICO, Princeton, NJ, USA) [37,38]. Yields of these gas-N were calculated by Equation (3):

$$Y_{gas-N} = \frac{c_n \times V}{m_{biomass} \times w_N} \times 100\% \tag{3}$$

where Y_{gas-N} means the yields of NH$_3$-N or HCN-N (wt%), c_n is the concentrations of NH$_3$-N or HCN-N in the absorption solutions (g/mL), and V represents the volume of absorption solution (200 mL).

3. Results and Discussion
3.1. Characteristics of Raw/Acid Leached Maize Straw

TG and derivative thermogravimetric (DTG) curves of RMA and AMS samples are shown in Figure 2. Similarly, both two biomass samples were found in three stages of mass loss. According to relevant researches, these three stages were summarized as water evaporation (below 200 °C), rapid decomposition of hemicellulose and cellulose (around 200–400 °C), and slow pyrolysis of lignin and other residual materials (above 400 °C) [39–41]. By comparison with the TG curve of RMS (Figure 2a), the second pyrolysis stage of AMS began later (at around 300 °C). This hysteresis was due to the fact that AAEMs in the ash could promote the decomposition of cellulose and hemicellulose under low temperature [42]. A shoulder peak at 305.67 °C was observed during the pyrolysis of acid-leached maize straw (Figure 2a). It is understood that the peak at around 300 °C originated from hemicellulose degradation [43]. RMS had a maximum degradation rate of 8.43%/min at 322.33 °C, which was lower than AMS's (10.60%/min at 356.83 °C) (Figure 2b). As the temperature increased to about 350 °C, the AMS samples continued to release volatiles at a rapid rate, while the weightlessness rate of RMS slowed. These results are consistent with those reported in other studies [42,44,45]. According to the description of Zareihassangheshlaghi et al. [23], the increased concentration of metal impurities on the surface of straw biomass ash could be found during the conversion, which could promote the formation of agglomeration and char, thereby hindering the following biomass decomposition. Also, acid-leaching treatment could increase the porosity of biomass and promote the conversion of cellulose and hemicellulose into sugars, leading to a large release of volatiles [20].

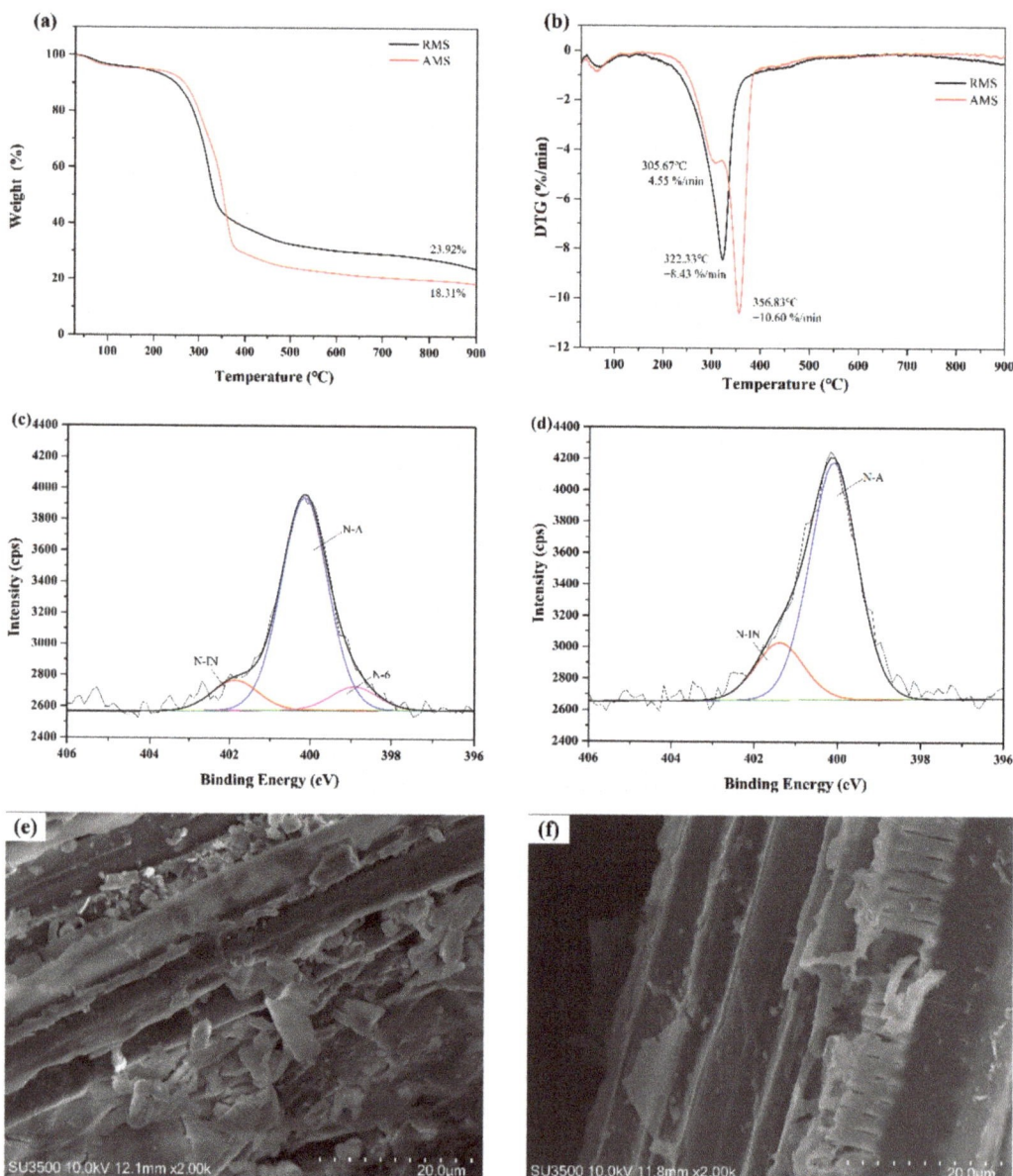

Figure 2. Curves of TG, DTG, and XPS analysis and SEM images of raw/acid leached maize straw: (**a**) TG curves; (**b**) DTG curves; (**c**) XPS analysis of raw maize straw (RMS); (**d**) XPS analysis of acid-leached maize straw (AMS); (**e**) SEM image of RMS; (**f**) SEM image of AMS.

The functional group characteristics of specific elements in different samples were investigated using XPS analysis. The predominant peak was observed in both biomass samples at 399.9 ± 0.2 eV, which can be verified as N-A (Figure 2c,d) [12]. Maize can absorb nitrogen elements from soil and chemical fertilizer for growth, the fixed nitrogen in biomass mainly existing in the forms of amine-N/amide-N/protein-N [46]. Some nitrogen in biomass exists in inorganic nitrogen such as NH_4^+-N, while quaternary-N groups are usually observed after heating treatment [47–49], thus the small peaks at 401.88 eV of RMS and 401.39 eV of AMS should be defined as N-IN in this study (Figure 2c,d) [36]. Moreover, the RMS spectrum contained the N-6 peak at 398.93 eV, which was missing in the AMS spectrum. These losses may be due to the high solubility of pyridine nitrogen and its reaction with acids during the pre-treatment. Moreover, a lot of floccules and small particles were observed in raw maize straw compared to acid leached straw (Figure 2e,f), which mainly belong to various salty, inorganic constituents in ash.

3.2. Distribution of Pyrolysis Products

In this study, solid products were represented by biochar, liquid products consisted of tar and condensed water, and the remaining non-condensable gases were referred to as gaseous products. According to the TG analysis reported above, the moisture in biomass was extracted before 200 °C, so the change of liquid products from 350 °C to 650 °C could be attributed to different tar productions. Under different temperatures, similar change trends in the pyrolysis product yields were obtained for the two biomass samples (Figure 3). Biochar was the predominant product during biomass samples pyrolysis at 350 °C, accounting for 48.55% for RMS and 44.81% for AMS. When the temperature increased to 450 °C, the char proportions decreased quickly for RMS (39.06%) and AMS (35.66%) due to the devolatilization. Gaseous product yields increased as the temperature increased from 350 °C to 450 °C, while liquid product yields increased from 450 °C to 550 °C and peaked at 550 °C. The highest yields of liquid products were 42.33% and 46.92% for RMS and AMS, respectively. During pyrolysis at low temperatures, the unstable chemical bonds of biomass break quickly and primarily produce non-condensable gases, while the ring cracking reactions in lignin and char were much stronger at higher temperatures and a large amount of tar are generated [50]. More surface folds and structural cracking were observed in SEM images from high-temperature pyrolysis (Figure 4). As the reaction temperature increased from 550 °C to 650 °C, the liquid product yields fell rapidly, which were attributed to the secondary decomposition of tar products. Moreover, higher solid product yields and lower liquid product yields were observed in RMS groups than that in AMS groups at the same pyrolysis temperature. According to Figure 4, ash agglomeration and sintering of particles resulted in larger ash particles coating the surface of char products from RMS. The acid-leaching pre-treatment could increase the porosity of the biomass and promoted the removal of volatiles [20].

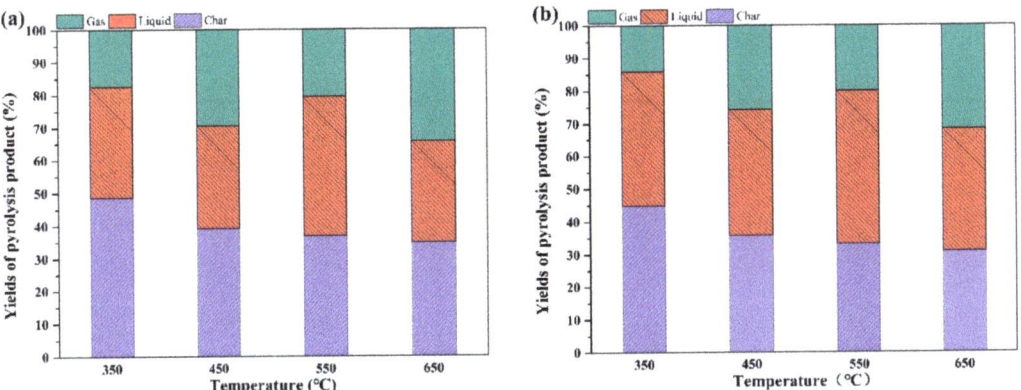

Figure 3. Pyrolysis products from RMS (**a**) and AMS (**b**) under different pyrolysis temperatures.

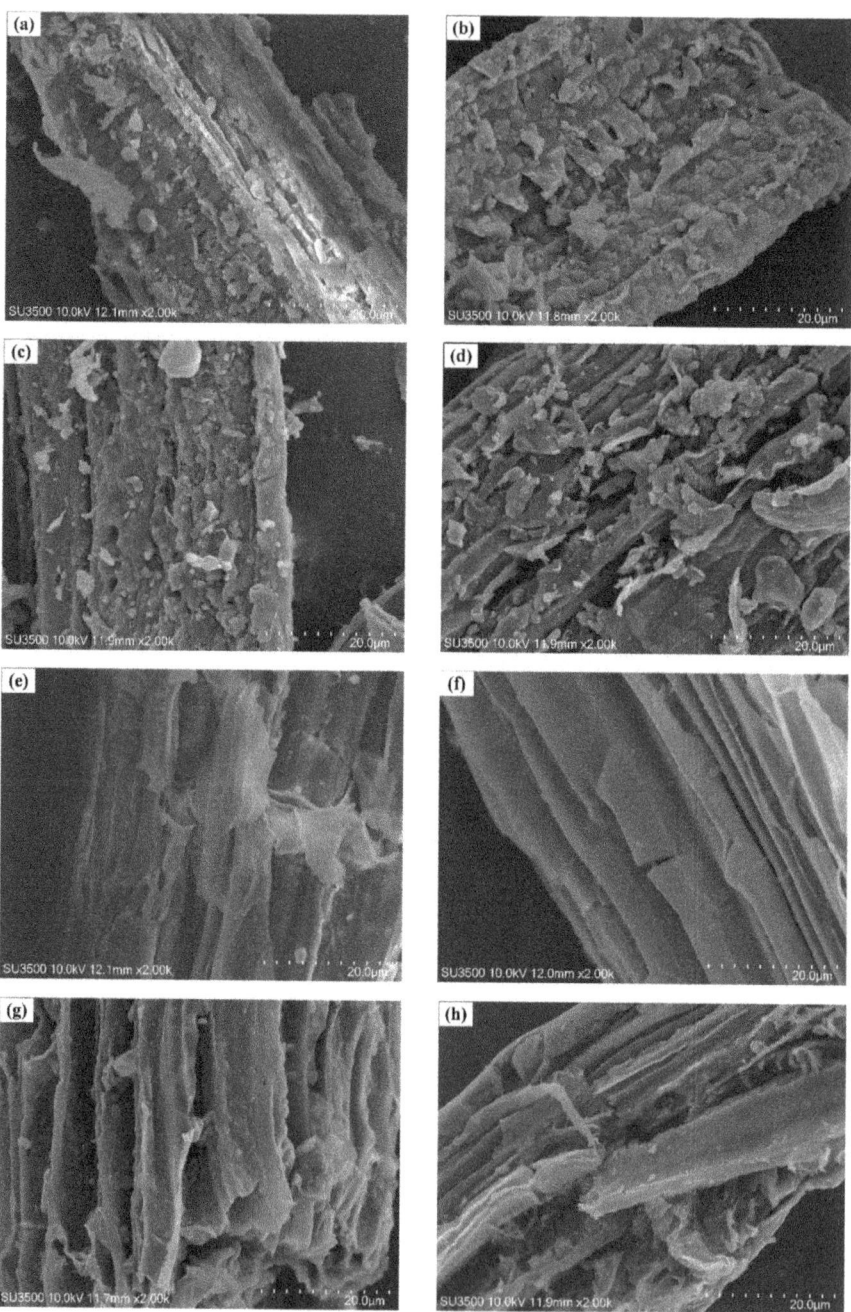

Figure 4. SEM images of the char products: (**a**) RMS$_{350}$; (**b**) RMS$_{450}$; (**c**) RMS$_{550}$; (**d**) RMS$_{650}$; (**e**) AMS$_{350}$; (**f**) AMS$_{450}$; (**g**) AMS$_{550}$; (**h**) AMS$_{650}$.

3.3. Distribution of Gaseous NO_x Precursors

As shown in Figure 5, NH_3 was the dominant NO_x gaseous precursors during pyrolysis of the two straw samples. According to the literature, when the fuel-N mostly existed as aromatic ring-N, the dominant gaseous NO_x precursors was HCN. When it existed in amine-N groups, NH_3 would become the key intermediate product. The rapid increase of NH_3 yield from 350 °C to 450 °C was primarily due to the decomposition of N-A, while the decomposition of char and tar could also produce NH_3 at the higher temperature. For HCN-N yield, it remained at a low level in the low-temperature pyrolysis for RMS and AMS, while a large increase was achieved at 650 °C. This meant that more heterocyclic-N was broken and HCN was released as the temperature rose.

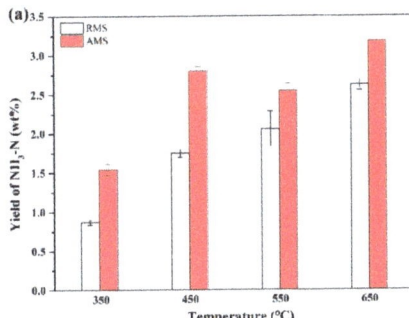 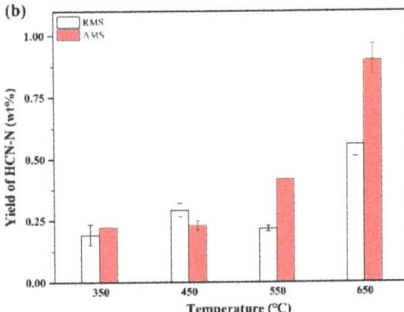

Figure 5. Yields of NH_3-N (**a**) and HCN-N (**b**) from RMS and AMS under different pyrolysis temperatures.

In the literature, it is reported that the removal of AAEM species from coal fuel before pyrolysis will reduce nitrogen conversion to NO_x gaseous precursors, but the situation is more complicated for biomass fuels [51]. Different amino acid structures have different nitrogen release characteristics during the chemical chain cracking. Potassium, for example, has been found to promote the NH_3-N production during aspartic acid pyrolysis, while inhibiting the HCN-N conversion from phenylalanine thermal decomposition [6]. By comparison with RMS, AMS released more NH_3-N at each pyrolysis experiment, while obtained much higher HCN-N yields in high-temperature pyrolysis. The highest NH_3-N and HCN-N yields were 3.17% ± 0.14% and 0.90% ± 0.07%, respectively, obtained from AMS pyrolysis at 650 °C. It's clear that the gas-N proportion was small for the total fuel-N. In the following part, we investigated the characteristics of char-N at various temperatures to further quantitatively confirm the nitrogen conversion from the two biomass samples pyrolysis.

3.4. Nitrogen Migration in Char-N during Biomass Pyrolysis

During pyrolysis, C, H, O, N, and S elements can be removed in various compound forms. However, since decarburization and denitrification take far longer than other reactions like dehydrogenation and deoxygenation, the increased carbon and nitrogen contents could be observed in biochar compared to the biomass sample (Table 2).

Table 2. Ultimate analysis (wt%, dry basis) of char products from raw/acid-leached maize straw.

Sample	C	H	S	O [1]	N
RMS_{350}	53.47 ± 0.19	3.71 ± 0.03	0.18 ± 0.03	40.70 ± 0.28	1.94 ± 0.04
RMS_{450}	54.56 ± 0.06	3.16 ± 0.04	0.29 ± 0.03	40.18 ± 0.18	1.82 ± 0.04
RMS_{550}	56.32 ± 0.16	2.45 ± 0.05	0.35 ± 0.04	39.19 ± 0.30	1.69 ± 0.04
RMS_{650}	57.92 ± 0.16	1.75 ± 0.03	0.29 ± 0.04	38.63 ± 0.28	1.41 ± 0.04

Table 2. Cont.

Sample	C	H	S	O [1]	N
AMS$_{350}$	58.56 ± 0.16	4.40 ± 0.03	0.65 ± 0.06	34.49 ± 0.29	1.91 ± 0.04
AMS$_{450}$	61.42 ± 0.14	3.87 ± 0.04	0.93 ± 0.04	31.78 ± 0.27	2.01 ± 0.04
AMS$_{550}$	63.93 ± 0.20	2.98 ± 0.08	0.64 ± 0.06	30.37 ± 0.38	2.09 ± 0.04
AMS$_{650}$	65.52 ± 0.20	2.36 ± 0.04	0.53 ± 0.05	29.83 ± 0.35	1.77 ± 0.06

[1] By difference.

The XPS spectra of biochar obtained at different pyrolysis temperatures are shown in Figure 6 and yields of nitrogen functional groups in the solid phase are shown in Figure 7. The biochar obtained from low-temperature pyrolysis kept some residual N-A groups and N-IN groups, which eventually vanish at 450 °C and higher temperatures. They were considered as the main source of NH_3-N in low-temperature pyrolysis [49]. N-6 and N-5 were mainly produced from N-A through direct cyclization, dimerization, and other reactions [52,53]. During the pyrolysis process, N-5 and N-6 dominated in nitrogen functional groups from solid products and accounted for more than 50% of biochar-N. The yields of N-6 and N-5 increased with the temperature in the range from 350 °C to 450 °C, but decreased at higher temperatures due to the secondary decomposition (Figure 7). The ring-opening reaction of N-5 was the important way of HCN formation [54]. The binding energies of N-IN and N-Q groups were very close (401.4 ± 0.2 eV). Considering the thermal stability, N-IN could easily decompose when heated, while N-Q was primarily formed by the cyclo-condensation and hydrogenation reaction of N-6 at high temperature [55]. Previous studies have reported that N-6 and N-Q conversions play an important role in NH_3-N generation at high pyrolysis temperature [36,56]. N-X stands for various nitrogen oxide functional groups, which can be observed in char products under each temperature. Zhan et al. [29] suggested that it could be produced by the combination of N-6 and oxygen functional groups.

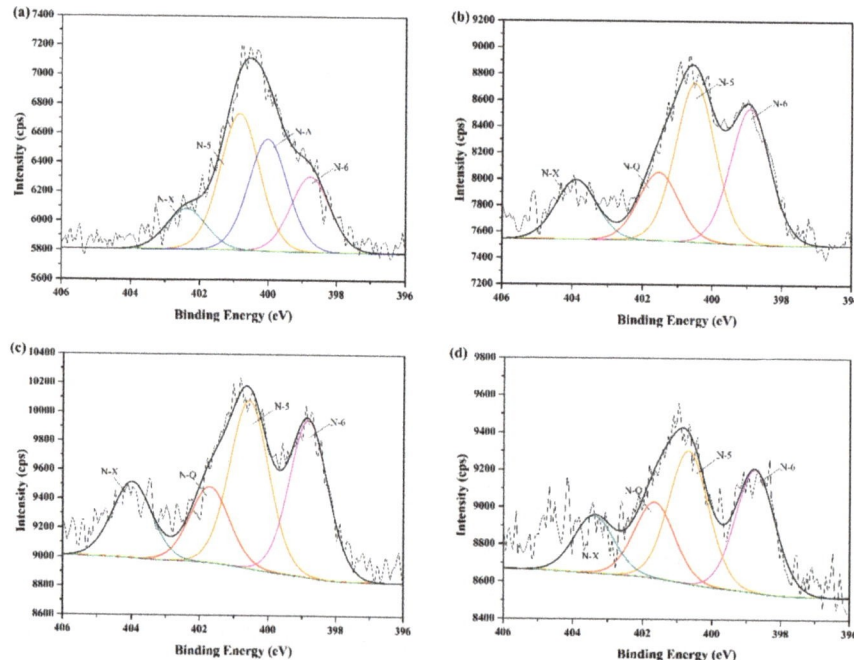

Figure 6. Cont.

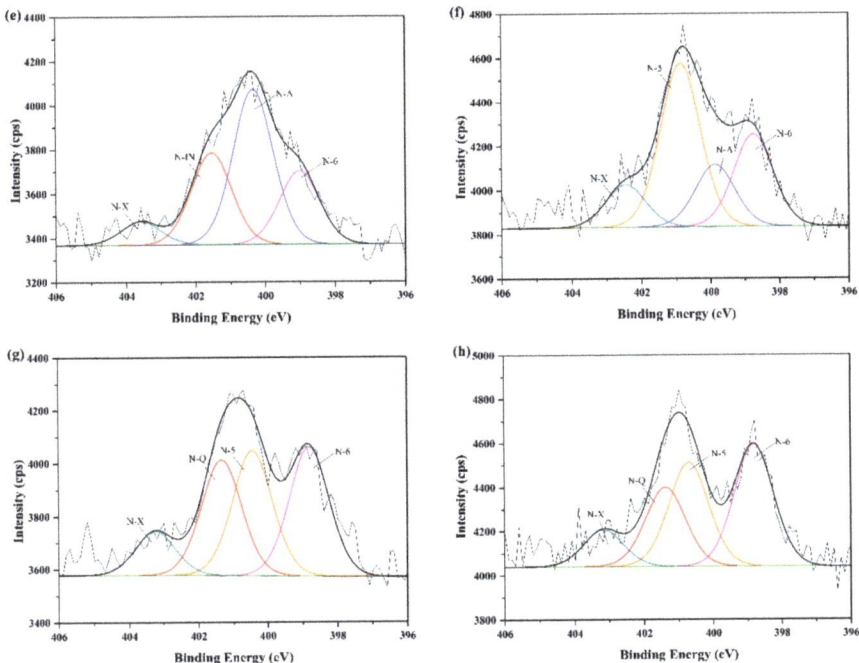

Figure 6. XPS analyses of the char products at various temperatures: (**a**) RMS$_{350}$; (**b**) RMS$_{450}$; (**c**) RMS$_{550}$; (**d**) RMS$_{650}$; (**e**) AMS$_{350}$; (**f**) AMS$_{450}$; (**g**) AMS$_{550}$; (**h**) AMS$_{650}$.

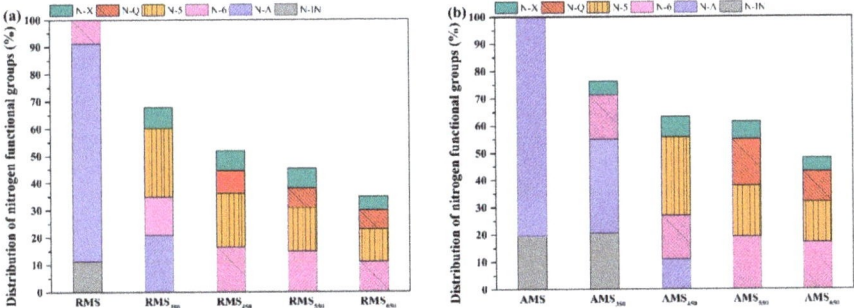

Figure 7. Yields of nitrogen functional groups in char products from RMS (**a**) and AMS (**b**) under different pyrolysis temperatures.

For nitrogen conversion in RMS pyrolysis, N-A yield dramatically dropped from 80.00 wt% to 20.82 wt% at 350 °C, while the corresponding gaseous-N (NH$_3$-N and HCN-N) only accounted for 1.06 wt%. It meant that in addition to the conversion of N-A to gaseous-N and N-5 and N-6 in char-N, a large part of N-A flowed into tar products. Moreover, by comparing the pyrolysis performance of RMA and AMS, delayed removals of N-IN and N-A were found in AMS pyrolysis with the same temperature. As the temperature increased from 350 °C to 450 °C, N-IN was totally converted, and N-A yield showed a decrease from 34.57 wt% to 10.94 wt%, which were responsible for the significant increase in NH$_3$-N yields in AMS pyrolysis (Figure 5a).

Based on the characteristic analysis of the pyrolysis products and related literature [12,36,57], nitrogen migration pathways during raw and acid-leached maize straw pyrolysis were illustrated in Figure 8. Thermal decomposition of N-A and N-IN was the main source of NH_3-N at low temperature, and the conversions of N-5 (char-N) and heterocyclic-N (tar-N) contributed to HCN-N yield at high temperature.

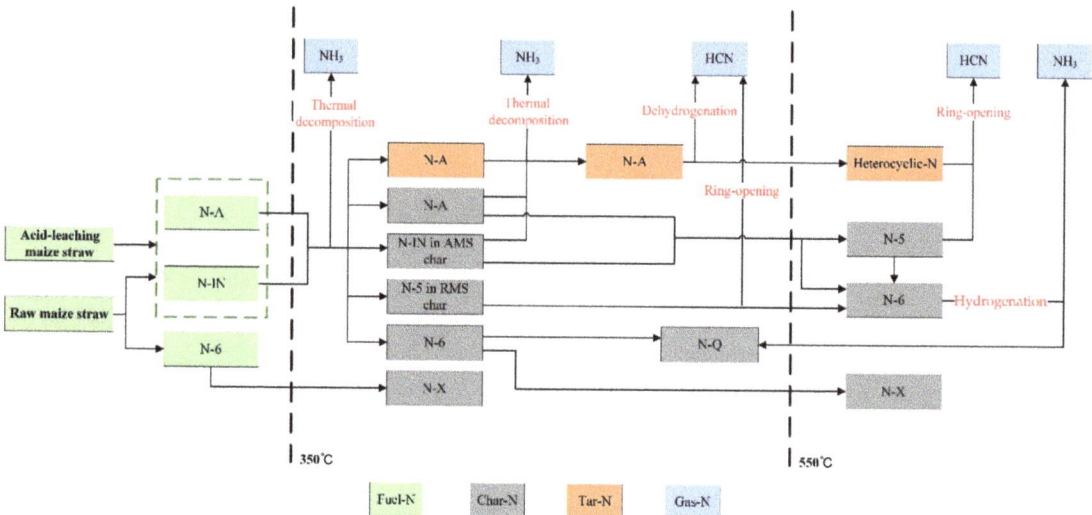

Figure 8. Possible fuel-N migration pathways during pyrolysis of RMA and AMS.

3.5. Nitrogen Distribution in Pyrolysis Products of Raw and Acid-Leached Maize Straw

In general, the distribution of nitrogen in each phase product from RMS and AMS were shown in Table 3. The yield of rest-N was detected by difference, of which tar-N accounted for more than 90%. During low temperature (350 °C, 450 °C) pyrolysis processes, more than half of the biomass-N was deposited in biochar products. As pyrolysis temperature increased, it can be seen that nitrogen elements gradually flowed from solid products to gas and tar. Among the three-phase products, gas-N took a small part of the total, and the majority of nitrogen was combined in the tar products during high-temperature pyrolysis. The proportion of gaseous NO_x precursors of sludge and some industrial waste biomass can be higher than that of straw biomass [49]. Zhan et al. [49] reported the gas-N yield (<8.5 wt%) and variation trend from straw biomass pyrolysis between 200 °C and 500 °C, which was close to the results in this study.

Table 3. Nitrogen distribution (wt%) in different pyrolysis products from RMS and AMS.

Sample	Char-N	Gas-N	Tar-N and Others [1]
RMS_{350}	67.80	1.06	31.14
RMS_{450}	51.87	2.04	46.09
RMS_{550}	45.23	2.28	52.49
RMS_{650}	34.98	3.18	61.84
AMS_{350}	76.39	1.76	21.85
AMS_{450}	63.32	3.03	33.65
AMS_{550}	61.53	2.96	35.51
AMS_{650}	48.13	4.07	47.80

[1] By difference.

The higher tar-N yield was obtained from RMS than AMS with the same temperature, AAEMs in the ash may promote the generation of tar-N in raw straw. N-rich tar products can be used in the production of high value-added chemicals, which is meaningful for industrial production. However, the complex nitrogen bonds in tar were labile and would cause severe NO_x emissions during fuel combustion. The slow nitrogen release from a stable char structure would support nitrogen control of straw fuel combustion. Although a slight increase of gas-N yield was found in AMS, the char-N yield was much higher than that of RMS. The char-N yield (76.39 wt%) was peaked at 350 °C from AMS pyrolysis, which was higher than that of other studies [34,49,58]. These results can provide some guidance for straw fuel modification and its low NO_x combustion.

4. Conclusions

The pyrolysis performance and nitrogen migration pathways of raw and acid leached straw biomass at various temperatures were investigated in this paper. The acid-leaching pretreatment effectively reduced the AAEMs in the ash of the straw. Thermogravimetric analysis showed that AMS released more volatiles than RMS during pyrolysis, but the maximum degradation rate of the former moved to a higher temperature than that of the latter. NH_3 dominated in gaseous NO_x precursor from slow pyrolysis of two straw biomass. The conversions of unstable N-IN and N-A groups in straw contributed to the NH_3 generation at low pyrolysis temperatures (350 °C and 450 °C), while more than half of the HCN was formed from the secondary reactions of N-5 in char and other heterocyclic-N in tar products. Particularly, higher char-N and gas-N yields were observed from AMS than RMS with the same temperature. The highest char-N yield of 76.39 wt% was obtained from AMS pyrolysis at 350 °C. By operating low pyrolysis temperature and acid-leaching treatment, the fuel nitrogen can be effectively stored in the stable char-N structure, rather than gaseous NO_x precursors or labile tar-N, which could provide support for NO_x emissions control from straw fuel combustion.

Author Contributions: Conceptualization, H.L., H.M. and Y.Z.; software, N.Z., Y.Y.; methodology, Q.H.; writing—original draft preparation, H.L.; writing—review and editing, M.P. and H.B.D.; supervision, Y.Z. and R.D. All authors have read and agreed to the published version of the manuscript.

Funding: This study was supported by the National Natural Science Foundation of China (Grant No. U20A2086; Grant No. 51806242), the Special Project on Innovation Methodology, Ministry of Science and Technology of China (No. 2020IM020900), the Inner Mongolia Autonomous Region Science and Technology Plan Project (Grant No. 2020GG0123), the Yantai Educational-Local Synthetic Development Project (Grant No. 2019XDRHXMXK25, No. 2019XDRHXMQT36), and DBFZ Deutsches Biomasseforschungszentrum Gemeinnützige GmbH.

Institutional Review Board Statement: Not applicable.

Informed Consent Statement: Informed consent was obtained from all subjects involved in the study.

Data Availability Statement: The data that supported the findings of this study are available from the corresponding author upon reasonable request.

Acknowledgments: We appreciate the support from the Key Laboratory of Clean Production and Utilization of Renewable Energy, Ministry of Agriculture and Rural Affairs, China; DBFZ Deutsches Biomasseforschungszentrum gemeinnützige GmbH, Department Thermo-chemical Conversion; National Center for International Research of BioEnergy Science and Technology, Ministry of Science and Technology, China; the National Energy R&D Center for Biomass, National Energy Administration of China; and Beijing Municipal Key Discipline of Biomass Engineering.

Conflicts of Interest: The authors declare no conflict of interest.

References

1. Zhou, Y.; Zhang, Z.; Zhang, Y.; Wang, Y.; Yu, Y.; Ji, F.; Ahmad, R.; Dong, R. A comprehensive review on densified solid biofuel industry in China. *Renew. Sustain. Energy Rev.* **2016**, *54*, 1412–1428. [CrossRef]
2. Houshfar, E.; Skreiberg, Ø.; Todorović, D.; Skreiberg, A.; Løvås, T.; Jovović, A.; Sørum, L. NO_x emission reduction by staged combustion in grate combustion of biomass fuels and fuel mixtures. *Fuel* **2012**, *98*, 29–40. [CrossRef]

3. EPA. Nitrogen Oxides (NOx), Why and How They Are Controlled. 1999. Available online: https://www3.epa.gov/ttncatc1/dir1/fnoxdoc.pdf (accessed on 26 March 2021).
4. He, F.; Li, Y.; Dou, S.; Yi, W. Design and Experiments on Residential Heating Furnace with Biomass Smoldering. *Nongye Jixie Xuebao/Trans. Chin. Soc. Agric. Mach.* **2006**, *37*, 75–78.
5. Susastriawan, A.A.P.; Badrawada, I.G.G.; Budi, D.P. An effect of primary air draft and flow rate on thermal perfor-mance and CO/CO_2 emission of the domestic stove fed with the briquette of coconut shell. *Biomass Convers. Biorefin.* **2020**, *10*, 1099–1104. [CrossRef]
6. Ren, Q.; Zhao, C. NO_x and N_2O precursors (NH_3 and HCN) from biomass pyrolysis: Interaction between amino acid and mineral matter. *Appl. Energy* **2013**, *112*, 170–174. [CrossRef]
7. Shah, I.A.; Gou, X.; Zhang, Q.; Wu, J.; Wang, E.; Liu, Y. Experimental study on NO_x emission characteristics of oxy-biomass combustion. *J. Clean. Prod.* **2018**, *199*, 400–410. [CrossRef]
8. Tian, F.; Li, B.; Chen, Y.; Li, C. Formation of NO_x precursors during the pyrolysis of coal and biomass. Part V. Pyrolysis of a sewage sludge. *Fuel* **2002**, *81*, 2203–2208. [CrossRef]
9. Ren, Q. NO_x and N_2O precursors from biomass pyrolysis. *J. Anal. Calorim.* **2014**, *115*, 881–885. [CrossRef]
10. Wang, Y.; Dong, B.; Fan, Y.; Hu, Y.; Zhai, X.; Deng, C.; Xu, Y.; Shen, D.; Dai, X. Nitrogen transformation during pyrolysis of oilfield sludge with high polymer content. *Chemosphere* **2019**, *219*, 383–389. [CrossRef] [PubMed]
11. Zaker, A.; Chen, Z.; Wang, X.; Zhang, Q. Microwave-assisted pyrolysis of sewage sludge: A review. *Fuel Process. Technol.* **2019**, *187*, 84–104. [CrossRef]
12. Tian, Y.; Zhang, J.; Zuo, W.; Chen, L.; Cui, Y.; Tan, T. Nitrogen conversion in relation to NH_3 and HCN during microwave pyrolysis of sewage sludge. *Environ. Sci. Technol.* **2013**, *47*, 3498–3505. [CrossRef]
13. Becidan, M.; Skreiberg, Ø.; Hustad, J.E. NO_x and N_2O precursors (NH_3 and HCN) in pyrolysis of biomass residues. *Energy Fuel* **2007**, *21*, 1173–1180. [CrossRef]
14. Hansson, K.; Samuelsson, J.; Tullin, C.; Åmand, L. Formation of HNCO, HCN, and NH_3 from the pyrolysis of bark and nitrogen-containing model compounds. *Combust. Flame* **2004**, *137*, 265–277. [CrossRef]
15. Cheng, L.; Ma, F.; Ranwala, D. Nitrogen storage and its interaction with carbohydrates of young apple trees in response to nitrogen supply. *Tree Physiol.* **2004**, *24*, 91–98. [CrossRef]
16. Ren, Q.; Zhao, C. Evolution of fuel-N in gas phase during biomass pyrolysis. *Renew. Sustain. Energy Rev.* **2015**, *50*, 408–418. [CrossRef]
17. Sher, F.; Pans, M.A.; Afilaka, D.T.; Sun, C.; Liu, H. Experimental investigation of woody and non-woody biomass combustion in a bubbling fluidised bed combustor focusing on gaseous emissions and temperature profiles. *Energy* **2017**, *141*, 2069–2080. [CrossRef]
18. Cheng, S.; Qiao, Y.; Huang, J.; Wang, W.; Wang, Z.; Yu, Y.; Xu, M. Effects of Ca and Na acetates on nitrogen transformation during sewage sludge pyrolysis. *Proc. Combust. Inst.* **2019**, *37*, 2715–2722. [CrossRef]
19. Ren, Q.; Zhao, C.; Wu, X.; Liang, C.; Chen, X.; Shen, J.; Tang, G.; Wang, Z. Effect of mineral matter on the formation of NO_x precursors during biomass pyrolysis. *J. Anal. Appl. Pyrol.* **2009**, *85*, 447–453. [CrossRef]
20. Dong, Q.; Zhang, S.; Zhang, L.; Ding, K.; Xiong, Y. Effects of four types of dilute acid washing on moso bamboo pyrolysis using Py–GC/MS. *Bioresour. Technol.* **2015**, *185*, 62–69. [CrossRef]
21. Long, J.; Song, H.; Jun, X.; Sheng, S.; Lun-shi, S.; Kai, X.; Yao, Y. Release characteristics of alkali and alkaline earth metallic species during biomass pyrolysis and steam gasification process. *Bioresour. Technol.* **2012**, *116*, 278–284. [CrossRef]
22. Wigley, T.; Yip, A.C.K.; Pang, S. Pretreating biomass via demineralisation and torrefaction to improve the quality of crude pyrolysis oil. *Energy* **2016**, *109*, 481–494. [CrossRef]
23. Zareihassangheshlaghi, A.; Beidaghy Dizaji, H.; Zeng, T.; Huth, P.; Ruf, T.; Denecke, R.; Enke, D. Behavior of metal impurities on surface and bulk of biogenic silica from rice husk combustion and the impact on ash-melting tendency. *ACS Sustain. Chem. Eng.* **2020**, *8*, 10369–10379. [CrossRef]
24. Oudenhoven, S.R.G.; Westerhof, R.J.M.; Kersten, S.R.A. Fast pyrolysis of organic acid leached wood, straw, hay and bagasse: Improved oil and sugar yields. *J. Anal. Appl. Pyrol.* **2015**, *116*, 253–262. [CrossRef]
25. AQSIQ. *Proximate Analysis of Solid Biofuels, GB/T 28731-2012*; General Administration of Quality Supervision, Inspection and Quarantine of the People's Republic of China: Beijing, China, 2012.
26. Zhao, N.; Li, B.; Chen, D.; Bahargul, T.; Wang, R.; Zhou, Y.; Annegarn, H.J.; Pemberton-Pigott, C.; Dong, R.; Ju, X. The effect of coal size on $PM_{2.5}$ and PM-bound polycyclic aromatic hydrocarbon (PAH) emissions from a domestic natural cross-draft stove. *J. Energy Inst.* **2019**, *93*, 542–551. [CrossRef]
27. Li, Q.; Zhao, Y.; Chen, H.; Hou, P.; Cheng, X. Effect of cornstalk ash on the microstructure of cement-based material under sulfate attack. IOP conference series. *Earth Environ. Sci.* **2019**, *358*, 52010.
28. Guo, J.; Cui, X.; Sun, H.; Zhao, Q.; Wen, X.; Pang, C.; Dong, R. Effect of glucose and cellulase addition on wet-storage of excessively wilted maize stover and biogas production. *Bioresour. Technol.* **2018**, *259*, 198–206. [CrossRef]
29. Zhan, H.; Yin, X.; Huang, Y.; Yuan, H.; Wu, C. NO_x precursors evolving during rapid pyrolysis of lignocellulosic industrial biomass wastes. *Fuel* **2017**, *207*, 438–448. [CrossRef]
30. Nefedov, V.I.; Salyn, Y.V.; Leonhardt, G.; Scheibe, R. A comparison of different spectrometers and charge corrections used in X-ray photoelectron spectroscopy. *J. Electron. Spectrosc.* **1977**, *10*, 121–124. [CrossRef]

31. Nesbitt, H.W.; Bancroft, G.M.; Tse, J.S.; Gao, X.; Skinner, W.; Zakaznova-Herzog, V.P. High-resolution valence-band XPS spectra of the nonconductors quartz and olivine. *Phys. Rev. B.* **2005**, *72*, 205113.
32. Sawyer, R.; Nesbitt, H.W.; Secco, R.A. High resolution X-ray Photoelectron Spectroscopy (XPS) study of K_2O–SiO_2 glasses: Evidence for three types of O and at least two types of Si. *J. Non Cryst. Solids* **2012**, *358*, 290–302. [CrossRef]
33. Stoch, J.; Ladecka, M. An XPS study of the KCl surface oxidation in oxygen glow discharge. *Appl. Surf. Sci.* **1988**, *31*, 426–436. [CrossRef]
34. Gao, P.; Guo, D.; Liang, C.; Liu, G.; Yang, S. Nitrogen conversion during the rapid pyrolysis of raw/torrefied wheat straw. *Fuel* **2020**, *259*, 116227. [CrossRef]
35. Wei, L.; Wen, L.; Yang, T.; Zhang, N. Nitrogen transformation during sewage sludge pyrolysis. *Energy Fuel* **2015**, *29*, 5088–5094. [CrossRef]
36. Zhan, H.; Zhuang, X.; Song, Y.; Yin, X.; Wu, C. Insights into the evolution of fuel-N to NO_x precursors during pyrolysis of N-rich nonlignocellulosic biomass. *Appl. Energy* **2018**, *219*, 20–33. [CrossRef]
37. MOEP. *Water Quality-Determination of Cyanide Volumetric and Spectrophotometry, HJ 484-2009*; Ministry of Environmental Protection of the People's Republic of China: Beijing, China, 2009.
38. MOEP. *Water Quality-Determination of Ammonia Nitrogen-Nessler's Reagent Spectrophotometry, HJ 535-2009*; Ministry of Environmental Protection of the People's Republic of China: Beijing, China, 2009.
39. Deng, S.; Wang, X.; Tan, H.; Mikulčić, H.; Yang, F.; Li, Z.; Duić, N. Thermogravimetric study on the co-combustion characteristics of oily sludge with plant biomass. *Thermochim. Acta* **2016**, *633*, 69–76. [CrossRef]
40. Huang, J.; Liu, J.; Chen, J.; Xie, W.; Kuo, J.; Lu, X.; Chang, K.; Wen, S.; Sun, G.; Cai, H.; et al. Combustion behaviors of spent mushroom substrate using TG-MS and TG-FTIR: Thermal conversion, kinetic, thermodynamic and emission analyses. *Bioresour. Technol.* **2018**, *266*, 389–397. [CrossRef] [PubMed]
41. Wang, T.; Peng, L.; Ai, Y.; Zhang, R.; Lu, Q. Pyrolytic behaviors of decocting residues of *Rhodiola rosea*. *J. Anal. Appl. Pyrol.* **2018**, *129*, 61–65. [CrossRef]
42. Eom, I.; Kim, K.; Kim, J.; Lee, S.; Yeo, H.; Choi, I.; Choi, J. Characterization of primary thermal degradation features of lignocellulosic biomass after removal of inorganic metals by diverse solvents. *Bioresour. Technol.* **2011**, *102*, 3437–3444. [CrossRef]
43. Zhang, S.; Zhang, H.; Su, Y.; Xu, D.; Zhu, S.; Liu, X. Effects of torrefaction and organic-acid leaching pretreatment on the pyrolysis behavior of rice husk. *Energy* **2018**, *149*, 804–813. [CrossRef]
44. Meng, A.; Zhou, H.; Qin, L.; Zhang, Y.; Li, Q. Quantitative and kinetic TG-FTIR investigation on three kinds of biomass pyrolysis. *J. Anal. Appl. Pyrol.* **2013**, *104*, 28–37. [CrossRef]
45. Yang, C.; Lu, X.; Lin, W.; Yang, X.; Yao, J. TG-FTIR Study on Corn Straw Pyrolysis-influence of Minerals. *Chem. Res. Chin. Univ.* **2006**, *22*, 524–532. [CrossRef]
46. Yuan, S.; Tan, Z.; Huang, Q. Migration and transformation mechanism of nitrogen in the biomass-biochar-plant transport process. *Renew. Sustain. Energy Rev.* **2018**, *85*, 1–13. [CrossRef]
47. Ma, D.; Zhang, G.; Areeprasert, C.; Li, C.; Shen, Y.; Yoshikawa, K.; Xu, G. Characterization of NO emission in combustion of hydrothermally treated antibiotic mycelial residue. *Chem. Eng. J.* **2016**, *284*, 708–715. [CrossRef]
48. Yuan, S.; Zhou, Z.; Li, J.; Wang, F. Nitrogen conversion during rapid pyrolysis of coal and petroleum coke in a high-frequency furnace. *Appl. Energy* **2012**, *92*, 854–859. [CrossRef]
49. Zhan, H.; Zhuang, X.; Song, Y.; Huang, Y.; Liu, H.; Yin, X.; Wu, C. Evolution of nitrogen functionalities in relation to NO precursors during low-temperature pyrolysis of biowastes. *Fuel* **2018**, *218*, 325–334. [CrossRef]
50. Wang, T.; Dong, X.; Jin, Z.; Su, W.; Ye, X.; Dong, C.; Lu, Q. Pyrolytic characteristics of sweet potato vine. *Bioresour. Technol.* **2015**, *192*, 799–801. [CrossRef]
51. Chang, L.; Zhao, Y.; Xie, K. Effect of inherent minerals on the release of fuel-nitrogen during coal pyrolysis. *Energy Sources* **2005**, *27*, 1019–1027. [CrossRef]
52. Choi, S.; Ko, J. Analysis of cyclic pyrolysis products formed from amino acid monomer. *J. Chromatogr. A* **2011**, *1218*, 8443–8455. [CrossRef]
53. Zhan, H.; Zhuang, X.; Song, Y.; Yin, X.; Cao, J.; Shen, Z.; Wu, C. Step pyrolysis of N-rich industrial biowastes: Regulatory mechanism of NO precursor formation via exploring decisive reaction pathways. *Chem. Eng. J.* **2018**, *344*, 320–331. [CrossRef]
54. Zhang, J.; Tian, Y.; Cui, Y.; Zuo, W.; Tan, T. Key intermediates in nitrogen transformation during microwave pyrolysis of sewage sludge: A protein model compound study. *Bioresour. Technol.* **2013**, *132*, 57–63. [CrossRef]
55. Li, C.; Tan, L.L. Formation of NO_x and SO_x precursors during the pyrolysis of coal and biomass. Part III. Further discussion on the formation of HCN and NH_3 during pyrolysis. *Fuel* **2000**, *79*, 1899–1906. [CrossRef]
56. Kambara, S.; Takarada, T.; Yamamoto, Y.; Kato, K. Relation between functional forms of coal nitrogen and formation of nitrogen oxide (NO_x) precursors during rapid pyrolysis. *Energy Fuel* **1993**, *7*, 1013–1020. [CrossRef]
57. Liu, H.; Yi, L.; Hu, H.; Xu, K.; Zhang, Q.; Lu, G.; Yao, H. Emission control of NO_x precursors during sewage sludge pyrolysis using an integrated pretreatment of Fenton peroxidation and CaO conditioning. *Fuel* **2017**, *195*, 208–216. [CrossRef]
58. Li, Y.; Hong, C.; Wang, Y.; Xing, Y.; Chang, X.; Zheng, Z.; Li, Z.; Zhao, X. Nitrogen migration mechanism during pyrolysis of penicillin fermentation residue based on product characteristics and quantum chemical analysis. *ACS Sustain. Chem. Eng.* **2020**, *8*, 7721–7740. [CrossRef]

Article

Effects of Ionic Liquid and Biomass Sources on Carbon Nanotube Physical and Electrochemical Properties

Kudzai Mugadza [1,2], Annegret Stark [2], Patrick G. Ndungu [3] and Vincent O. Nyamori [1,*]

1. School of Chemistry and Physics, University of KwaZulu-Natal, Westville Campus, Private Bag X54001, Durban 4000, South Africa; mugadzakudzie@gmail.com
2. SMRI/NRF SARChI Research Chair in Sugarcane Biorefining, School of Engineering, University of KwaZulu-Natal, Durban 4041, South Africa; starka@ukzn.ac.za
3. Energy, Sensors and Multifunctional Nanomaterials Research Group, Department of Chemical Sciences, University of Johannesburg, Doornfontein, Johannesburg 2028, South Africa; pndungu@uj.ac.za
* Correspondence: nyamori@ukzn.ac.za

Abstract: The ongoing research toward meeting global energy demands requires novel materials from abundant renewable resources. This work involves an investigation on nitrogen-doped carbon nanotubes (N-CNTs) synthesized from relatively low-cost and readily available biomass as carbon precursors and their use as electrodes for supercapacitors. The influence of the ionic liquid 1-butyl-3-methylimidazolium chloride, or its combination with either sugarcane bagasse or cellulose (IL-CNTs, ILBag-CNTs, and ILCel-CNTs, respectively), in the synthesis of N-CNTs and the resultant effect on their physical and electrochemical properties was studied. Systematic characterizations of the N-CNTs employing transmission electron microscopy (TEM), thermogravimetric analysis, X-ray photoelectron spectroscopy (XPS), elemental analysis, nitrogen sorption analysis, cyclic voltammetry, and electrochemical impedance spectroscopy were performed. TEM data analysis showed that the mean outer diameters decreased, in the order of IL-CNTs > ILBag-CNTs > ILCel-CNTs. The N-CNTs possess only pyridinic and pyrrolic nitrogen-doping moieties. The pyridinic nitrogen-doping content is lowest in IL-CNTs and highest in ILCel-CNTs. The N-CNTs are mesoporous with surface areas in the range of 21–52 m^2 g^{-1}. The ILCel-CNTs had the highest specific capacitance of 30 F g^{-1}, while IL-CNTs has the least, 10 F g^{-1}. The source of biomass is beneficial for tuning physicochemical properties such as the size and surface areas of N-CNTs, the pyridinic nitrogen-doping content, and ultimately capacitance, leading to materials with excellent properties for electrochemical applications.

Keywords: carbon nanotubes; biomass; cellulose; sugarcane bagasse; capacitance

Citation: Mugadza, K.; Stark, A.; Ndungu, P.G.; Nyamori, V.O. Effects of Ionic Liquid and Biomass Sources on Carbon Nanotube Physical and Electrochemical Properties. *Sustainability* 2021, 13, 2977. https://doi.org/10.3390/su13052977

Academic Editor: Asterios Bakolas

Received: 8 February 2021
Accepted: 2 March 2021
Published: 9 March 2021

Publisher's Note: MDPI stays neutral with regard to jurisdictional claims in published maps and institutional affiliations.

Copyright: © 2021 by the authors. Licensee MDPI, Basel, Switzerland. This article is an open access article distributed under the terms and conditions of the Creative Commons Attribution (CC BY) license (https://creativecommons.org/licenses/by/4.0/).

1. Introduction

Across the globe, the quest for sustainable energy supply has seen societies moving from the extensive usage of petroleum to renewable sources of energy, with active research still ongoing [1]. Mainly, the energy from the sun, which is in abundance, is being highly considered, amongst other alternative reliable sources of energy. The challenge is to effectively harness and store solar energy [2]. Therefore, firstly, photovoltaics are successfully implemented in the conversion of energy from the sun directly into electricity [3]. Secondly, the power converted from the sun is amassed using energy storage devices such as batteries and supercapacitors.

What distinguishes supercapacitors from batteries is their integral constituents and the mechanisms associated with their charge and discharge processes [4]. Supercapacitors possess higher energy-storage and power-delivery capability than batteries [5]. Thus, they are suitable for use in devices where there is a great power demand at relatively short periods. In addition, supercapacitors are associated with a longer life cycle, fast charge and discharge rates coupled with low internal resistance [5]. Hence, they are suitable for applications that require storing high current transient electrical surges to improve the efficiency and energy of a system. Supercapacitors are currently gaining popularity in diverse applications, including portable electronic devices, hybrid electric vehicles, and commercial electric utilities [6]. Since there has been an increase in energy demand and reliability, supercapacitors are continuously involved in modification with new electrode materials. These materials allow for operation over a wider potential window with a rapid current response to voltage reversal, which offers high specific capacitances, thus enhancing their storage abilities [7,8].

Mesoporous carbon nanostructured materials (CNMs) are prominent electrode materials for supercapacitors, and this is due to their low cost, exceptional cycle stability, and wide voltage operating windows [9]. Research on the use of biomass-derived CNMs as an electrode material for supercapacitors exists [7,8]. However, to the best of our knowledge, this is the first report on the effect of the precursor composition (IL or IL + biomass) on the properties and electrochemical performance of the resulting carbon nanotubes (CNTs).

The unique one-dimensional geometry of CNTs and the associated physicochemical properties such as mechanical strength and excellent electrical properties afford CNTs to be suitable for charge storage and numerous other applications [10]. For achieving a high specific capacitance, a characteristic carbon material should possess a high electroactive surface area and suitable pore sizes that match the size of electrolyte ions [10].

Material engineering by the inclusion of a heteroatom during doping of the carbon material should provide favorable surface characteristics that promote ion intercalation. Of interest is the introduction of heteroatoms such as sulfur, boron, or nitrogen in the graphitic carbon network [11]. N-enriched carbon structures can establish a high capacitance, and this is due to the Faradaic redox reactions, i.e., pseudo-capacitance effects, provided by the nitrogen functional groups [11,12]. Additionally, improved electron donor capability and enhanced surface wetting by the electrolyte can increase the charge storage [12].

Our current work focuses on studying the electrochemical performance of CNTs synthesized using renewable raw materials, namely sugarcane bagasse or cellulose, in combination with ferrocene and an ionic liquid, namely, 1-butyl-3-methylimidazolium chloride ([C_4MIM]Cl). The primary objectives were to perform capacitance measurements of the N-CNTs and interpret the results with respect to the nitrogen-doping effect, surface area, and tube diameter, amongst others, emanating from the impact of sources. This work contributes advantageously to material beneficiation from available, renewable, and abundant waste biomass for the fabrication of electronic devices such as supercapacitors.

2. Materials and Methods

All materials and methods employed are stated, and experimental details are provided in subsequent sections under sample preparation, characterization, and electrochemical measurements.

2.1. Preparation of the Samples

H_2SO_4 (98%) was obtained from Merck, South Africa, HNO_3 (70%) was obtained from Associated Chemical Enterprises, South Africa. Ferrocene, 98% (AR), and cellulose Avicel®, PH 101, 50 µm were purchased from Sigma Aldrich, Germany. Sugarcane bagasse was obtained frozen from the Sugar Milling Research Institute (SMRI). Sugarcane bagasse was shredded to small pieces, sieved, and dried at 100 °C for 10 h (weight constant). After drying, the bagasse was ground with mortar and pestle to fine particles.

The ionic liquid (IL), 1-butyl-3-methylimidazolium chloride ([C_4MIM]Cl), was prepared as previously described [13].

The synthesis of CNTs was conducted with either sugarcane bagasse (Bag) or cellulose (Cel) with ferrocene and [C_4MIM]Cl (IL), following the previously reported procedure (Scheme 1) [13].

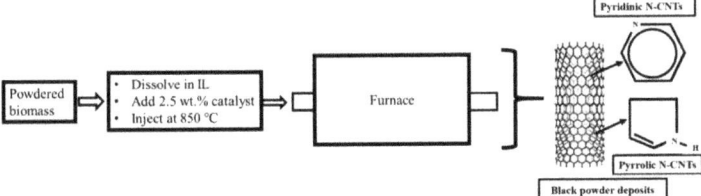

Scheme 1. Schematic presentation of the synthesis of nitrogen-doped carbon nanotubes (N-CNTs).

In short, ferrocene (0.25 g) and either IL-Bag or IL-Cel solution (containing 10% of either biomass; 9.75 g) were mixed (to make 2.5 wt % of catalyst) in an ultra-sound sonicator (Digital ultrasonic heater UD150SH-6L) for approximately 20 min. Then, the mixture was transferred into a 10 mL plastic syringe and set for injection using a purpose-built injection system. The injection system had the requisite gas fittings and was designed to deliver the mixture directly into the centre of a quartz tube located within a horizontally aligned tube furnace.

A horizontally aligned tube furnace, fitted with a main zone temperature controller, was set to the desired temperature using a ramp rate of 10 °C/min. The carrier gas, 10% H_2 balanced in Ar, was switched on at 600 °C, and a flow rate of 100 mL/min was applied. When the respective desired temperature was reached, injection of the sonicated mixture into the chemical vapour deposition (CVD) system at a set injection rate of 0.8 mL/min followed. The reaction was set for 30 min, and after that, the furnace was allowed to cool down to room temperature. All the samples collected were purified. The purification process involved calcining all samples using a muffle furnace at 300 °C in air for 3 h, followed by ultrasonication in H_2SO_4/HNO_3 (3:1 by volume) for 40 min using an ultrasonic heater. Thereafter, the mixture was refluxed at 100 °C for 24 h at a constant stirring speed of 300 rpm. After refluxing, the acid was neutralized with a solution of NaOH, and the mixture was ultrasonicated for 1 h. After filtration and washing with deionized water (until the pH was 7), the materials were dried at 120 °C for 3 h. The CNTs synthesized from IL and ferrocene are denoted IL-CNTs, while those from the IL-bagasse and IL-cellulose mixtures are denoted as ILBag-CNTs and ILCel-CNTs, respectively.

2.2. Characterization

Characterization of the carbon nanostructures utilized TEM (JEOL JEM 1010) with the aid of Image J software for structural dimension determination. Elemental analysis was performed using an LECO CHNS-932 elemental analyzer standardized with acetanilide. X-ray photoelectron spectroscopy (XPS) analysis was performed with a Thermo ESCAlab 250Xi equipped with monochromatic Al Kα (1486.7 eV) X-rays. An X-ray power of 300 W, with an X-ray spot size of 900 µm, pass energy (Survey) of 100 eV, pass energy (Hi-resolution) of 20 eV, and pressure of 10^{-8} mbar were used. The thermogravimetric analysis utilized the TA instruments Q seriesTM Thermal Analyzer DSC/TGA (Q600) with a TA instruments universal analysis 2000 software for data acquisition. The samples were run in oxygen at a 100 mL/min flow rate and from ambient temperatures to 1000 °C at 10 °C/min. Nitrogen sorption studies were performed with a Micromeritics TriStar 3020 V1.03 (V1.03) instrument. Surface area and porosity were calculated with the use of Brunauer–Emmett–Teller (BET) and Barrett–Joyner–Halenda (BJH) methods, respectively.

2.3. Electrochemical Measurements

The electrochemical tests were performed using cyclic voltammetry (CV) and electrochemical impedance spectroscopy (EIS) on an electrochemical CHI 660E work station, CH Instruments, Austin, Texas, USA. All the electrochemical measurements were achieved by the use of a three-electrode system. The counter and reference electrodes were Pt and Ag/AgCl, respectively. The working electrode was prepared by drop-casting the sample mixed with Nafion onto a glassy carbon electrode (3 mm diameter). The deposited mixture, on the electrode, was dried at room temperature. The electrolyte, H_2SO_4 (0.1 M), was prepared by diluting 98% solution with double deionized water. The electrodes used for EIS were likewise prepared.

3. Results and Discussions

Effects of the precursor on the physicochemical properties of the synthesized CNTs and the electrochemical performance are reported herein.

3.1. Morphological Analysis

The TEM image analysis of the three samples shows hollow tubular structures with some bamboo compartments (Figure 1), which resemble nitrogen-doped CNTs [14,15]. The outer diameters (ODs) of the CNTs after purification are 101, 63, and 38 nm for IL-CNTs, ILBag-CNTs, and ILCel-CNTs, respectively. The differences in the size of CNT diameters were attributed to the presence of oxygen in the biomass containing precursors. Oxygen affects the activation of the catalyst, the decomposition of the carbon precursor, and also acts as an etchant of amorphous carbon [16]. Additionally, the diameters of the purified samples are smaller than those from the crude samples (172, 104, and 112 nm for IL-CNTs, ILBag-CNTs, and ILCel-CNTs, respectively). Similarly, the reduction in sizes was attributed to the etching effect of the purification processes.

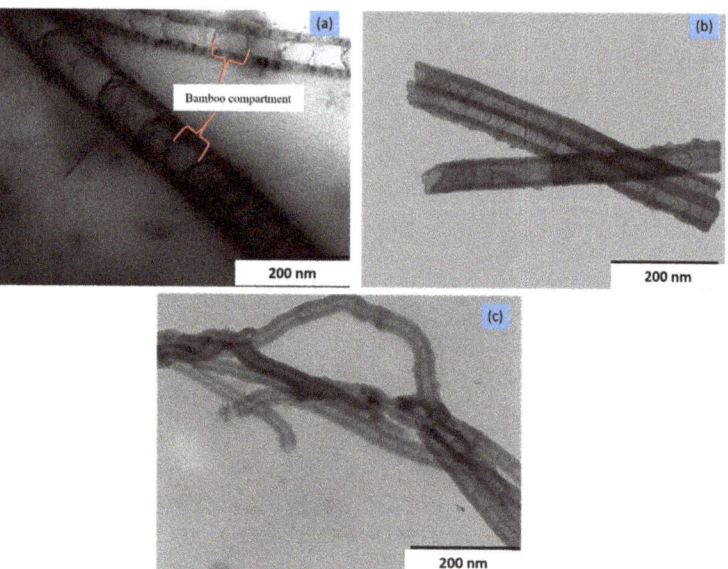

Figure 1. TEM images for the synthesized structures: (**a**) IL-CNTs, (**b**) ILBag-CNTs and (**c**) ILCel-CNTs. IL-CNTs, ILBag-CNTs, and ILCel-CNTs: ionic liquid 1-butyl-3-methylimidazolium chloride, or its combination with either sugarcane bagasse or cellulose.

Apart from being smaller, the ILCel-CNTs are notably more bent than the IL-CNTs and ILBag-CNTs (Figure 1). The inward bent structure defects suggest that pentagons were formed because of nitrogen doping [13].

3.2. Chemical Composition Analysis

Elemental analysis (EA) was initially used for qualitative analysis and showed the presence of carbon, nitrogen, hydrogen, and oxygen in the samples (Supplementary Information, Figure S1). IL-CNTs have a higher content of doped N, followed by ILBag-CNTs and ILCel-CNTs. This concurs with the TEM observations where IL-CNTs were found to have more pronounced bamboo compartments. Further elemental quantitative analysis through XPS shows a trend of atomic nitrogen composition similar to that from EA, i.e., thus confirming a higher percentage of nitrogen in the IL-CNTs sample (Figure 2). The atomic nitrogen content for IL-CNTs, ILBag-CNTs, and ILCel-CNTs are 7.2%, 7.1%, and 6.2%, respectively (Figure 2).

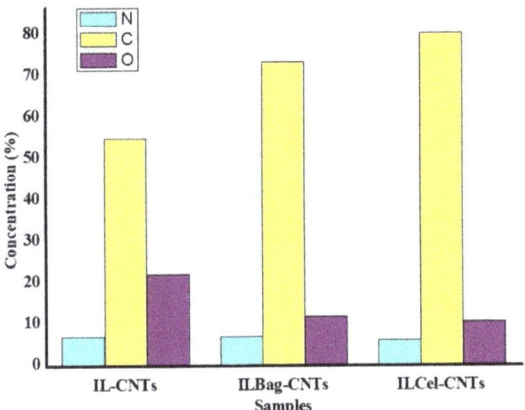

Figure 2. Elemental composition of C, N, and O in all samples from XPS.

The N1s peak show a broad peak between 397.7 and 400.4 eV (Figure 3a–c), which was deconvoluted to reveal the presence of two major nitrogen-doped species in the carbon lattices, namely pyridinic and pyrrolic around 397.6–398.5 and 399.6–400.5 eV, respectively [17–20]. These peaks confirm the presence of nitrogen doping in the samples, which concurs with the presence of bent structures and bamboo compartments as observed using TEM (Figure 1).

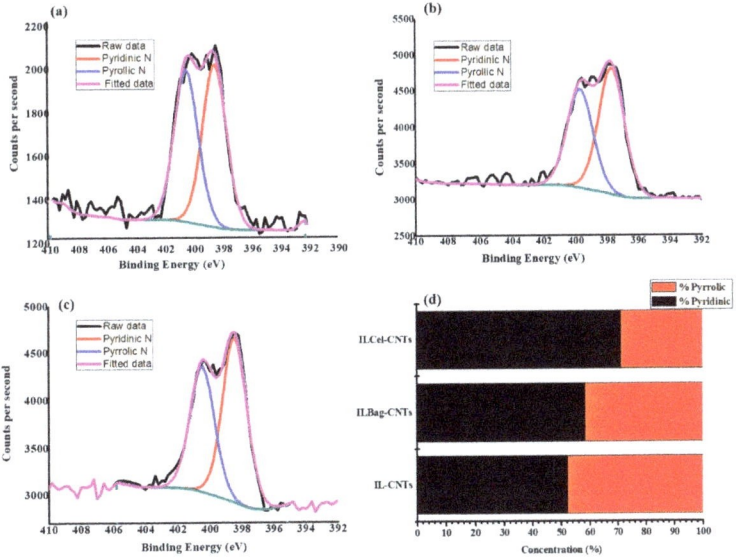

Figure 3. XPS nitrogen species in samples (a) IL-CNTs, (b) ILBag-CNTs, (c) ILCel-CNTs, and (d) composition of these deconvoluted N species in the materials.

The deconvoluted N1s species ratio was used to determine the percentage composition of the identified species (Figure 3d). In all samples, the percentage of the pyridinic nitrogen is higher than the pyrrolic nitrogen, i.e., 52%, 58%, and 71% IL-CNTs, ILBag-CNTs, and ILCel-CNTs, respectively. Interestingly, ILCel-CNT, which shows bent structures in the TEM analysis (Figure 1c), has the lowest pyrrolic-nitrogen content, and thus, the inclusion of the unsymmetrical pyrrole ring is not responsible for this phenomenon. Hence, these results suggest that the more pronounced bent structures of ILCel-CNTs are due to the smaller tube sizes.

3.3. Thermal Analysis

The thermogravimetric analysis (TGA) profiles of the CNTs show minimal water loss (≈2 wt %), which is expected to occur below 200 °C (Figure 4). There is no noticeable decrease in weight in the range of 200–400 °C, corroborating TEM analysis and inferring the absence of amorphous carbon. Significant weight loss occurs from around 500 °C, with a weight loss of ≈95 wt % from both ILCel-CNTs and IL-CNTs, at 630 and 611 °C, respectively, and ≈97 wt % from ILBag-CNTs samples at 618 °C. This was assigned to the decomposition of N-CNTs, which is attributed to the disintegration of a single type of sp^2 carbon structures. This is further supported by TEM analysis where no other structures such as spheres, fibers, or amorphous carbons were observed (Figures 1 and 4). In addition, the XPS analysis further supports the findings from the TGA analysis.

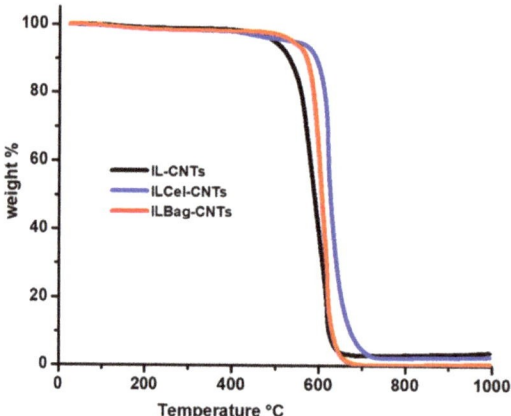

Figure 4. Thermal analysis profiles of IL-CNTs, ILCel-CNTs, and ILBag-CNTs samples.

The thermogram in Figure 4 indicates that the order of material tolerance to heat is IL-CNTs < ILBag-CNTs < ILCel-CNTs, i.e., T_{max} DTA of 611, 618, and 630 °C, respectively. These thermal stability trends agree with the amount of doped N determined by XPS. The thermal decomposition temperatures are close to each other, and this is due to the similar relationship of nitrogen contents, as seen in Figure 2. IL-CNTs show the highest amount of doped N within the graphitic structure; hence, they have relatively higher compromised C-C bond strengths and poorer heat tolerance. Some residual Fe remains in the materials, as evidenced by the incomplete decomposition. The sample containing ILBag-CNTs has almost zero residual Fe content while ILCel-CNTs and IL-CNTs have a residual Fe content of around 3 wt %. This could be due to non-dissolved Fe after the purification process since the concentrated H_2SO_4/HNO_3 mixture has the ability to dissolve the metal catalyst, separate the CNTs, and remove carbonaceous impurities.

3.4. Surface Area and Porosity Analysis

Nitrogen sorption and desorption data can be used to determine the surface area, pore size distribution, as well as the ratio of micro- to mesopores. Surface area and porosity were calculated with the BET and BJH methods, respectively. The samples exhibit type IV sorption−desorption isotherms with an H2 hysteresis loop observed in the range of 0.45–1.0 P/P$_0$ (Figure 5a).

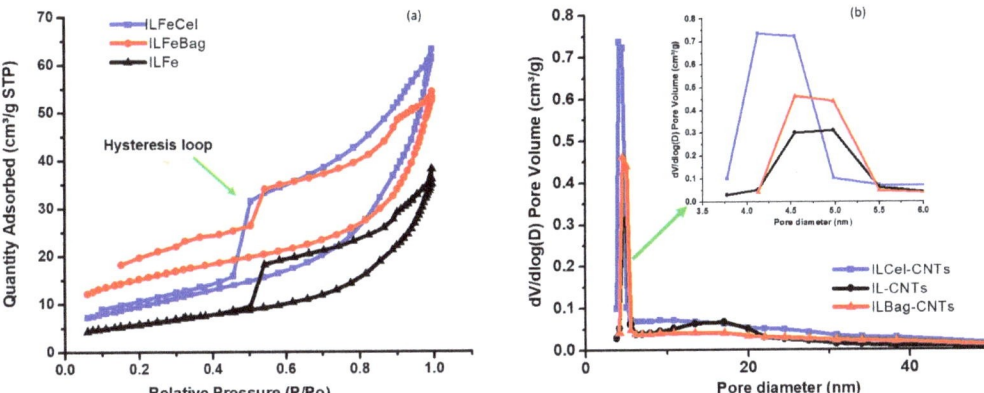

Figure 5. (**a**) Nitrogen adsorption and desorption isotherms; and (**b**) pore size distribution.

The H2 hysteresis loop suggests the presence of pores with narrow mouths (ink-bottle pores), comparatively constant channel-like pores, and pore network-linking effects [21], therefore suggesting the samples are mesoporous materials. The hysteresis loop of ILCel-CNTs is slightly broader than the other two, suggesting the presence of more porous structures with possibly greater surface area and pore volume. This observation correlates to the highest BET surface area of 52 m^2 g^{-1} for ILCel-CNTs, which is followed by ILBag-CNTs, 35 m^2 g^{-1}, and IL-CNTs with the lowest surface area of 21 m^2 g^{-1}. These surface areas correspond to the increase in ODs of the CNTs, i.e., 38, 63, and 101 nm, respectively, hence suggesting that the smaller the OD, the greater the surface area. The pore size distribution of samples is displayed in Figure 5b, and the insert is for the size region of 3.5–5.5 nm. The order of population frequency of these pores was IL-CNTs < ILBag-CNTs < ILCel-CNTs, respectively. The most probable reason for this is the different dimensions and composition of nitrogen-doping types leading to altered structures and textural characteristics. The pore distribution indicates that the CNTs are mostly composed of mesopores (2–50 nm).

3.5. Cyclic Voltammetry

Cyclic voltammetry (CV) is an effective and popular electrochemical technique frequently utilized to study the electrochemical performance of materials. The charge accumulation at electrodes was determined using a scan rate of 10 mV s^{-1} in a potential range of 0 to 0.8 V. The specific capacitance (Cs) was calculated from the CV curves according to Equation (1) [22]:

$$Cs = \frac{\int IdV}{2v\Delta Vm} \quad (1)$$

where Cs is the specific capacitance (F g^{-1}), I is the current response (A), ΔV is the potential window (V), v is the scan rate (V/s), and m is the mass of electrode material (g). The capacitive performance of the CNTs was obtained in 0.1 M H$_2$SO$_4$ at room temperature. The fabricated electrodes from all the N-CNT materials display some pseudo-capacitance traits and rectangular-shaped CV curves (Figure 6). This means that all the N-CNT materials investigated show an excellent rapid current response to voltage reversal and excellent electrochemical capacitive behavior. A pair of reversible redox peaks at 0.43 and 0.53 V was attributed to the presence of N introducing Faradic reactions between nitrogen atoms in the N-CNT matrix and the electrolyte. In acidic medium, the pseudo-Faradaic effects resulting from CNTs with pyridinic nitrogen functionality can be characterized according to Equation (2) [18,23]:

$$C^* = N + e^- + H^+ \leftrightarrow C^* NH \quad (2)$$

where C* stands for the carbon network. There were no oxidation and reduction peaks of Fe observed in CV curves, and this observation suggests that the effect of the residual Fe nanoparticles from synthesis is negligible [24,25].

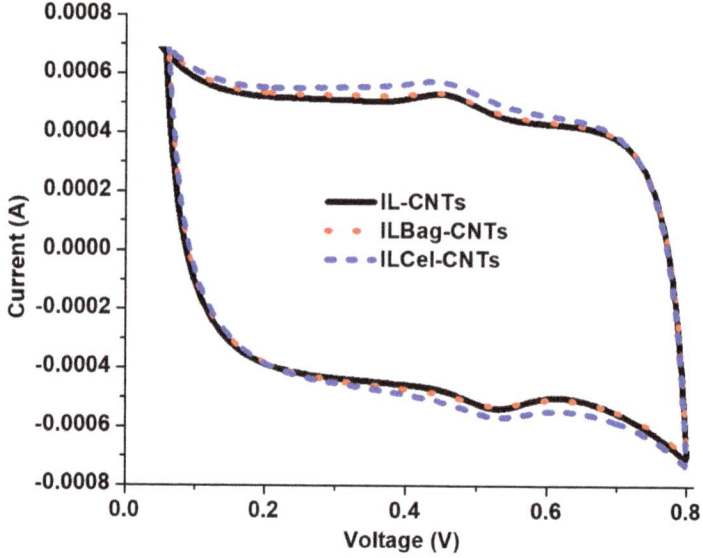

Figure 6. Cyclic voltammetry (CV) curves of all nanostructures at a scan speed of 10 mV s^{-1}.

The existence of nitrogen as a dopant in CNTs results in enhanced capacitance due to the modified electronic properties [26]. However, an increase in pyrrolic nitrogen-doping in N-CNTs has been reported to cause severe deterioration regarding both the electrochemical capacitor (EC) quality and the current [27]. However, in the present work, the EC quality did not deteriorate with an increase in pyrrolic nitrogen doping (Figure 6).

Furthermore, introducing nitrogen, an electron-rich atom, in the form of pyridinic moieties in the graphitic carbon network adds a lone pair of electrons into the delocalized π system [19]. The function of the lone pair of electrons is to act as a charge carrier, which improves the charge density and, in turn, leads to increased electrical conductivity, therefore resulting in increased capacitance [28]. It is also important to note that the substitution of carbon with nitrogen provides natural carbon materials with other redox reactions for pseudocapacitors [29]. On the other hand, pyrrolic nitrogen is known to induce a high concentration of defects within the graphitic structure since five-membered rings are formed instead of six-membered rings [30], which compromise the associated electrical conductivity of carbon and cause a loss of capacitance.

The synthesized N-CNTs exhibit Cs of 10, 26, and 30 F g^{-1} for IL-CNTs, ILBag-CNTs, and ILCel-CNTs, respectively. The Cs is comparable to the reported values of commercially available doped N-CNTs [31]. The total nitrogen content in the samples is very similar (7.2%, 7.1%, and 6.2% for IL-CNTs, ILBag-CNTs, and ILCel-CNTs), but the amount of pyrrolic N-doping decreases in the series (48%, 42%, and 29% (Figure 3d)) and is hence in line with the literature. The data indicate that the performance of the supercapacitors can be tuned by pyridinic nitrogen-doping composition.

Another reason for the increase in specific capacitance in that series could be the different porosities and specific surface areas [23]. Sufficiently large surface areas must be accessible to the electrolyte ions for adsorption and ultimately forming a double layer. In the above series, both the surface area and the average pore volume increase, indicating indirectly that the surface area is available for electrosorption. This contrasts to other reports [32,33], where no relationship between EC quality and performance with BET surface area was found. This work indicates that the physical properties, namely, diameter, porosity, and surface area of N-CNTs, are suitably tailored by introducing cellulose or bagasse as feedstock combined with the ionic liquid.

Cycle stability studies were performed on all three samples at 50 mV s^{-1} for 50 cycles (Figure 7a).

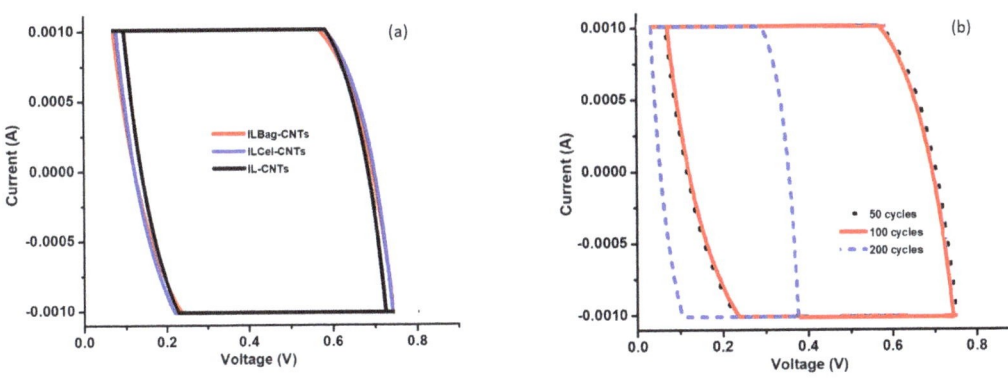

Figure 7. Cycle stability of (**a**) all samples at 50 mV s^{-1}; (**b**) ILCel-CNTs at 50, 100, and 200 cycles at 50 mV s^{-1}.

It is observed that all samples are stable, with ILCel-CNTs being slightly more stable based on the voltage window. Therefore, the most stable sample, ILCel-CNTs, which also features the best capacitive performance, was further studied for long-term cycle stability studies, i.e., 50, 100, and 200 cycles, as shown in Figure 7b. The voltammogram maintained its quasi-rectangular shape at 50 and 100 cycles, at a scan rate of 50 mV s^{-1}, indicating quick and efficient charge transfer and excellent capacitive behavior [34]. However, it deteriorated severely between the 100th and 200th cycles, as shown by the smaller voltage window (Figure 7b).

3.6. Electrochemical Impedance Spectroscopy (EIS)

Nyquist plots display the frequency responses of the samples, which demonstrate the associated impedance properties of the CNTs [35]. The high-frequency region shows negligible charge transfer resistance for all samples (Figure 8). Ideally, in cases where there is charge transfer resistance, the Nyquist plot may show a distorted semicircle in the high-frequency region. However, in this case, the semicircle is absent, indicating that the resistance is negligible, and hence, there is a remarkable charge reversal at voltage change, leading to the rectangular shapes shown in Figure 6 for the CV analysis.

The lower frequency region of the three CNT materials shows a straight line, which is attributed to the low ion diffusion resistance in the electrolyte, suggesting a nearly ideal charge storage device [35]. The slope for ILCel-CNTs is slightly steeper than those for ILBag-CNTs and IL-CNTs, suggesting faster ion diffusion and better EC quality.

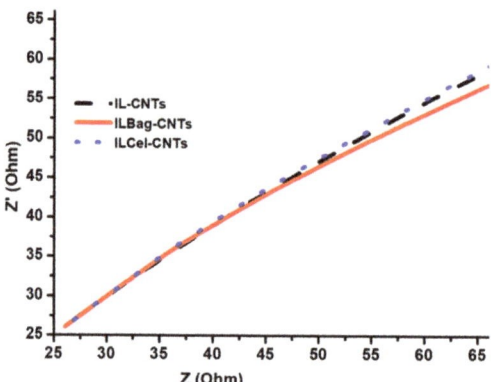

Figure 8. Nyquist plot.

4. Conclusions

In conclusion, the results have shown that excellent electrochemical properties are imparted to N-CNTs when these are prepared from 1-butyl-3-methylimidazolium chloride or a mixture of this ionic liquid with either bagasse or cellulose. Hence, the developed protocol allows fabricating electronic devices such as supercapacitors from renewable resources. Depending on the starting material composition, the physical (dimensions, pore size distribution, surface area), chemical (elemental composition, level of N-doping, ratio of pyridinic and pyrollic nitrogen, thermal stability) and electrochemical properties electrochemical capacitor, EC, (EC quality, specific capacitance, impedance) of the material are tailored. In particular, the N-CNTs synthesized from cellulose display excellent performance.

Supplementary Materials: The following are available online at https://www.mdpi.com/2071-1050/13/5/2977/s1.

Author Contributions: Conceptualization, K.M.; methodology, K.M.; validation, K.M., A.S., P.G.N. and V.O.N.; formal analysis, K.M.; investigation, K.M.; data curation, K.M.; writing—original draft preparation, K.M.; writing—review and editing, A.S., P.G.N., and V.O.N.; supervision, A.S., P.G.N. and V.O.N. All authors have read and agreed to the published version of the manuscript.

Funding: This work is based on the research supported in part by grant numbers 116610 (KM), 103979 (VON), 115465 (AS) by the National Research Foundation of South Africa.

Institutional Review Board Statement: Not applicable.

Informed Consent Statement: Not applicable.

Data Availability Statement: Data is contained within the article or Supplementary Materials.

Acknowledgments: The authors would like to thank the Synthetic and Medicinal Chemistry Research Group (SMCRG) at the University of KwaZulu-Natal for the use of the CHI instruments. Kudzai Mugadza wishes to thank the Sugar Milling Research Institute (SMRI) and the National Research Foundation-The World Academy of Science (NRF-TWAS) for the supply of the feedstock and funding, respectively.

Conflicts of Interest: The authors declare no conflict of interest.

References

1. Dresselhaus, M.S.; Thomas, I.L. Alternative energy technologies. *Nature* **2001**, *414*, 332–337. [CrossRef] [PubMed]
2. Lewis, N.S.; Nocera, D.G. Powering the planet: Chemical challenges in solar energy utilization. *Proc. Natl. Acad. Sci. USA* **2006**, *103*, 15729–15735. [CrossRef] [PubMed]
3. Green, M.A.; Bremner, S.P. Energy conversion approaches and materials for high-efficiency photovoltaics. *Nat. Mater.* **2016**, *16*, 23–34. [CrossRef] [PubMed]

4. Dubal, D.P.; Ayyad, O.; Ruiz, V.; Gomez-Romero, P. Hybrid energy storage: The merging of battery and supercapacitor chemistries. *Chem. Soc. Rev.* **2015**, *44*, 1777–1790. [CrossRef]
5. Chen, H.; Cong, T.N.; Yang, W.; Tan, C.; Li, Y.; Ding, Y. Progress in electrical energy storage system: A critical review. *Prog. Nat. Sci.* **2009**, *19*, 291–312. [CrossRef]
6. Kötz, R.; Carlen, M. Principles and applications of electrochemical capacitors. *Electrochim. Acta* **2000**, *45*, 2483–2498. [CrossRef]
7. Wang, H.; Yang, Y.; Guo, L. Nature-inspired electrochemical energy-storage materials and devices. *Adv. Energy Mater.* **2017**, *7*, 1601709. [CrossRef]
8. Li, B.; Dai, F.; Xiao, Q.; Yang, L.; Shen, J.; Zhang, C.; Cai, M. Nitrogen-doped activated carbon for a high energy hybrid supercapacitor. *Energy Environ. Sci.* **2016**, *9*, 102–106. [CrossRef]
9. Bairi, P.; Shrestha, R.G.; Hill, J.P.; Nishimura, T.; Ariga, K.; Shrestha, L.K. Mesoporous graphitic carbon microtubes derived from fullerene C 70 tubes as a high performance electrode material for advanced supercapacitors. *J. Mater. Chem. A* **2016**, *4*, 13899–13906. [CrossRef]
10. Zhang, Q.; Huang, J.-Q.; Qian, W.-Z.; Zhang, Y.-Y.; Wei, F. The road for nanomaterials industry: A review of carbon nanotube production, post-treatment, and bulk applications for composites and energy storage. *Small* **2013**, *9*, 1237–1265. [CrossRef]
11. Fic, K.; Frackowiak, E.; Béguin, F. Unusual energy enhancement in carbon-based electrochemical capacitors. *J. Mater. Chem.* **2012**, *22*, 24213–24223. [CrossRef]
12. Wickramaratne, N.P.; Xu, J.; Wang, M.; Zhu, L.; Dai, L.; Jaroniec, M. Nitrogen enriched porous carbon spheres: Attractive materials for supercapacitor electrodes and CO_2 adsorption. *Chem. Mater.* **2014**, *26*, 2820–2828. [CrossRef]
13. Mugadza, K.; Ndungu, P.G.; Stark, A.; Nyamori, V.O. Ionic liquids and cellulose: Innovative feedstock for synthesis of carbon nanostructured material. *Mater. Chem. Phys.* **2019**, *234*, 201–209. [CrossRef]
14. Labulo, A.H.; Ngidi, N.P.; Omondi, B.; Nyamori, V.O. Physicochemical properties of nitrogen-doped carbon nanotubes from metallocenes and ferrocenyl imidazolium compounds. *J. Organomet. Chem.* **2018**, *868*, 66–75. [CrossRef]
15. Steinmetz, M.; Lima, D.; Machado RR, L.; Sundararaj, U.; Arjmand, M.; da Silva, A.B.; Santos, J.P.; Pessôa, C.A.; Wohnrath, K. Nitrogen-doped carbon nanotubes towards electrochemical sensing: Effect of synthesis temperature. *Diam. Relat. Mater.* **2020**, *110*, 108093. [CrossRef]
16. Mori, S.; Suzuki, M. Effect of oxygen and hydrogen addition on the low-temperature synthesis of carbon nanofibers using a low-temperature CO/Ar DC plasma. *Diam. Relat. Mater.* **2008**, *17*, 999–1002. [CrossRef]
17. Hassan, F.M.; Chabot, V.; Li, J.; Kim, B.K.; Ricardez-Sandoval, L.; Yu, A. Pyrrolic-structure enriched nitrogen-doped graphene for highly efficient next-generation supercapacitors. *J. Mater. Chem. A* **2013**, *1*, 2904–2912. [CrossRef]
18. Biemolt, J.; Denekamp, I.M.; Slot, T.K.; Rothenberg, G.; Eisenberg, D. Boosting the supercapacitance of nitrogen-doped carbon by tuning surface functionalities. *ChemSusChem* **2017**, *10*, 4018–4024. [CrossRef]
19. Zhu, Y.; Chen, M.; Zhang, Y.; Zhao, W.; Wang, C. A biomass-derived nitrogen-doped porous carbon for high-energy supercapacitor. *Carbon* **2018**, *140*, 404–412. [CrossRef]
20. Li, H.F.; Wang, C.; Liu, L.F.; Xie, N.; Pan, M.; Wu, P.; Wang, X.D.; Zeng, Z.; Deng, S.; Dai, G.P. Facile one-step synthesis of N-doped carbon nanotubes/N-doped carbon nanofibers hierarchical composites by chemical vapor deposition. *J. Nanoparticle Res.* **2020**, *22*, 10. [CrossRef]
21. Sing, K.; Williams, R. Physisorption hysteresis loops and the characterization of nanoporous materials. *Adsorp. Sci. Technol.* **2004**, *22*, 773–782. [CrossRef]
22. Shahrokhian, S.; Mohammadi, R.; Asadian, E. One-step fabrication of electrochemically reduced graphene oxide/nickel oxide composite for binder-free supercapacitors. *Int. J. Hydrog. Energy* **2016**, *41*, 17496–17505. [CrossRef]
23. Frackowiak, E.; Lota, G.; Machnikowski, J.; Vix-Guterl, C.; Béguin, F. Optimisation of supercapacitors using carbons with controlled nanotexture and nitrogen content. *Electrochim. Acta* **2006**, *51*, 2209–2214. [CrossRef]
24. Kim, Y.-S.; Kumar, K.; Fisher, F.T.; Yang, E.-H. Out-of-plane growth of CNTs on graphene for supercapacitor applications. *J. Nanotechnol.* **2011**, *23*, 015301. [CrossRef]
25. Mombeshora, E.T.; Ndungu, P.G.; Jarvis, A.L.L.; Nyamori, V.O. Oxygen-modified multiwalled carbon nanotubes: Physicochemical properties and capacitor functionality. *Int. J. Energy Res.* **2017**, *41*, 1182–1201. [CrossRef]
26. Qian, X.; Li, N.; Imerhasan, M.; Wang, W. Conversion of low molecular weight hydrogel to nitrogen-doped carbon materials and its application as supercapacitor. *Colloids Surf. A Physicochem. Eng. Asp.* **2019**, *573*, 255–261. [CrossRef]
27. Mombeshora, E.T.; Ndungu, P.G.; Jarvis, A.L.L.; Nyamori, V.O. The physical and electrochemical properties of nitrogen-doped carbon nanotube- and reduced graphene oxide-titania nanocomposites. *Mater. Chem. Phys.* **2018**, *213*, 102–112. [CrossRef]
28. Han, X.; Jiang, H.; Zhou, Y.; Hong, W.; Zhou, Y.; Gao, P.; Ding, R.; Liu, E. A high performance nitrogen-doped porous activated carbon for supercapacitor derived from pueraria. *J. Alloys Compd.* **2018**, *744*, 544–551. [CrossRef]
29. Yang, M.; Zhou, Z. Recent breakthroughs in supercapacitors boosted by nitrogen-rich porous carbon materials. *Adv. Sci.* **2017**, *4*, 1600408. [CrossRef]
30. Ewels, C.; Glerup, M. Nitrogen doping in carbon nanotubes. *J. Nanosci. Nanotechnol.* **2005**, *5*, 1345–1363. [CrossRef]
31. Zhao, J.-R.; Jun, H.; Jiang-Feng, L.; Ping, C. N-doped carbon nanotubes derived from waste biomass and its electrochemical performance. *Mater. Let.* **2020**, *261*, 127146. [CrossRef]
32. Misnon, I.I.; Zain, N.K.M.; Aziz, R.A.; Vidyadharan, B.; Jose, R. Electrochemical properties of carbon from oil palm kernel shell for high performance supercapacitors. *Electrochim. Acta* **2015**, *174*, 78–86. [CrossRef]

33. Zhang, G.; Chen, Y.; Chen, Y.; Guo, H. Activated biomass carbon made from bamboo as electrode material for supercapacitors. *Mater. Res. Bull.* **2018**, *102*, 391–398. [CrossRef]
34. Liou, T.-H.; Wu, S.-J. Characteristics of microporous/mesoporous carbons prepared from rice husk under base- and acid-treated conditions. *J. Hazard. Mater* **2009**, *171*, 693–703. [CrossRef] [PubMed]
35. Xu, X.; Wang, M.; Liu, Y.; Lu, T.; Pan, L. Metal-organic framework-engaged formation of a hierarchical hybrid with carbon nanotube inserted porous carbon polyhedra for highly efficient capacitive deionization. *J. Mater. Chem. A* **2016**, *4*, 5467–5473. [CrossRef]

Article

Alkaline Fractionation and Subsequent Production of Nano-Structured Silica and Cellulose Nano-Fibrils for the Comprehensive Utilization of Rice Husk

Hyun Jin Jung [1,2], Hyun Kwak [2], Jinyoung Chun [3] and Kyeong Keun Oh [1,2,*]

1. Department of Chemical Engineering, Dankook University, Yongin, Gyeonggi-do 16890, Korea; hjjung@sugaren.co.kr
2. R&D Center, SugarEn Co., Ltd., Yongin, Gyeonggi-do 16890, Korea; hkwak@sugaren.co.kr
3. Korea Institute of Ceramic Engineering and Technology (KICET), Jinju, Gyeongsangnam-do 52851, Korea; jchun@kicet.re.kr
* Correspondence: kkoh@dankook.ac.kr; Tel.: +82-31-8005-3549

Abstract: The parameters of the alkaline fractionation process were investigated and optimized using a statistical analysis method to simultaneously remove hemicellulose and ash from rice husk (RH) concomitantly. After the alkaline fractionation process, the residual solid contained high cellulose, and the recovery yield of hemicellulose was enhanced in the fractionated liquid hydrolyzate. The hemicellulosic sugar recovery yield (71.6%), de-ashing yield (>99%), and lignin removal (>80%) were obtained at the reaction conditions of 150 °C of temperature, 40 min of reaction time, and 6% (w/v) of NaOH concentration. Subsequently, nano-structured silica was synthesized using black liquor obtained as a by-product of this fractionation process. For the production of nano-structured silica, it was observed that the pH of a black liquor and the heat treatment temperature significantly influenced the textural properties of silica product. In addition, the two-stage bleaching of solid residue followed by colloid milling for the production of high value-added CNF with was attempted. As a result, in addition to 119 g of fermentable sugar, 143 g of high-purity (>98%) silica with a surface area of 328 m^{2}g^{-1} and 273.1 g of high-functional CNF with cellulose content of 80.1% were simultaneously obtained from 1000 g of RH.

Keywords: biorefinery; multi-objectives RSM; nano-silica; de-ashing; cellulose crystals

Citation: Jung, H.J.; Kwak, H.; Chun, J.; Oh, K.K. Alkaline Fractionation and Subsequent Production of Nano-Structured Silica and Cellulose Nano-Fibrils for the Comprehensive Utilization of Rice Husk. *Sustainability* **2021**, *13*, 1951. https://doi.org/10.3390/su13041951

Academic Editor: Dirk Enke

Received: 3 December 2020
Accepted: 8 February 2021
Published: 11 February 2021

Publisher's Note: MDPI stays neutral with regard to jurisdictional claims in published maps and institutional affiliations.

Copyright: © 2021 by the authors. Licensee MDPI, Basel, Switzerland. This article is an open access article distributed under the terms and conditions of the Creative Commons Attribution (CC BY) license (https://creativecommons.org/licenses/by/4.0/).

1. Introduction

In response to growing concerns about environmental pollution caused by excessive consumption of fossil fuels, many researchers are pursuing alternative bio-based resources that can be applied to related industries. Among various candidates, lignocellulosic biomass, including agro-industrial and agricultural wastes, has attracted an intensive interest in the production of biofuels and chemicals, which are desirable alternatives to conventional petroleum-based fuels and chemicals. The biodegradability, reproducibility, and abundance of lignocellulosic biomass make it a promising resource for the production of these bio-based products [1]. Thus, lignocellulosic biomass has emerged as a potential renewable resource due to its annual reproducibility and abundance.

To secure the economic feasibility of biorefinery, a biorefinery system for producing value-added chemicals, i.e., an integrated biomass conversion process, must be conceived. To operate the biorefining process efficiently, each component of biomass must be separated sequentially by an effective and technically feasible fractionation process [2,3]. Rice is one of the most cultivated crops in the world; in 2018, approximately 661 million tons of rice were cultivated worldwide. Therefore, rice husk (RH) is a major side product of the agricultural industry, because its weight accounts for approximately 20 wt.% of rice plant [4]. Since RH is obtained through the milling process, it has already been processed first, suitable for further chemical and physical treatment without any additional drying or grinding process. In Korea, domestic RPCs (rice process complexes) perform rice milling according to demand, so they can produce relatively uniform amounts of RH throughout the year [5]. Nevertheless, most of the current RH is either burned in the field or used as compost in the soil as a quick and easy disposal method [6]. However, these methods of disposal are a major source of air pollution from smoke and greenhouse gases. In particular, RH combustion is evaluated as socially unacceptable due to the widespread air pollution increasing organic carbon emission [7]. The inclusion of some RH in the soil is a common management practice to improve soil fertility [8] and to prevent the harmful effects of long-term agricultural activities [9]. However, composting of RH increases emission of methane gas, one of the major greenhouse gases [10].

There has been a considerable amount of research on the use of RH over the past few years. Since RH is composed of a large amount of lignocellulose and ash, numerous studies have been focused on the preparation of high value-added products such as lignin, silica, and fermentable sugar to utilize RH as a raw material in various bio-based industries [11–13]. For examples, hemicellulose is hydrolyzed into sugars (mainly xylose) and substituted sugar oligomers [14,15], cellulose is hydrolyzed into glucose and then fermented into ethanol [16,17], and ash is used for preparation of pure fine silica [18,19]. In particular, the silicified surface of RH acts as a protective barrier, specifically conferring an outstanding recalcitrance toward biological threats, which leads to inefficient hydrolysis of cellulose and hemicellulose. Accordingly, the release of fermentable sugar from the tightly woven structure of cell wall is considered a prerequisite to obtaining bio-based products through the biological conversion process [20].

Interestingly, silica (SiO_2) accounts for 80–95% of the inorganic components in RH [21,22]. Therefore, after extraction of the cellulosic components through fractionation, silica-based products can be obtained from the byproduct liquor (black liquor). The morphology and purity of the silica products can be changed through post-treatment of black liquor. Therefore, the development of an appropriate post-treatment process with optimized conditions can lead to the production of high purity/grade nano-structured silica from black liquor. Currently, a large amount of silica is widely used in various industries. In particular, nano-structured silica with unique characteristics is required for various modern applications such as biosensors [23], drug delivery systems [24], wastewater treatment [25], and superhydrophobic coatings [26], which are expected to grow rapidly [27]. Accordingly, the facile preparation of high-purity nano-structured silica from RH is an important research theme, because the production of high value-added silica from a low-cost raw material and the effective treatment of waste biomass can be achieved simultaneously. Fernandes et al. [28] suggested that the effective production of silica from RH is generally based on a combination of appropriate chemical and heat treatment conditions.

On the other hand, due to the specific chemical and physical properties of biomass-derived cellulose, the interest in their research and development is rapidly increasing. However, cellulose is still limited in its use in industrial applications due to its unique properties such as water insolubility, hygroscopicity, and non-melting properties. Nevertheless, until recently, many researchers have found that when biomass cellulose is subjected to mechanical shear or controlled hydrolysis, elongated fibrils or defect-free rod-shaped crystal particles are obtained in the nanoscale range, and that they were named nano-cellulose [29]. Nano-cellulose has many advantages such as high mechanical strength, large aspect ratio, low density, high specific surface area, and excellent biocompatibility. In particular, by modifying the surface chemical properties of nano-cellulose through various development strategies, it is possible to change the properties to meet various industrial needs [30]. Thus, nano-cellulose can be a versatile source of various products in the form of commercial cellulose derivatives through chemical modification [31]. Typically, nano-cellulose is divided into two main types. That is, it can be broadly classified into cellulose nanocrystals (CNC) obtained from acid treatment and cellulose nano-fibrils (CNF) mainly produced by mechanical disintegration. CNF has become a more attractive material for industrial use, and as CNF can be produced on a commercial scale, this material has become readily available through the market. Thus, CNF has received more attention in industrial applications. In relation to CNF from lignocellulosic biomass, many R&Ds have been proposed to develop and optimize production technologies or to impart new properties to materials of special and high industrial value [32–36].

In our previous work to effectively utilize the lignocellulosic biomass, glucose and xylose were produced as the raw materials for production of biofuel by hydrolyzing cellulose and hemicellulose through optimized fractionation process. In addition, ball milling combined in the presence of chemical catalyst has been shown to be an effective fractionation method because of increasing mixing effectiveness, reducing particle size and increasing larger accessible surface area, resulting in significantly improved fractionation of herbaceous biomass [37–39]. However, unavoidable substantial byproduct formation after recovery of the fermentable sugars incurs additional costs for the associated waste treatment, which can increase the production cost. Therefore, it is speculated that a conversion strategy for the utilizing by-products produced in the fractionation process is required [40].

In the current study, alkaline fractionation of RH has been attempted to selectively separate the components from RH. The objective of this study was to recover hemicellulosic sugar and ash (mainly silica), while retaining most of the cellulosic component in the solid residue. Numerous studies, however, have been conducted to produce bioethanol using xylose in liquid phase and cellulose in solid phase. Therefore, the focus of the research is on the multi-purpose optimization of the alkali fractionation process, which can simultaneously meet the recovery of hemicellulosic sugar (mainly xylose) and the de-ashing yield from RH. In addition, the studies were conducted to secure bio-based materials from rice husk other than biofuels. After fractionation of the RH, with the aim of producing high value-added bio-based materials from fractionation residues, new synthetic methods for producing high-purity nano-silica from liquid residues were investigated, and a series of experiments were conducted to produce CNF from solid residues.

2. Materials and Methods

2.1. Feedstock Preparation

RH was obtained as a by-product from local rice process complex (RPC); the rice was harvested from Gimpo, Gyeonggi-do, Korea, in the fall of 2017. Prior to the experiment, RH was dried for 48 h at 45 ± 5 °C using a convection drying oven (FC-PO-1500, Lab House, Seoul, Korea) and stored in an automatic dehumidification desiccator. The moisture content of the RH was 4.6% based on its oven-dried weight.

2.2. Compositional Analysis of Raw and Fractionated RH

The carbohydrate components of the fractionated or untreated RH were subjected to a two-stage acid hydrolysis extraction process, which was standardized according to the procedure specified in the corresponding Laboratory Analytical Procedure of the National Renewable Energy Laboratory (NREL-LAP) [41]. In addition, in the analysis of liquid samples obtained during the fractionation, glucose and hemicellulosic sugar (xmg: xylose, mannose, and galactose) were analyzed by high-performance liquid chromatography (HPLC; 1260 Infinity, Agilent Technologies Inc., Santa Clara, CA, USA). The analytical column and detector were an Aminex HPX-87H organic acid column (Bio-Rad) and a refractive index detector (1260 RID, Agilent Technologies Inc.), respectively. The operating conditions for the HPLC column were 65 °C and a mobile-phase (sulfuric acid) flow rate of 0.6 mL/min.

2.3. Experimental Setup and Operation of Bench-Scale Fractionation

A tubular reactor was used to optimize the lab-scale alkaline fractionation process. The reactor was made of stainless steel (SS-316L) and had an internal diameter of 10.7 mm and a length of 150 mm, with an internal volume of 13.5 mL. Five hundred milligrams of oven-dried RH was placed in the reactor, and sufficient alkaline solution was added to give a solid-to-liquid ratio of 1:10. To increase the temperature of the reactor to the target temperature within approximately 1.0 min, it was first immersed in the first heating bath (molten salt) set at 240 °C. When the temperature of the reactor reached the target temperature, it was quickly transferred to the second bath (silicone oil) set to the target reaction temperature (116–184 °C). After the reaction was conducted for a fixed period of time, the reactor was quenched in an ice-water bath to quickly stop the reaction.

The bench-scale (30 L) combined ball milling reactor used for the production of hemicellulosic sugar and black liquor via an alkaline fractionation of RH [42]; this reactor was designed to be capable of operating at 60 rpm at a pressure of 20 kg_f cm^{-2} and a temperature of 200 °C (Sugaren Co. Ltd., Yongin, Korea). Alumina balls (HD, sphere type, 10 mm diameter, and 3.6 g cm^{-3} density) were additionally placed in the reactor to improve the mixing efficiency. The ball/biomass/alkaline solution ratio was 30:1:10 ($w/w/v$). Upon completion of the reaction, the solid and liquid inside the reactor were separated. The remaining solid samples were washed and then used for glucose production by enzymatic hydrolysis.

2.4. Optimization of the Process Variables Using Response Surface Methodology (RSM)

A second-order model was used to fit the data individually for the response of hemicellulosic sugar extraction yield and de-ashing yield by the general model with three variables in the alkaline fractionation of rice husk: reaction temperature, reaction time, and NaOH concentration [43]. The range of process variables in the central composite design (CCD) are shown in Table 1.

Table 1. Process variables and their levels for 3^3 central composite design (CCD).

Process Variables	Coded Levels and Actual Conditions				
	−1.68	−1	0	+1	+1.68
Temperature (°C)	116	130	150	170	184
Time (min)	6	20	40	60	74
NaOH Conc. (%)	1	2	3.5	5	6

To achieve high hemicellulosic sugar extraction yield and de-ashing yield from RH, the alkaline fractionation conditions were optimized by response surface methodology (RSM) based on the 3^3 factorial CCD. To determine the optimal conditions for extracting hemicellulosic sugar and de-ashing, analysis of variance (ANOVA) and multiple regression analysis were performed using Design Expert (Ver 8, Stat-Ease, Minneapolis, MN, USA).

2.5. Preparation of Nano-Structured Silica Powders from Black Liquor

Nano-structured silica powders (NSP series) were obtained from a black liquor. A block diagram according to the preparing sequence and conditions is shown in Figure 1. In a typical preparation of NSP-1, 450 mL of the black liquor diluted with 50 mL of distilled water was used. First, the pH of the black liquor was adjusted to ~7.0 by addition of acetic acid. After the resulting solution was stirred for 12 h at room temperature (~25 °C), it was centrifuged at 5000 rpm for 5 min; the supernatant was then decanted away. The remaining products were washed several times with distilled water and dried overnight at 80 °C. Finally, the dried powders were heat-treated at 600 °C for 2 h under air atmosphere. Through this process, we obtained ~15.5 g of silica powders per liter of black liquor. The overall synthesis procedure of NSP-2 and NSP-3 was similar to that of NSP-1, except that the pH of the black liquor was adjusted to ~6.0 and ~8.0, respectively.

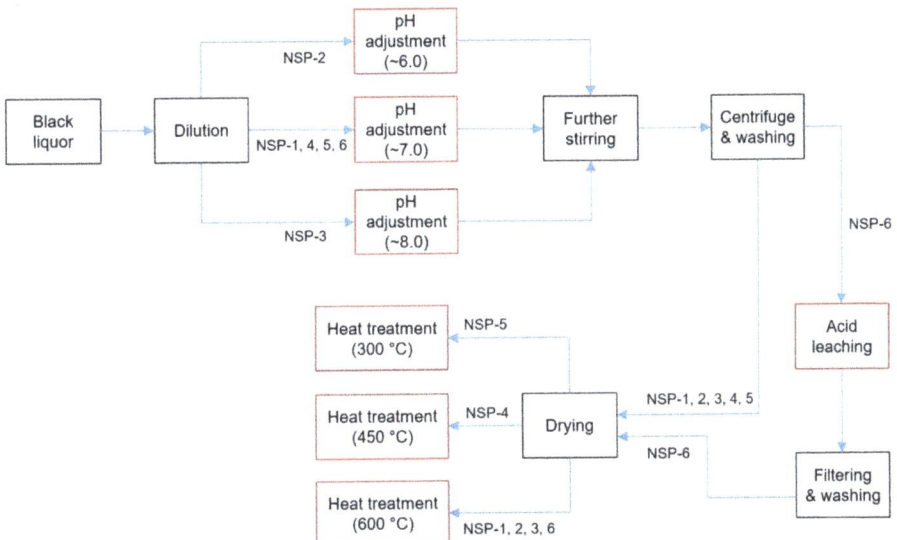

Figure 1. Block diagram depicting the preparation of nanostructured silica powder (NSP) from black liquor under different conditions.

In the case of NSP-4 and NSP-5, the overall synthetic procedure was the same as that of NSP-1, except that the heat-treatment temperatures at the last step ere 450 °C and 300 °C, respectively. NSP-6 was prepared by a procedure similar to that used for NSP-1, except that the dried powders at 80 °C were further treated with HCl solution before the heat treatment at 600 °C; specifically, the dried powders were placed in 100 mL of 10 wt% HCl solution and then stirred at 90 °C for 2 h. After reaction, the powders were filtered and washed with distilled water. After drying overnight at 80 °C, the final products were obtained by the heat treatment at 600 °C for 2 h under air atmosphere.

2.6. Preparation of Cellulose Nano-Fibrils from Fractionated Residual Solid

The two-stage bleaching process was performed using hydrogen peroxide (H_2O_2) and chlorine dioxide (ClO_2) to remove residual lignin from fractionated solid phase. In the first step, the solid/liquid ratio was set to 10% using 5% (v/v) H_2O_2, and reacted at 100 °C for 60 min. In a second step using 5% (v/v) ClO_2, the pH was controlled to 3 to 4 using CH_3COOH, reacted at 70 °C for 90 min, then washed with distilled water and dried at 45 °C. A 2 wt.% (w/v) dry solid suspension was prepared using distilled water, and CNF was prepared by grinding with a colloid milling device, Super Mass Colloider (MKCA6-5J, Masuko Sangyo Co., Kawaguchi, Japan). A non-porous grinder (MKGA10–80, Masuko Sangyo Co., Kawaguchi, Japan) that prevents contamination of bacteria and can add catalysts was used, for which the grinder interval was 150 μm and the rotation speed was 1600 rpm.

2.7. Material Characterization of Nano-Structured Silica Powders and Cellulose Nano-Fibril

X-ray diffraction (XRD) patterns were obtained using a Rigaku D/Max 2500/PC diffractometer (Japan). The material morphologies were investigated by scanning electron microscopy (SEM; X-Max T50, Oxford Instruments, Abingdon, UK) and transmission electron microscopy (TEM; JEM-2000EX, JEOL Ltd., Tokyo, Japan). The nitrogen (N2) physisorption isotherms were obtained at 77 K using a Tristar II 3020 system (Micromeritics Inc., Norcross, GA, USA). The surface areas of materials were calculated from the measured isotherms according to the Brunauer–Emmett–Teller (BET) method, and the pore volumes were taken at the $P/P_0 \approx 0.995$ single point. The pore size distributions of materials were calculated by the Barrett–Joyner–Halenda (BJH) method from the adsorption branches of the isotherms. The inorganic chemical composition and carbon content of silica were determined using a sequential X-ray fluorescence (XRF) spectrometer (ZSX100, Rigaku, Akishima, Japan) and an elemental analyzer (Elementar Analysensysteme GmbH, vario MICRO cube, Langenselbold, Germany), respectively.

3. Results

3.1. Compositional Analysis of Raw and Fractionated RH

Table 2 summarizes the chemical constituents of the raw and NaOH-fractionated RH. The composition of carbohydrates in the raw RH was measured to be 35.6 ± 0.8% of cellulose, 13.6 ± 0.4% of hemicellulose, 0.8 ± 0.1% of galactan, and 1.7 ± 0.0% of arabinan; thus, the total carbohydrates account for 51.7 ± 1.3%. In addition, the noncarbohydrate component was analyzed to consist of 23.4 ± 0.2% lignin, 4.2 ± 0.1% extractives, 3.2% protein, 0.5% lipid, and 15.7 ± 0.2% ash. Therefore, the mass closure of each constituent of raw RH was reached at 99.6 ± 1.6% for oven-dried biomass. The results of the component analysis of the NaOH-fractionated RH in the second row in Table 2 were used to compare the changes in content of each component.

Table 2. Chemical composition of raw and NaOH-fractionated rice husk based on an oven-dry biomass.

Components	Carbohydrates			Lignin		Extractives		Protein [2]	Lipid [3]	Ash	Mass Closure
	Glucan	XMG [2]	Arabinan	Insoluble Lignin	Soluble Lignin	Water	Ethanol				
Raw rice husk	35.6 ± 0.8	13.6 ± 0.4	0.8 ± 0.1	22.7 ± 0.0	0.7 ± 0.2	3.5 ± 0.1	0.7 ± 0.0	3.2	0.5	15.7 ± 0.2	97.9
Fractionated rice husk [1]	73.9 ± 1.1	13.4 ± 0.6	1.1 ± 0.2	11.5 ± 0.3							99.9

[1] NaOH-fractionated at optimized reaction conditions; reaction temperature of 150 °C, reaction time of 45 min, and NaOH. concentration of 6%(w/v) with solid remaining of 35.5%. [2] XMG: Major component of hemicellulose (Xylan, Mannan, and Galactan). [3], [4] Protein and lipid contents were analyzed by KFRI, Korea Food Research Institute. N-factor = 5.95.

Only carbohydrate and lignin contents are shown for fractionated rice husk, because most of the noncarbohydrate components, including ash, were estimated to have been extracted into liquid by-product during the fractionation, resulting in no analysis. Similar studies have reported that alkaline fractionation facilitates the degradation of lignin and hemicellulose because of saponification of the intermolecular crosslinking between hemicellulose and lignin. In addition, significant changes in physical properties such as increased porosity, cellulose expansion and destruction of carbohydrate–lignin complexes can also be the cause [44,45].

The cellulose content of the NaOH-fractionated RH increased from 35.6% to 73.9%, and the hemicellulosic sugar content was found to be very slightly reduced. Meanwhile, the lignin content was greatly reduced from 22.7% to 11.5%. These values are relative, because the noncarbohydrate fraction and some of the hemicellulose and lignin fractions were solubilized during the fractionation; given the amount of residual solids, the hemicellulose and lignin fractions correspond to 26.2% cellulose, 4.8% hemicellulose, and 4.1% lignin based on the raw RH. These results are in good agreement with those of Shahabazuddin et al., who reported that the cellulose content increased from 32.7% to 51.7% and that the hemicellulose and lignin contents decreased from 18.1% to 16.1% and 26.7% to 12.3%, respectively [46]. Compositional analysis was established from three replicate measurements performed independently; error values are expressed as standard deviations.

3.2. Effects of Process Variables for Alkaline Fractionation

The effects of independent variables, reaction temperature, reaction time, and NaOH concentration, were evaluated to maximize the extraction of lignin, ash, and hemicellulosic sugar. In the 17 experimental conditions investigated in this study, the hemicellulosic sugar extraction yields ranged from 21.7% to 77.4%, and the de-ashing yields ranged from 69.2 to 95.6% (Table S1). In examining the influence of each independent variable, we observed that hemicellulosic sugar extraction and de-ashing yields tended to increase proportionally as the level of all of the independent variables increased. The effects of each independent variable on hemicellulosic sugar extraction and de-ashing yields were evaluated based on ANOVA (Table S2). The effect of NaOH concentration on the hemicellulosic sugar extraction yield was found to be greatest, followed by the effect of reaction time and temperature. By contrast, in the case of the de-ashing yield, significance was confirmed in the order of NaOH concentration, reaction temperature, and reaction time. The coefficients of determination (R^2), which measure the fitness of the second-order regression equations for hemicellulosic sugar extraction and de-ashing yield, as derived using the CCD, were evaluated as 0.9803 and 0.8372, respectively. The p-values were 0.0001 and 0.0406, respectively, indicating that the confidence intervals were within 95%. The highest yields of hemicellulosic sugar extraction and de-ashing at 150 °C, 40 min of reaction time, and 6% (w/v) NaOH concentration were 77.4% and 95.6%, respectively. These results are consistent with previous reports that lignin and ash were extracted as black liquor with increasing NaOH concentration.

3.3. Multi-Objectives Optimization of the Alkaline Fractionation of RH

The second-order polynomial equations related to hemicellulosic sugar extraction yield and de-ashing yield were obtained from the regression analysis and given by Equations (1) and (2).

$$Y_{xmg} = 49.35 + 2.14 \times X_1 - 0.34 \times X_2 + 14.56 X_3 - 2.18 X_1 X_2 - 2.84 X_1 X_3 - 1.17 X_2 X_3 - 6.20 X_1^2 - 1.60 X_2^2 + 0.076 X_3^2 \qquad (1)$$

$$Y_{de\text{-}ashing} = 93.65 + 1.25 X_1 + 1.68 X_2 + 5.47 X_3 - 0.10 X_1 X_2 + 0.13 X_1 X_3 + 0.25 X_2 X_3 + 0.083 X_1^2 - 1.19 X_2^2 - 3.37 X_3^2 \qquad (2)$$

where X_1: reaction temperature; X_2: reaction time; X_3: NaOH concentration.

The coefficients of determination (R^2), which measure the fitness of the second-order regression equations for hemicellulosic sugar extraction and de-ashing yield, were evaluated as 0.9803 and 0.8372, respectively, which further indicated that the model was suitable for adequately representing the real relationships among the selected reaction variables. The statistical significance of the model was evaluated by the F-test that showed that the regression was statistically significant. For the case of hemicellulosic sugar extraction yield, the "Prob > F" represented as p-value for the model was <0.0001, which indicated that the model was statistically significant with a confidence level of 99.99%.

The predicted maximum yield of hemicellulosic sugar extraction from the regression equation of dependent variable hemicellulosic sugar extraction yield was 74.7% at 148.1 °C reaction temperature, 27.0 min reaction time, and 5.9% NaOH concentration, respectively. Furthermore, the predicted maximum de-ashing yield was 96.1% at 142.8 °C reaction temperature, 60.6 min reaction time, and 4.9% NaOH concentration. The conditions for highest yields of hemicellulosic sugar extraction and de-ashing were determined by the analysis of the statistical model generated.

Figure 2 attempts to represent the change of dependent variables by overlaid contour lines for the interactions between the independent variables; therefore, the gray-toned contours represent the changes in the de-ashing yield, and the black-toned contours were expressed for the changes in the hemicellulosic sugar extraction yield. Since a relatively high de-ashing yield was obtained in most of the experimental range, the optimization criterion was given priority to the range in which the hemicellulosic sugar extraction yield could be obtained more than 70%. It was determined that the de-ashing yield was high, and the overlapping portion appeared to be suitable for the purpose of the processing, which was indicated in grayscale. Figure 2a shows that over 72% hemicellulosic sugar extraction yield and 94% de-ashing yield can be obtained in the range of reaction temperature of about 146 to 158 °C and reaction time of about 41 to 55 min. Likewise, in the case of reaction temperature, there was a possibility to obtain an hemicellulosic sugar extraction yield of 70% or more and a de-ashing yield of 93% or more in a wide range of about 140 to 160 °C, but the NaOH concentration to meet these yields was determined to be possible only in a narrow experimental range of about 5.8–6.3% (Figure 2b).

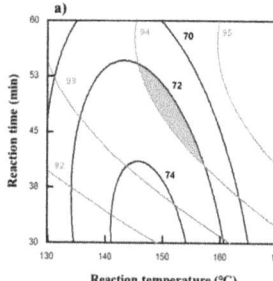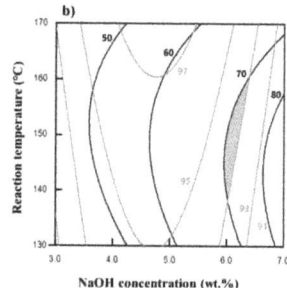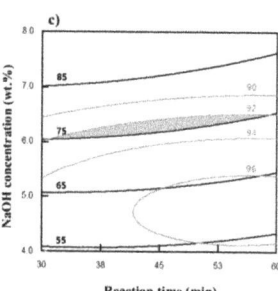

Figure 2. Multi-objective optimization on hemicellulosic sugar recovery and ash removal as a function of reaction temperature, reaction time and NaOH concentration, where, (**a**) NaOH concentration; 3.5 (w/v)%, (**b**) Reaction time; 40 min., (**c**) Reaction temperature; 150 °C.

It can be seen in Figure 2c that NaOH concentration greatly influences the sugar recovery and ash removal in the alkaline fractionation process of RHs. The reaction range capable of simultaneously matching the hemicellulosic sugar yield of 75% and the de-ashing yield of 92% or higher was possible in the entire range of reaction time. However, the effect of NaOH concentration was significant, as it was found to be possible only in the range of about 6.0% to 6.3%.

By overlapping the response surfaces of the resultant dependent variables, the optimized reaction conditions were estimated with a reaction temperature of 149.8 °C, a reaction time of 45.1 min, and a NaOH concentration of 5.9 wt.%, under which the hemicellulosic sugar extraction and de-ashing yields were estimated to be 73.2% and 93.9%, respectively. When the RH was fractionated under the optimized conditions; 150 °C, a reaction time of 45 min, and a NaOH concentration of 6.0 wt.%, it was confirmed that the extraction yield of hemicellulosic sugar was 71.6% and the de-ashing yield was greater than 99%, which are in good agreement with the estimated values obtained from the optimized conditions.

3.4. Synthesis of Nano-Structured Silica from the Residual Black Liquor

Residual black liquor is a type of basic silicate solution that contains both hemicellulosic sugar and lignin components. To obtain silica powders from black liquor preventing excessive precipitation of other components (e.g., lignin), the pH of black liquor was adjusted to neutral using a certain amount of acid solution. In this study, we used acetic acid instead of sulfuric acid for pH adjustment, because the former is weaker and less harmful than the latter. After purification of precipitated silica by centrifugation and washing, trace organic components were removed by heat treatment under an air atmosphere.

To prepare an initial test sample (NSP-1), the pH of black liquor was adjusted to approximately 7.0 using acetic acid, followed by heat treatment at a high temperature (600 °C) to completely remove the organic components. The XRD pattern of NSP-1 exhibited a broad diffraction near 20°, which is typical for amorphous silica (Figure 3a). It was seen that silica content among the inorganic components of NSP-1 was as high as 98.0%, and a small amount (1.76%) of sodium compound was also measured. SEM and TEM images of NSP-1 showed that the small nano-particles aggregated into large particles with sizes of a few hundred nanometers (Figure 3b,c). These types of hierarchical nano-structures impart NSP-1 with a large surface area of 175 m^2 g^{-1} and a pore volume of 0.81 cm^3 g^{-1} (Table 3).

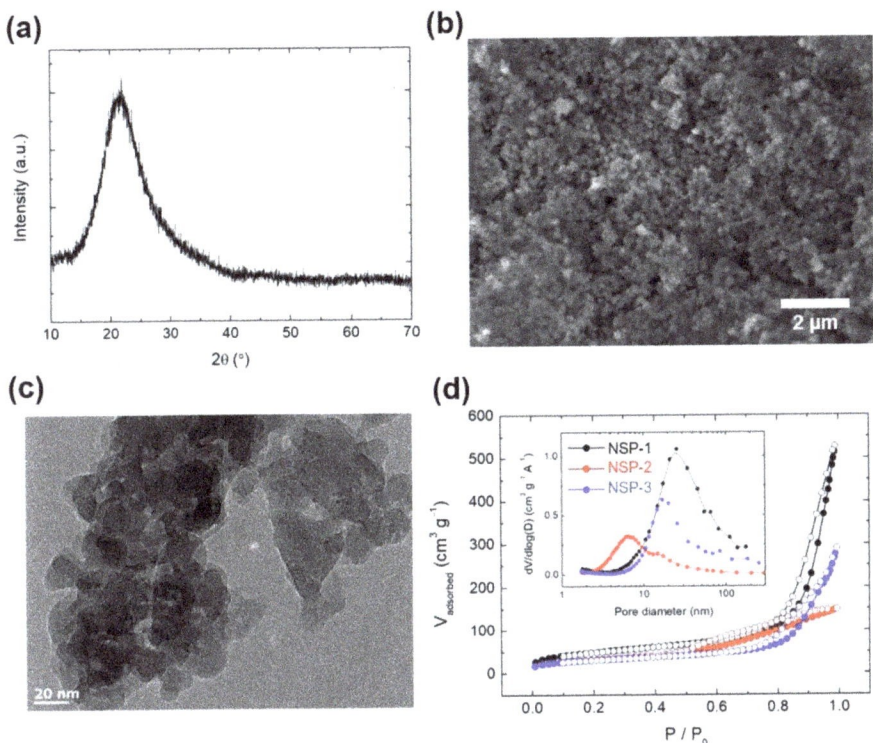

Figure 3. (**a**) XRD pattern, (**b**) SEM image, and (**c**) TEM image of NSP-1. (**d**) N2 physisorption isotherms and pore size distributions (inset) of NSP-1, NSP-2, and NSP-3.

Table 3. Experimental conditions and characterization results of NSP series.

	NSP-1	NSP-2	NSP-3	NSP-4	NSP-5	NSP-6
pH control	~7.0	~6.0	~8.0	~7.0	~7.0	~7.0
Acid leaching	–	–	–	–	–	O
Heat treatment temperature (°C)	600	600	600	450	300	600
Surface area (m^2 g^{-1})	175	113	109	273	328	392
Pore volume (cm^3 g^{-1})	0.81	0.23	0.45	0.57	0.61	0.65
Main pore size (nm)	~25	~7	~20	~10	~10	~10
SiO$_2$ (%) among the inorganic components	98.0	95.5	97.2	98.2	98.1	99.8

In the synthesis procedure, it was confirmed that the pH of black liquor and the heat-treatment temperature affected the textural properties of silica products substantially. NSP-2 and NSP-3 were obtained using the same procedure used for NSP-1, except that the pH levels of the black liquor were adjusted to approximately 6.0 and 8.0, respectively. Although the pH levels of NSP-2 and NSP-3 only differed from that of NSP-1 by approximately ±1.0 pH unit, their surface areas and pore volumes were noticeably small. NSP-2 exhibited a surface area of 113 m^2 g^{-1} and a pore volume of 0.23 cm^3 g^{-1}. A similar tendency was observed for NSP-3 (Table 3), which exhibited a surface area of 109 m^2 g^{-1} and a pore volume of 0.45 cm^3 g^{-1}. We speculated that these differences in textural properties resulted from the different rates of condensation reactions. After the formation of silicic acid by the addition of acetic acid, the condensation reaction of silicic acid was strongly dependent on the pH of the reactant solution [47]. Since the formation of the silica framework was affected by the condensation reaction rate, the textural properties of NSP powder were also altered by the pH of the solution.

3.5. Optimization for the Synthesis of Nano-Structured Silica with Large Surface Area

NSP-4 and NSP-5 were prepared following the same synthesis procedure used for NSP-1, except that the heat-treatment temperature at the last step was decreased to 450 °C and 300 °C, respectively. Because of the lower heat-treatment temperatures, the surface area of the silica substantially increased (Table 3). Between NSP-4 and NSP-5, NSP-5 showed a surface area of 328 $m^2\ g^{-1}$, which was nearly two times larger than that of NSP-1.

The main pore sizes of NSP-4 and NSP-5 were approximately 10 nm, whereas that of NSP-1 was approximately 25 nm (Figure 4a). This phenomenon might be induced by the residual sodium contents in the NSP powders, because the melting points of sodium silicates are substantially lower than that of pure silica [48]. Therefore, a small sodium content could reduce the melting point of the silica. In addition, because the particle size decreased to the nano-scale, the melting point of the material also decreased [49]. Accordingly, the nanosized silica framework of the NSP series could be easily melted and aggregated even at 600 °C. This process leads to the generation of large mesopores through the collapse of small mesopores, resulting in the decrease of surface area. Inevitably, because of the low heat-treatment temperature, the carbon content (0.61 wt.%) of NSP-5 was greater than that (0.06 wt.%) of NSP-1. However, we found that nano-structured silica with a high surface area could be readily prepared through a simple optimization of the established process.

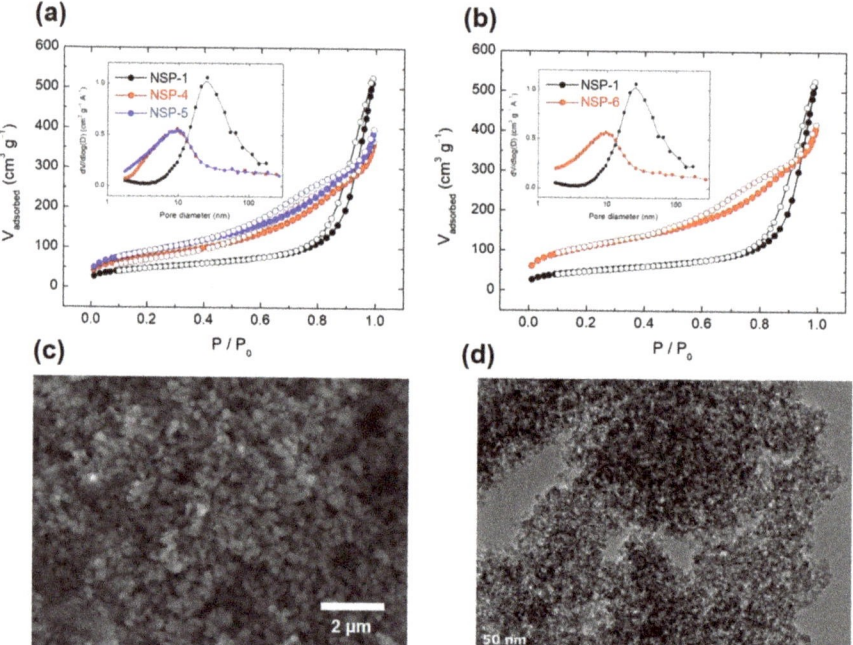

Figure 4. (a) N_2 physisorption isotherms and pore size distributions (inset) of NSP-1, NSP-4, and NSP-5. (b) N2 physisorption isotherms and pore size distributions (inset) of NSP-1 and NSP-6. (c) SEM and (d) TEM images of NSP-6.

It was also found that high purity silica can be obtained without compromising textural properties through additional acid leaching. After centrifugation and washing, dried powders were further treated with HCl solution before heat treatment at high temperature.

Through this acid leaching process, residual sodium compounds could be completely removed. XRF analysis results showed that the purity of the silica product (NSP-6) was as high as 99.8%, but that of the sodium compound was less than 0.2%. Interestingly, although NSP-6 was obtained through heat treatment at 600 °C, it retained a high surface area of 392 m^2 g^{-1} with a main pore size of ~10 nm (Table 3 and Figure 4b). In comparison with the TEM image in Figure 4c, the high-magnification TEM image of NSP-6 (Figure 4d) clearly showed the preservation of small-sized nanopores without agglomeration of the framework. This result supports the fact that the residual sodium content strongly affects the surface area of silica by deforming the silica framework during heat treatment at high temperatures. Because of its high treatment temperature, NSP-6 contained practically no carbon (0.06 wt.%). Therefore, high-purity silica with a large surface area could be successfully obtained from a black liquor. In particular, the surface area of NSP-6 is much greater than that of RH-derived silica prepared in previous studies [50,51], despite the absence of the use of additional surfactants or block copolymers.

3.6. Characteristics of CNF Prepared from Fractionated Residual Solid

Figure 5 shows the changes in chemical composition of raw RH and sequentially treated RH fibers. The untreated RH showed component content of cellulose 35.6%, hemicellulose 13.6%, lignin 23.4%, and ash 15.7%. The relative composition of the residual solid fractionated by NaOH showed 70.8% of cellulose, 12.3% of hemicellulose, 12.2% of lignin, and 0% of ash, confirming that all ash and a significant amount of lignin were removed through the optimized NaOH fractionation, whose conditions were 150 °C, a reaction time of 45 min, and a NaOH concentration of 6.0 wt.%. As a result of the first bleaching stage (H$_2$O$_2$) on the NaOH fractionated residual solid, the relative content of the 1st bleached solid was shown as cellulose of 81.1%, hemicellulose of 12.0%, and lignin of 6.1%, representing that some of the lignin was removed. In the case of the second bleached (ClO$_2$) RH, the relative composition of cellulose 85.2%, hemicellulose 10.4%, and lignin 0.6% were shown, so that most of the lignin was removed, resulting in highly pure cellulose.

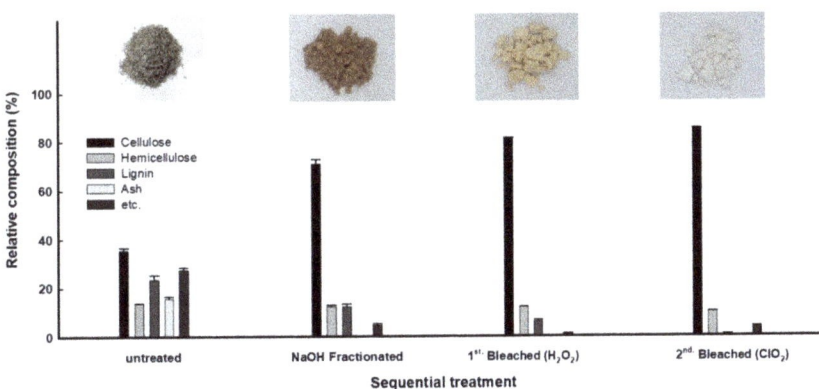

Figure 5. Changes in chemical composition and whiteness of rice husk fiber according to sequential treatment: untreated RH, NaOH fractionated RH fiber, first bleached RH fiber, second bleached RH fiber.

In addition, figures of residual solids for each sequential bleaching were intended to be shown at the top of Figure 5. After the second-stage bleaching process, the residual solid had a white color, which was believed to be due to the removal of lignin and other impurities as the bleaching proceeds sequentially. The change in the colors of the residual solids in each sequence well explained the change in the chemical composition as the bleaching progresses.

Figure 6a shows the results of XRD analysis to compare the change in crystallinity of RH and RH fibers according to each sequence. RH raw material, NaOH-treated RH fiber, and first-stage bleached RH fiber showed Cellulose I form, which is the crystal structure of lignocellulose that exists in nature. On the other hand, CNF obtained after second-stage bleaching showed a Cellulose II form. It is believed that this is because the number of hydrogen bonds between molecules increases as hydroxyl groups are exposed due to mechanical shear and friction forces, resulting in a more stable crystal structure. The crystallization index (CrI) at each stage of bleaching was the lowest at 37.3% in RH raw samples, 51.6% in NaOH-treated RH fiber, and 60.5% in first-stage bleached RH fiber. On the other hand, the crystallinity of the CNF obtained after proceeding to the second-stage bleaching was measured to be 57.1%, which was slightly lowered. This result is in good agreement with the results of Liu et al., which suggested that the crystallinity of CNF increased as the proportion of cellulose increases due to the removal of amorphous components such as lignin or hemicellulose during chemical bleaching step [52].

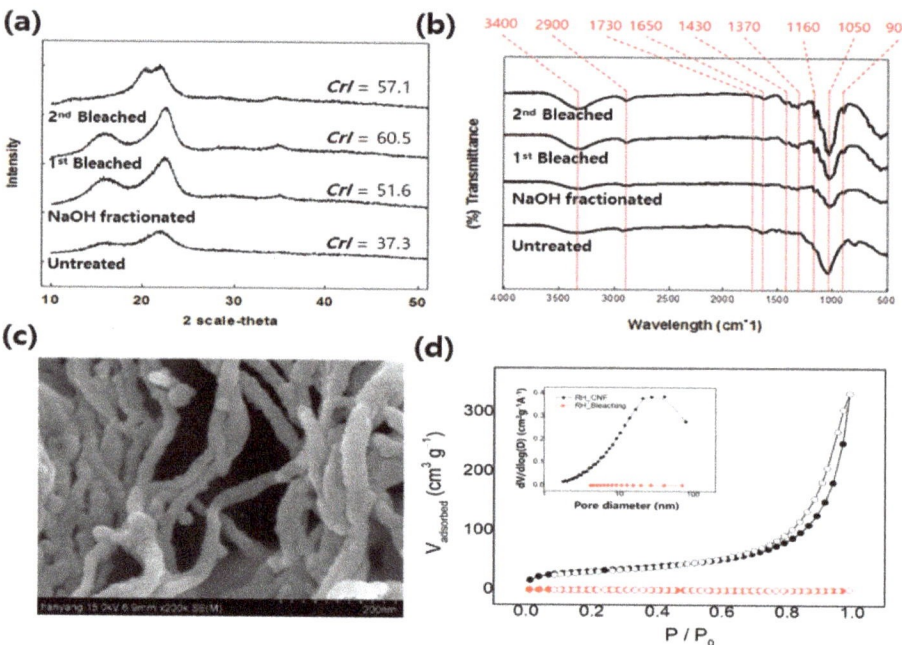

Figure 6. (a) XRD patterns, (b) FT-IR patterns, (c) SEM image of CNF, (d) N_2 physisorption isotherms and pore size distributions (inset) of NaOH + bleaching treated RH, CNF.

Figure 6b shows the FT-IR spectra of raw RH and RH fibers obtained by chemical bleaching for each stage. All samples exhibited major absorption areas in the range of 900–1600 cm^{-1}, 2900–3400 cm^{-1}. Compared to the raw RH raw material, the narrow and high peak of 3335–3344 cm^{-1} in the chemically treated RH fiber indicates that it contains more -OH groups. The 1159–1108 cm^{-1} peak, 1027–1051 cm^{-1} peak, and 896–897 cm^{-1} peak represent typical β-(1-4) glycosidic bonds in cellulose including C-O-C stretching and pyranose ring skeleton vibration. As the bleaching progressed, more significant effects were observed, which is believed to be due to the increase in the proportion of cellulose due to the removal of non-cellulosic substances by chemical bleaching. The peak at 1734 cm^{-1} in the spectrum of the RH raw material (a) is due to C=O bonding, and it was found that most of the hemicellulose and lignin components were removed as the peak disappeared in the treated RH fiber. In the CNF spectrum (d), characteristic peaks were 1429 and 1370 cm^{-1}, which were related to CH$_2$ bending and O-H bending, respectively.

As a result of the FE-SEM analysis, it was confirmed that the nano-fibrillation of the milled RH fibril was properly performed compared to the unmilled RH fiber (image is not shown) as shown in Figure 6c. Before colloid milling, the diameter of RH fiber was observed to be 10–50 μm, and after colloid milling, the diameter of RH fibril was observed to be 19–21 nm, confirming that the nano-sized fibril was properly formed.

In order to observe the structural characteristics of CNF, BET analysis was performed (Figure 6d), and the results of analysis of the specific surface area, pore volume, and pore size of RH fibrils before and after colloid milling are summarized in Table 4. The surface area of RH fibril before milling was 0.79 m^2 g^{-1}, the pore volume was 0.0026 cm^3 g^{-1}, and the pore size was 14.81 nm. After milling, the surface area, pore volume, the pore size of CNF were all increased; i.e., 110.52 m^{2}g^{-1}, 0.52 cm^{3}g^{-1}, and 15.82 nm, respectively. It was confirmed that the aspect ratio and specific surface area were greatly improved during colloid milling, and colloid milling can be used as an effective CNF manufacturing method.

Table 4. Effects of colloid milling treatment on characteristics of CNF.

Sample	BET Surface Area (m^2 g^{-1})	Pore Volume (cm^3)	Pore Size (nm)
Before milling treatment	0.79	<0.01	14.81
After milling treatment	110.52	0.52	15.82

3.7. Overall Mass Balance on the Comprehensive Utilization of RH

Figure 7 summarizes the simplified flow diagram and overall mass balance for each sequential process aimed at producing high value-added bio-based materials from RH. Based on 1000 g of RH input, the cellulose and hemicellulose, lignin, and ash contents were 356, 136, 227, and 157 g, respectively. Under the optimized fractionation conditions, approximately 64.5% of the input biomass mass was solubilized into the black liquor, with 26.2% cellulose, 64.7% hemicellulose, and nearly 100% of ash from the raw RH.

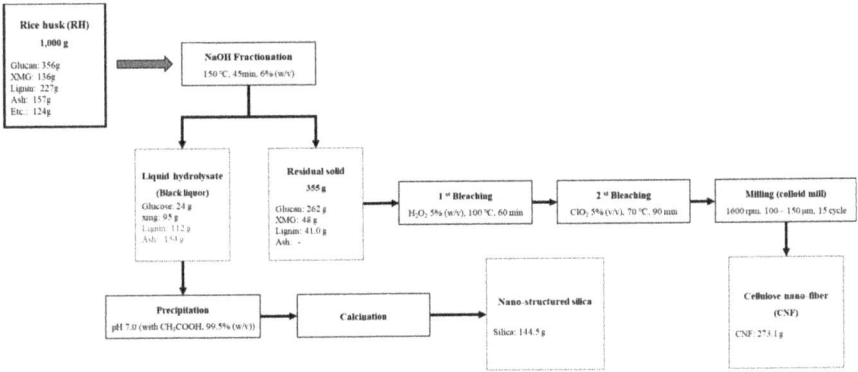

Figure 7. Simplified flow diagram of the RH process including the overall mass balance of RH fractionation process for cellulose nano-fibril and nano-structured silica.

Black liquor from alkaline fractionation contained 24 g of glucose, 95 g of hemicellulosic sugar, and 152 g of ash, and 262 g of cellulose and 48 g of hemicellulose were retained unhydrolyzed in 355 g of the residual solid based on 1 kg of raw RH. During the alkaline fractionation process, the undetected 79 g of glucose, which corresponded to 20.3% of cellulose, was presumed to be in the sugar filtrate as a cello-oligomer, because small amounts of degradation products from glucose, such as HMF and levulinic acid, were found in the analysis.

The black liquor resulting from the alkaline fractionation was subjected to consecutive precipitation and calcination, yielding 143 g of high-purity nano-structured silica.

In order to produce the CNF from the fractionated solid, a two-stage bleaching process using hydrogen peroxide and chlorine dioxide followed by colloid milling was performed, which resulted in 273.1 g of CNF (80.1% of cellulose content). After the bleaching process, 41.6 g (12.2%) of lignin was almost removed, and the whiteness of the cellulose could be enhanced. As a result, 96.3% of 283.5 g of cellulose was converted to CNF throughout the colloid milling process.

4. Discussion

For production of high value-added materials from lignocellulosic biomass, there have been many previous reports on the cellulose nano-crystals [53] and silica [54]. In addition, recent studies have focused on the opportunity to simultaneously fractionate and recover the useful components of biomass through unique fractionation methods using various catalytic solutions [55,56].

In this work, an alkaline fractionation was selected for the fractionation of the lignocellulosic materials, because the NaOH solutions are widely employed to remove lignin from wood and non-woody lignocellulosic materials [57,58]. It should be noted that the alkaline fractionation is also able to effectively solubilize hemicellulose, residual extractives, and ashes [59]. It is also important that the large amount of ash confers RH an outstanding recalcitrance toward biological and environmental threats. It was observed that the 30 L-bench scale alkaline fractionation completely removed ash from RH, in addition to the effective production of hemicellulosic sugars. We have already confirmed that the integration of mechanical and chemical pretreatment can be implemented in a large-scale reactor system [37].

RH is a naturally ash-rich agricultural by-product, which has more than 90% of this inorganic fraction as a pure silica [60], and the natural silica obtained from it represents an economically viable raw material that can be effectively used in life science and health care industries including cosmetics and pharmaceuticals. In this study, it was verified that silica can be recovered from the fractionated liquid phase generated and there was an advantage in recovering cellulose nanofibrils (CNF) in solid phase. CNF are attracting much attention due to their high nano-sized surface area, low density and high mechanical strength, as well as their solubility and biodegradability properties. In these days, CNF is promising to be used in a variety of applications in fields such as paper, composites, packaging, coatings, biomedicine, and automobiles due to its outstanding properties [61].

5. Conclusions

One of the most impeding negatives for the biorefinery industry is how to improve low economics. To overcome this problem, it is important to achieve effective utilization of lignocellulosic biomass and improve overall process economy by simultaneous production of high value-added materials. In our study, we have demonstrated the excellent properties, physical properties, and purity of the fractions produced through the fractionation method, which can maximize the added value of each component. In conclusion, it was confirmed that rice husk is a resource with very high economic potential. For this reason, this study attempted to increase the added value of by-products for the development of efficient biorefinery process technology, and is currently conducting research on technologies and applications such as separation/purification of high value-added silica and production of cellulose fiber.

Supplementary Materials: The following are available online at https://www.mdpi.com/2071-1050/13/4/1951/s1: Figure S1. Central composite experimental design and responses obtained for the alkaline fractionation of rice husk on the yields of xmg extraction and de-ashing, Figure S2. Analysis of variance (ANOVA) for the adjusted quadratic model for alkaline fractionation of rice husk on the xmg extraction yield and de-ashing yield.

Author Contributions: Conceptualization: H.J.J., J.C. and H.K.; methodology: H.J.J. and H.K.; validation: H.K.; formal analysis: H.J.J. and J.C.; investigation: H.J.J., J.C. and K.K.O.; resources: H.J.J.; data curation: H.J.J., J.C. and H.K.; writing—original draft preparation: K.K.O. and J.C.; writing—review and editing: K.K.O. and J.C.; visualization: H.J.J.; supervision: K.K.O.; project administration: K.K.O.; funding acquisition: K.K.O. and J.C. All authors have read and agreed to the published version of the manuscript.

Funding: This work was supported by the R&D program of Korea Institute of Energy Technology Evaluation and Planning (KETEP) grant funded by the Ministry of Trade, Industry, and Energy (MOTIE), the Republic of Korea (No. 20183030091950).

Institutional Review Board Statement: Not applicable.

Informed Consent Statement: Not applicable.

Data Availability Statement: Not applicable.

Conflicts of Interest: The authors declare that they have no known competing financial interests or personal relationships that could have appeared to influence the work reported in this paper.

References

1. Jiang, Z.; Hu, D. Molecular mechanism of anionic dyes adsorption on cationized rice husk cellulose from agricultural wastes. *J. Mol. Liq.* **2019**, *276*, 105–114. [CrossRef]
2. Mansilla, H.D.; Baeza, J.; Urzua, S.; Maturana, G.; Villase, J.; Duran, N. Acid catalysed hydrolysis of rice hull: Evaluation of furfural production. *Bioresour. Technol.* **1998**, *66*, 189–193. [CrossRef]
3. Menon, V.; Rao, M. Trends in bioconversion of lignocellulose: Biofuels, platform chemicals and biorefinery concept. *Prog. Energy Combust. Sci.* **2012**, *38*, 522–550. [CrossRef]
4. Barana, D.; Salanti, A.; Orlandi, M.; Ali, D.S.; Zoia, L. Biorefinery process for the simultaneous recovery of lignin, hemicelluloses, cellulose nanocrystals and silica from rice husk and Arundo donax. *Ind. Crops Prod.* **2016**, *86*, 31–39. [CrossRef]
5. Johar, N.; Ahmad, I.; Dufresne, A. Extraction, preparation and characterization of cellulose fibres and nanocrystals from rice husk. *Ind. Crops Prod.* **2012**, *37*, 93–99. [CrossRef]
6. Hanafi, E.; Khadrawy, H.; Ahmed, W.; Zaabal, M. Some observations on rice straw with emphasis on updates of its management. *World Appl. Sci. J.* **2012**, *16*, 354–361.
7. Arai, H.; Hosen, Y.; van Nguyen, P.H.; Thi, N.T.; Huu, C.N.; Inubushi, K. Greenhouse gas emissions from rice straw burning and straw-mushroom cultivation in a triple rice cropping system in the Mekong Delta. *Soil Sci. Plant Nutr.* **2015**, *61*, 719–735. [CrossRef]
8. Liu, C.; Lu, M.; Cui, J.; Li, B.; Fang, C.M. Effects of straw carbon input on carbon dynamics in agricultural soils: A meta-analysis. *Glob. Chang. Biol.* **2014**, *20*, 1366–1381. [CrossRef] [PubMed]
9. Ray, D.K.; Ramankutty, N.; Mueller, N.D.; West, P.C.; Foley, J.A. Recent patterns of crop yield growth and stagnation. *Nat. Commun.* **2012**, *3*, 1293–1300. [CrossRef] [PubMed]
10. Conrad, R. Microbial ecology of methanogens and methanotrophs. *Adv. Agron.* **2007**, *96*, 1–63. [CrossRef]
11. Salanti, A.; Zoia, L.; Orlandi, M.; Zanini, F.; Elegir, G. Structural characterization and antioxidant activity evaluation of lignins from rice husk. *J. Agric. Food. Chem.* **2010**, *58*, 10049–10055. [CrossRef] [PubMed]
12. Liou, T.H. Preparation and characterization of nano-structured silica from rice husk. *Mater. Sci. Eng. A.* **2004**, *364*, 313–323. [CrossRef]
13. Lemons e Silva, C.F.; Schirmer, M.A.; Maeda, R.N.; Barcelos, C.A.; Pereira, N. Potential of giant reed (Arundo donax L.) for second generation ethanol production. *Electron. J. Biotechnol.* **2015**, *18*, 10–15. [CrossRef]
14. Wei, G.Y.; Lee, Y.J.; Kim, Y.J.; Jin, I.H.; Lee, J.H.; Chung, C.H.; Lee, J.W. Kinetic study on the pretreatment and enzymatic saccharification of rice hull for the production of fermentable sugars. *Appl. Biochem. Biotechnol.* **2010**, *162*, 1471–1482. [CrossRef] [PubMed]
15. Yu, J.; Zhang, J.B.; He, J.; Liu, Z.D.; Yu, Z.N. Combinations of mild physical or chemical pretreatment with biological pretreatment for enzymatic hydrolysis of rice hull. *Bioresour. Technol.* **2009**, *100*, 903–908. [CrossRef]
16. Megawati, M.; Sediawan, W.B.; Sulistyo, H.; Hidayat, M. Kinetics of sequential reaction of hydrolysis and sugar degradation of rice husk in ethanol production: Effect of catalyst concentration. *Bioresour. Technol.* **2011**, *102*, 2062–2067. [CrossRef] [PubMed]

17. Banerjee, S.; Sen, R.; Pandey, R.A.; Chakrabarti, T.; Satpute, D.; Giri, B.S.; Mudliar, S. Evaluation of wet air oxidation as a pretreatment strategy for bioethanol production from rice husk and proces s optimization. *Biomass Bioenergy* **2009**, *33*, 1680–1686. [CrossRef]
18. Li, Y.; Ding, X.F.; Guo, Y.P.; Rong, C.G.; Wang, L.L.; Qu, Y.N.; Ma, X.Y.; Wang, Z.C. A new method of comprehensive utilization of rice husk. *J. Hazard. Mater.* **2011**, *186*, 2151–2156. [CrossRef]
19. Carmona, V.B.; Oliveira, R.M.; Silva, W.T.L.; Mattoso, L.H.C.; Marconcini, J.M. Nanosilica from rice husk: Extraction and characterization. *Ind. Crops Prod.* **2013**, *43*, 291–296. [CrossRef]
20. Kim, S.J.; Kim, T.H.; Oh, K.K. Deacetylation Followed by Fractionation of Yellow Poplar Sawdust for the Production of Toxicity-Reduced Hemicellulosic Sugar for Ethanol Fermentation. *Energies* **2018**, *11*, 404. [CrossRef]
21. Santana Costa, J.A.; Paranhos, C.M. Systematic evaluation of amorphous silica production from rice husk ashes. *J. Clean. Prod.* **2018**, *192*, 688–697. [CrossRef]
22. Karera, A.; Nargis, S.; Patel, M. Silicon-based materials from rice husk. *J. Sci. Ind. Res.* **1986**, *45*, 441–448.
23. Walcarius, A. Silica-based electrochemical sensors and biosensors: Recent trends. *Curr. Opin. Electrochem.* **2018**, *10*, 88–97. [CrossRef]
24. Bharti, C.; Nagaich, U.; Pal, A.K.; Gulati, N. Mesoporous silica nanoparticles in target drug delivery system. *Int. J. Pharm. Investig.* **2015**, *5*, 124–133. [CrossRef] [PubMed]
25. Burakov, A.E.; Galunin, E.V.; Burakova, I.V.; Kucherova, A.E.; Agarwal, S.; Tkachev, A.G.; Gupta, V.K. Adsorption of heavy metals on conventional and nanostructured materials for wastewater treatment purposes. *Ecotoxicol. Environ. Saf.* **2018**, *148*, 702–712. [CrossRef] [PubMed]
26. Xu, L.; Zhu, D.; Lu, X.; Lu, Q. Transparent, thermally and mechanically stable superhydrophobic coating prepared by an electrochemical template strategy. *J. Mater. Chem. A* **2015**, *3*, 3801–3807. [CrossRef]
27. Mebert, A.M.; Baglole, C.J.; Desimone, M.F.; Maysinger, D. Nanoengineered silica: Properties, applications and toxicity. *Food Chem. Toxicol.* **2017**, *109*, 753–770. [CrossRef]
28. Fernandes, I.J.; Calheiro, D.; Sanchez, F.A.L.; Camacho, A.L.D.; de Rocha, T.L.A.C.; Moraes, C.A.M.; de Sousa, V.C. Characterization of silica produced from rice husk ash: Comparison of purification and processing methods. *Mater. Res.* **2017**, *20*, 512–518. [CrossRef]
29. Habibi, Y. Key advances in the chemical modification of nanocelluloses. *Chem. Soc. Rev.* **2014**, *43*, 1519–1542. [CrossRef]
30. Zhu, G.; Lin, N. Surface Chemistry of Nanocellulose. In *Nanocellulose: From Fundamentals to Advanced Materials*; Wiley-VCH Verlag GmbH & Co: Weinheim, Germany, 2019; pp. 115–153.
31. Klemm, D.; Heublein, B.; Fink, H.P.; Bohn, A. Cellulose: Fascinating biopolymer and sustainable raw material. *Angew. Chem. Int. Ed.* **2005**, *44*, 3358–3393. [CrossRef] [PubMed]
32. Moon, R.J.; Martini, A.; Nairn, J.; Simonsen, J.; Youngblood, J. Cellulosenanomaterials review: Structure, properties and nanocomposites. *Chem. Soc.Rev.* **2011**, *40*, 3941–3994. [CrossRef] [PubMed]
33. Lavoine, N.; Desloges, I.; Dufresne, A.; Bras, J. Microfibrillated cellulose—itsbarrier properties and applications in cellulosic materials: A review. *Carbohydr. Polym.* **2012**, *90*, 735–764. [CrossRef] [PubMed]
34. Isogai, A. Wood nanocelluloses: Fundamentals and applications as newbio-based nanomaterials. *J. Wood Sci.* **2013**, *59*, 449–459. [CrossRef]
35. Osong, S.H.; Norgren, S.; Engstrand, P. Processing of wood-basedmicrofibrillated cellulose and nanofibrillated cellulose, and applications relating to papermaking: A review. *Cellulose* **2016**, *23*, 93–123. [CrossRef]
36. Chauve, G.; Bras, J. Industrial point of view of nanocellulose materials and their possible applications. In *Handbook of Green Materials*; World Scientific: Singapore, 2014; pp. 233–252. [CrossRef]
37. Kim, T.H.; Im, D.; Oh, K.K.; Kim, T.H. Effects of Organosolv Pretreatment Using Temperature-Controlled Bench-Scale Ball Milling on Enzymatic Saccharification of Miscanthus × giganteus. *Energies* **2018**, *11*, 2657. [CrossRef]
38. Lee, J.Y.; Ryu, H.J.; Oh, K.K. Acid-catalyzed hydrothermal severity on the fractionation of agricultural residues for xylose-rich hydrolyzates. *Bioresour. Technol.* **2013**, *132*, 84–90. [CrossRef]
39. Kim, D.Y.; Kim, Y.S.; Kim, T.H.; Oh, K.K. Two-stage, acetic acid-aqueous ammonia, fractionation of empty fruit bunches for increased lignocellulosic biomass utilization. *Bioresour. Technol.* **2016**, *199*, 121–127. [CrossRef]
40. Zhang, H.; Ding, X.; Chen, X.; Ma, Y.; Wang, Z.; Zhao, X. A new method of utilizing rice husk: Consecutively preparing d-xylose, organosolv lignin, ethanol and amorphous superfine silica. *J. Hazard. Mater.* **2015**, *291*, 65–73. [CrossRef]
41. Sluiter, A.; Hames, B.; Ruiz, R.; Scarlata, C.; Sluiter, J.; Templeton, D. *Determination of Sugars, Byproducts, and Degradation Products in Liquid Fraction Process Samples*; Laboratory Analytical Procedure (LAP) NREL/TP-510-42623; National Renewable Energy Laboratory: Golden, CO, USA, 2008.
42. Kim, S.J.; Um, B.H.; Im, D.J.; Lee, J.H.; Oh, K.K. Combined Ball Milling and Ethanol Organosolv Pretreatment to Improve the Enzymatic Digestibility of Three Types of Herbaceous Biomass. *Energies* **2018**, *11*, 2457. [CrossRef]
43. Kim, T.H.; Oh, K.K.; Ryu, H.J.; Lee, K.H.; Kim, T.H. Hydrolysis of hemicellulose from barley straw and enhanced enzymatic saccharification of cellulose using acidified zinc chloride. *Renew. Energy* **2014**, *65*, 56–63. [CrossRef]
44. Sindhu, R.; Kuttiraja, M.; Binod, P.; Sukumaran, R.K.; Pandey, A. Physicochemical characterization of alkali pretreated sugarcane tops and optimization of enzymatic saccharification using response surface methodology. *Renew. Energy* **2014**, *62*, 362–368. [CrossRef]

45. Nikzad, M.; Movagharnejad, K.; Talebnia, F.; Aghaiy, Z.; Mighani, M. Modeling of alkali pretreatment of rice husk using response surface methodology and artificial neural network. *Chem. Eng. Commun.* **2015**, *202*, 728–738. [CrossRef]
46. Shahabazuddin, M.; Chandra, T.S.; Meena, S.; Sukumaran, R.K.; Shetty, N.P.; Mudliar, S.N. Thermal assisted alkaline pretreatment of rice husk for enhanced biomass deconstruction and enzymatic saccharification: Physico-chemical and structural characterization. *Bioresour. Technol.* **2018**, *263*, 199–206. [CrossRef] [PubMed]
47. Schlomach, J.; Kind, M. Investigations on the semi-batch precipitation of silica. *J. Colloid Interface Sci.* **2004**, *277*, 316–326. [CrossRef] [PubMed]
48. Kracek, F.C. The system sodium oxide-silica. *J. Phys. Chem.* **2002**, *34*, 1583–1598. [CrossRef]
49. Sun, J.; Simon, S.L. The melting behavior of aluminum nanoparticles. *Thermochim. Acta* **2007**, *463*, 32–40. [CrossRef]
50. Lee, J.H.; Kwon, J.H.; Lee, J.W.; Lee, H.S.; Chang, J.H.; Sang, B.I. Preparation of high purity silica originated from rice husks by chemically removing metallic impurities. *J. Ind. Eng. Chem.* **2017**, *50*, 79–85. [CrossRef]
51. Azat, S.; Korobeinyk, A.V.; Moustakas, K.; Inglezakis, V.J. Sustainable production of pure silica from rice husk waste in Kazakhstan. *J. Clean. Prod.* **2019**, *217*, 352–359. [CrossRef]
52. Liu, Q.; Lu, Y.; Aguedo, M.; Jacquet, N.; Ouyang, C.; He, W.; Yan, C.; Bai, W.; Guo, R.; Goffin, D.; et al. Isolation of high-purity cellulose nanofibers from wheat straw through the combined environmentally friendly methods of steam explosion, microwave-assisted hydrolysis, and microfluidization. *ACS Sustain. Chem. Eng.* **2017**, *5*, 6183–6191. [CrossRef]
53. Rosa, S.M.L.; Rehman, N.; de Miranda, M.I.G.; Nachtigall, S.M.B.; Bica, C.I.D. Chlorine-free extraction of cellulose from rice husk and whisker isolation. *Carbohydr. Polym.* **2012**, *87*, 1131–1138. [CrossRef]
54. Chun, J.Y.; Gu, Y.M.; Hwang, J.K.; Oh, K.K.; Lee, J.H. Synthesis of ordered mesoporous silica with various pore structures using high-purity silica extracted from rice husk. *J. Ind. Eng. Chem.* **2020**, *81*, 135–143. [CrossRef]
55. Kaparaju, P.; Serrano, M.; Thomsen, A.B.; Kongjan, P.; Angelidaki, I. Bioethanol, biohydrogen and biogas production from wheat straw in a biorefinery concept. *Bioresour. Technol.* **2009**, *100*, 2562–2568. [CrossRef] [PubMed]
56. Choi, C.H.; Oh, K.K. Application of a continuous twin screw-driven process for dilute acid pretreatment of rape straw. *Bioresour. Technol.* **2012**, *110*, 349–354. [CrossRef] [PubMed]
57. Park, Y.C.; Kim, J.S. Comparison of various alkaline pretreatment methods of lignocellulosic biomass. *Energy* **2012**, *47*, 31–35. [CrossRef]
58. Stoklosa, R.J.; Hodge, D.B. Fractionation and improved enzymatic deconstruction of hardwoods with alkaline delignification. *Bioenergy Res.* **2015**, *8*, 1224–1234. [CrossRef]
59. Singh, D.P.; Trivedi, R.K. Acid and alkaline pretreatment of lignocellulosic biomass to produce ethanol as biofuel. *Int. J. Chem. Tech. Res.* **2013**, *5*, 727–734.
60. Park, B.; Wi, S.; Lee, K.; Singh, A.; Yoon, T.; Kim, Y. Characterization of anatomical features and silica distribution in rice husk using microscopic and micro-analytical techniques. *Biomass Bioenergy* **2003**, *25*, 319–327. [CrossRef]
61. Nechyporchuk, O.; Belgacem, M.N.; Bras, J. Production of cellulose nanofibrils: A review of recent advances. *Ind. Crops Prod.* **2016**, *93*, 2–25. [CrossRef]

Article

The Production of Biogenic Silica from Different South African Agricultural Residues through a Thermo-Chemical Treatment Method

Ncamisile Nondumiso Maseko [1], Denise Schneider [2], Susan Wassersleben [3], Dirk Enke [1,3], Samuel Ayodele Iwarere [4,*], Jonathan Pocock [1] and Annegret Stark [1,5]

1. Discipline of Chemical Engineering, University of KwaZulu-Natal, 238 Mazisi Kunene Road, Glenwood, Durban 4041, South Africa; masekon@ukzn.ac.za (N.N.M.); dirk.enke@uni-leipzig.de (D.E.); pocockj@ukzn.ac.za (J.P.); starka@ukzn.ac.za (A.S.)
2. 3P Instruments GmbH & Co. KG, Rudolf-Diesel-Str. 12, 85235 Odelzhausen, Germany; denise.schneider@3P-instruments.com
3. Institute of Chemical Technology, Universität Leipzig, Linnéstr. 3, 04103 Leipzig, Germany; susan.wassersleben@uni-leipzig.de
4. Department of Chemical Engineering, University of Pretoria, Lynnwood Road, Hatfield, Pretoria 0028, South Africa
5. SMRI/NRF SARChI Research Chair in Sugarcane Biorefining, Durban 4001, South Africa
* Correspondence: samuel.iwarere@up.ac.za; Tel.: +27-12-420-4111

Abstract: A thermo-chemical treatment method was used to produce biogenic amorphous silica from South African sugarcane and maize residues. Different fractions of South African sugarcane (leaves, pith, and fiber) were processed for silica production. The biomass samples were leached with either 7 wt% citric acid or 7 wt% sulfuric acid at 353 K for 2 h prior to being rinsed, dried and combusted using a four-step program ranging from room temperature to 873 K in a furnace. The characterization of the pre-treated biomass samples was conducted using thermogravimetric analysis (TG/DTA), X-ray fluorescence analysis (XRF) and elemental analysis (CHN), while the final products were characterized by XRF, X-ray diffraction (XRD), elemental analysis, nitrogen physisorption and scanning electron microscopy (SEM). Citric acid pre-treatment proved to be an attractive alternative to mineral acids. Amorphous biogenic silica was produced from sugarcane leaves in good quality (0.1 wt% residual carbon and up to 99.3 wt% silica content). The produced biogenic silica also had great textural properties such as a surface area of up to 323 m^2 g^{-1}, average pore diameter of 5.0 nm, and a pore volume of 0.41 cm^3 g^{-1}.

Keywords: biogenic amorphous silica; green chemistry; maize leaves; sugarcane fiber; sugarcane leaves; sugarcane pith

Citation: Maseko, N.N.; Schneider, D.; Wassersleben, S.; Enke, D.; Iwarere, S.A.; Pocock, J.; Stark, A. The Production of Biogenic Silica from Different South African Agricultural Residues through a Thermo-Chemical Treatment Method. *Sustainability* **2021**, *13*, 577. https://doi.org/10.3390/su13020577

Received: 5 December 2020
Accepted: 7 January 2021
Published: 9 January 2021

Publisher's Note: MDPI stays neutral with regard to jurisdictional claims in published maps and institutional affiliations.

Copyright: © 2021 by the authors. Licensee MDPI, Basel, Switzerland. This article is an open access article distributed under the terms and conditions of the Creative Commons Attribution (CC BY) license (https://creativecommons.org/licenses/by/4.0/).

1. Introduction

Plants can either be classified as silica accumulators or silica non-accumulators. Sugarcane, maize, and other types of *Gramineae* (grasses) fall under silica accumulators. Besides water, agricultural residues such as maize leaves or sugarcane leaves, pith, and fiber consist of cellulose, hemicellulose, lignin, as well as inorganic matter that is generally referred to as ash. Extensive research has been conducted to extract biogenic silica from different agricultural residues such as rice husk [1,2], wheat straw and cereal remnant [3], rice straw, oat husk and spelt husk [4], and corn cob [5].

Biogenic silica can be obtained, e.g., from diatomite. However, this procedure utilizes complex thermal and mechanical treatments that result in biogenic silica with impurities like CaO, FeO$_3$, and Al$_2$O$_3$ [6]. Hence the production of high-quality porous silica from silica accumulating plants has become an area of interest to most researchers [1–5]. For the production of silica from biomass, pre-treatment of the biomass by leaching prior to combustion is essential as it improves the purity of the ash by increasing the SiO_2 content while decreasing the share of all the other inorganic constituents (Cl^-, K_2O, P_2O_5, MgO, etc.) [2]. Acid leaching disrupts the lignin-carbohydrate matrix in order to enable cellulose hydrolysis [7]. Inorganic acids (e.g., sulfuric and hydrochloric acid) and bases (e.g., sodium hydroxide and ammonium hydroxide) have been used for biomass leaching in previous studies [1,8,9]. Inorganic acid leaching protocols, however, are environmentally unfriendly. They have several drawbacks as they require a lot of water to rinse the biomass post leaching, are not economical since they require the usage of expensive corrosive resistant processing equipment and in addition, a special disposal treatment of the used acids is required. Several researchers have therefore opted for the use of carboxylic acids as are bio-based and economical with no special equipment needed. During the leaching, the carboxylic group is involved in the successful removal of inorganic impurities via a chelating reaction with the metal ions [10].

The importance of the incineration temperature has also been emphasized in literature [3]. Silica can exist in an amorphous form or in a variety of crystalline forms: cristobalite, tridymite, and quartz. Amorphous silica can be partially converted to crystalline silica at approximately 1280 K in sugarcane-related biomass [11]. Higher temperatures cause the cellular microstructure to collapse, and that results in the merging of the small pores [12]. Consequently, a crystalline phase will form, leading to a reduced specific surface area [12]. The phase-transition temperature, however, is dependent on the presence of impurities in the sample; i.e., the higher the impurity content in a sample, the lower the temperature at which amorphous silica converts to crystalline silica. This makes the incineration protocol an important aspect in removing organics while retaining amorphous silica from biomasses.

Sugarcane (scientifically known as *Saccharum officinarum*) is one of the principal agricultural crops that are cultivated in tropical countries. The annual world production of sugarcane in 2018/2019 was approximately 1904 million tons (of which approximately 19 million tons are produced in South Africa [13]. From the produced volume, approximately 279 million metric tons of biomass residues are generated worldwide [14], with South Africa responsible for about 2 million metric tons [15]. Sugarcane residues mainly consist of leaves and bagasse, where the bagasse can be further separated into fiber and pith. Maize (*Zea mays* L.) is the most important grain crop in South Africa, which is produced throughout the country. Approximately 8 million tons of maize are produced in South Africa annually, and a significant percentage of leaves are produced as trash [16].

In South Africa, 90% of the sugarcane is burnt prior to harvest to facilitate cane cutting and to reduce fungus as well as other plant disease growth on the soil, while 10% is harvested green [17]. The burning of the sugarcane leaves prior to harvest is not environmentally friendly as it causes pollution and health-related issues [11]. Fiber and pith can be produced from excess sugarcane bagasse, which is otherwise used in boilers in the sugarcane mills to generate steam, giving an opportunity to process them to value added materials. Maize leaves are used to feed livestock in rural areas, while they have no significant use in most urban areas. Silica has been exploited in several applications across many fields such as electronics, image sensing, solar power conversion and as a precursor for silicon production (the birth of semiconductor evolution) [18]. Silica has been used in desiccant packets for specimen storage [19], the manufacturing of toothpaste [20], and to synthesize amorphous silica nanowires [21].

Several studies that utilize citric acid as a biomass leaching agent have been conducted by various researchers [2,4,10,22]. Selective results obtained from the literature are reported in Table 1. All the presented results in Table 1 are based on pre-treatment methods that employed only carboxylic/organic acids. Rice husk seems to be the most utilized biomass when it comes to the separation of biogenic silica from agricultural residues.

Table 1. Comparison of different citric acid pre-treatment conditions for the production of biogenic silica from different biomasses.

Type of Biomass	Pre-Treatment Acid	Pre-Treatment Temperature (K)	Silica Content (wt%)	BET Surface Area ($m^2\ g^{-1}$)	References
Rice husk	10% (v/v) citric acid	423	98.8	n.d.	[22]
Rice husk	5 wt% citric acid	323	99.1	n.d.	[10]
Rice husk	5 wt% citric acid	353	97.7	313	[2]
Rice Straw		323	99.1	264	[4]
Spelt husk	3.25 M citric acid		95.8	185	
Oat husk			99.1	248	
Horsetail			91.6	301	

n.d: not determined.

There is no substantial existing literature that focuses on the usage of either sugarcane or maize leaves residues. This study focuses on the valorization of South African maize (maize leaves) and sugarcane (sugarcane leaves, pith, and fiber) residues for the first time to produce amorphous biogenic silica using a thermo-chemical method that utilizes an organic acid. This study aimed to demonstrate the effectiveness of citric acid pre-treatment on the utilized South African biomass residues in terms of purity and textural properties despite the nature of the biomass. Hot water pre-treatment and inorganic acid (sulfuric acid) pre-treatment was also carried out for comparison sake. Despite the different physical appearance of these biomasses, they were all handled/utilized the same way in order to highlight the effectiveness of the employed thermochemical method using citric acid as a leaching agent to produce biogenic silica with high silica yield and impressive textural properties. The results obtained from all the utilized biomasses were presented based on ash purity, morphology, textural and structural characteristics.

2. Materials and Methods

2.1. Materials

Sugarcane and maize leaves were supplied by farmers in Durban, South Africa. Sugarcane fiber and pith were supplied by the Sugar Milling Research Institute (SMRI) in Durban, South Africa. Upon collection, the biomasses were dried in an oven at 375 K overnight to reduce the moisture content. After drying, the leaves were cut into pieces of about 1.5 cm long prior to any treatment, while the pith and the fiber were used as supplied.

2.2. Methods

2.2.1. Preparation of Biogenic Silica

In separate experiments, about 50 g of each agricultural residue biomass was washed with 1200 mL distilled water while being agitated for 2 h at ambient temperature and then dried in the oven at 373 K overnight prior to leaching with either 1200 mL hot distilled water, 7 wt% citric acid or 7 wt% sulfuric acid (in the case of sugarcane leaves) at 353 K for 2 h with continuous stirring. After leaching, the biomass was rinsed with deionized water to remove excess acid, followed by drying at 323 K for 24 h.

The dried, leached biomass was subjected to sequential combustion at a heating rate of 10 K min^{-1} in a furnace. The incineration program was as follows: 30 min at 583 K, 60 min at 723 K, 210 min at 783 K, and lastly 30 min at 873 K. After cooling of the ash, the resulting samples were kept at room temperature for characterization.

The samples were labeled sugarcane (SC) leaves-raw (for dried but unprocessed sugarcane leaves), SC leaves ash -HW (for the hot distilled water leached sugarcane leaves), SC leaves ash-CA (for citric acid leached sugarcane leaves), SC leaves ash-SA (for sulfuric acid leached sugarcane leaves), SC pith-raw (for dried unprocessed sugarcane pith), SC pith ash-CA (for citric acid leached sugarcane pith), SC fiber-raw (for dried but unprocessed sugarcane fiber), SC fiber ash-CA (for citric acid leached sugarcane fiber), Maize leaves-raw (for dried but unprocessed maize leaves) and Maize leaves ash-CA (for citric acid leached maize leaves).

2.2.2. Characterization of the Biogenic Silica

A Vario EL micro analyzer system (Heraeus, Hanau, Germany) was used for carbon content analysis, approximately 3 mg of the biomass sample was used. The elemental composition of the specimen was carried out using X-ray fluorescence (XRF) analysis (S4 Explorer, WDXRF Bruker, Karlsruhe, Germany). For this purpose, 0.75 g of the sample was mixed with 0.25 g of a wax powder. The mixture was ground to form a fine powder prior to being pressed with a hydraulic press (PerkinElmer, Germany) at 10 tons for 2 min to produce a round pellet with 20 mm diameter.

The phase identification was carried out using a Seifert XRD 7 apparatus that is equipped with Ni-filtered, Cu-Kα radiation (λ = 1.54 Å). Thermal analysis of the biomass samples was performed using a Thermo Gravimetric/Differential Thermal analyzer TG/DTA equipment (DST2960 simultaneous apparatus, TA Instrument, New Castle, DE, USA) that had a supplement air flow rate of 150 mL min^{-1} and a heating rate of 10 K min^{-1}. Approximately 7 mg of each sample was used for the analysis. The operating temperature range was between 293 and 1073 K, while the heating rate was 10 K min^{-1}. The surface morphology was studied by Scanning Electron Microscopy (SEM) using an Ultra 55 (Zeiss, Jena, Germany) that was operated at 10 keV. Nitrogen sorption analysis was used to determine the textural properties of the biomass samples. This was carried out using an ASAP 2010, Micromeritics, Nocross, GA, USA). Prior to measurements, the biomass samples were degassed for 12 h at 523 K under ultra-high vacuum. The relative pressure (p/p_0) of 0.995 was used to calculate the total pore volume. The Brunauer-Emmett-Teller (BET) model was used in the relative pressure of the range of (p/p_0) between 0.05 and 0.25 to evaluate the specific surface area. The value of 0.162 nm^2 was used as cross-sectional area of a nitrogen molecule. The adsorption branch of the isotherm was used to determine the pore size distribution by applying the Barret-Joyner-Halenda (BJH) method [23].

3. Results and Discussion

3.1. Thermal Analysis and Assessment of the Combustion Protocol

The elemental analysis (CHN) of the untreated biomass and produced silica ash samples is given in Table 2.

Table 2. Elemental analysis (CHN) for sugarcane and maize-based biomass, as well as the ash derived from the respective biomass samples.

Biomass	Nitrogen (%)	Carbon (%)	Hydrogen (%)
SC leaves—Raw	0.40 ± 0.00	39.50 ± 0.00	5.42 ± 0.10
SC leaves Ash-HW	0.031 ± 0.01	0.19 ± 0.00	0.42 ± 0.01
SC leaves Ash-CA	0.02 ± 0.00	0.14 ± 0.01	0.31 ± 0.00
SC leaves Ash-SA	0.01 ± 0.01	0.02 ± 0.01	0.12 ± 0.01
SC Pith-Raw	0.10 ± 0.01	45.30 ± 0.10	5.80 ± 0.05
SC Pith Ash-CA	0.01 ± 0.01	0.04 ± 0.01	0.38 ± 0.04
SC Fiber—Raw	0.03 ± 0.00	45.20 ± 0.10	5.90 ± 0.00
SC Fiber Ash-CA	0.01 ± 0.00	0.03 ± 0.00	0.14 ± 0.03
Maize leaves—Raw	0.10 ± 0.00	45.30 ± 0.10	5.80 ± 0.10
Maize leaves Ash-CA	0.03 ± 0.01	0.05 ± 0.01	0.37 ± 0.04

In order to assess the combustion method for the biomasses, thermogravimetric analysis (TGA) was conducted (air, 10 K min^{-1}, max. temperature 1073 K). Figure 1 shows the TG/DTA profiles of the utilized biomass samples, indicating similar decomposition patterns irrespective of the type of biomass. The first observed peak was at 339 K, 352 K, 334 K and 355 K with a weight loss of 3.7, 7.1, 6.3 and 5.2 wt% for sugarcane leaves, pith, maize leaves, and sugarcane fiber, respectively. This peak is due to the evaporation of moisture and other volatiles. The second observed peak (first degradation peak) occurred at 575 K, 578 K, 562 K and 554 K, with a weight loss of 59.8, 54.1, 60.6 and 58.7 wt% for sugarcane leaves, sugarcane pith, maize leaves and sugarcane fiber, respectively. This peak is due to the thermal degradation of hemicellulose [4].

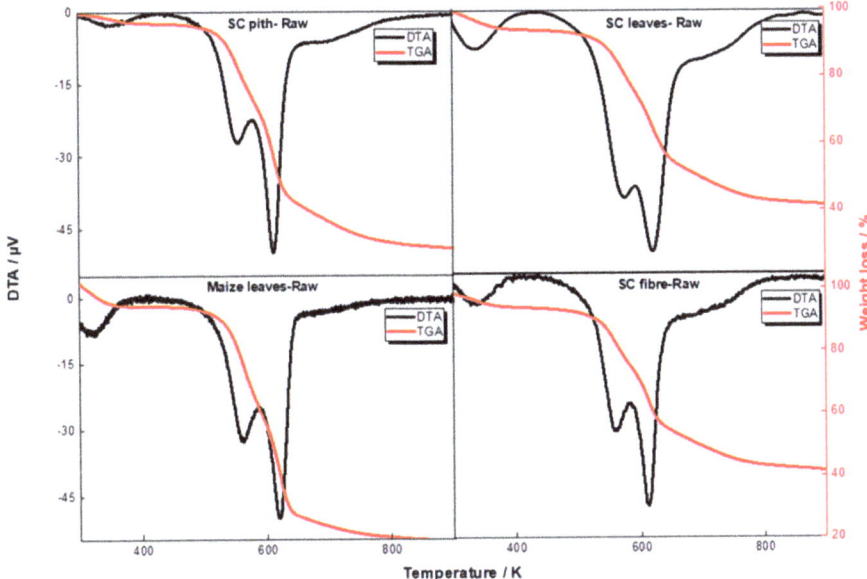

Figure 1. TG/DTA profiles of SC pith-Raw, SC leaves-Raw, Maize leaves-Raw and SC fiber-Raw.

Pure cellulose has been reported to thermally degrade at 628 K [10]. In the thermograms of the materials investigated herein, the third observed peak (2nd degradation peak) was at 619 K, 619 K, and 617 K with a weight loss of 29.9, 37.7, and 28.9 wt% for sugarcane leaves, pith, and maize leaves, respectively. Sugarcane fiber exhibited a somewhat lower degradation temperature of 607 K with a corresponding weight loss of 35.6 wt%. According to Alyosef et al. [2], the presence of high concentrations of potassium and other inorganic species found in biomass samples reduces the thermal stability of cellulose, in line with the results reported in this work: all the biomass samples contain potassium (see Table 3), with the highest potassium concentration found for sugarcane fiber.

Table 3. Chemical analysis of inorganics from sugarcane leaves, pith and fiber, as well as maize leaves, before and after pre-treatment with hot water or acids (determined by XRF).

Constituent	SC Leaves-Raw (wt%)	SC Leaves Ash-HW (wt%)	SC Leaves Ash-CA (wt%)	SC Leaves Ash-SA (wt%)	SC Pith-Raw (wt%)	SC Pith Ash-CA (wt%)	SC Fiber-Raw (wt%)	SC Fiber Ash-CA (wt%)	Maize Leaves-Raw (wt%)	Maize Leaves Ash-CA (wt%)
SiO_2	63.8	87.6	95.4	99.3	39.2	94.8	33.7	94.3	51.4	93.0
P_2O_5	1.4	0.7	0.4	0.1	1.3	1.1	4.0	2.6	5.3	3.3
K_2O	3.8	1.4	0.2	<LOD	2.4	0.1	12.0	0.1	5.8	0.3
CaO	14.9	7.5	2.6	0.2	19.0	0.4	14.9	0.6	21.4	1.3
MgO	1.5	1.2	0.4	<LOD	2.2	0.3	2.9	0.9	2.0	0.4
SO_3	4.5	0.9	0.8	0.3	4.4	0.3	7.3	0.4	5.5	0.4
Fe_2O_3	2.3	0.2	0.1	<LOD	5.7	0.9	2.5	0.2	2.4	0.3
Al_2O_3	3.1	0.1	0.1	0.1	11.7	0.8	12.4	0.2	1.7	0.6
Cl	1.7	<LOD	<LOD	<LOD	5.6	<LOD	6.2	<LOD	0.8	<LOD
Others [a]	3.0	0.4	0.0	0.0	8.5	1.3	4.1	0.7	3.7	0.4

Others: [a] Other metal oxides; LOD: limit of detection = 0.1 ppm.

Kumar et al. [24] and Yang et al. [25] reported the degradation of lignin to take place over a wide temperature range of 453–1173 K, concurrent with both the 1st and 2nd degradation peaks.

Hence, for the sequential combustion protocol, the first programmed temperature was kept at 583 K for 30 min to allow for the removal of residual water from the sample. The second step was performed at 723 K for 60 min. This focused on thermal degradation of organic components and the transformation of cellulose, hemicellulose, and lignin into carbon/CO_2, which can be noticed by the formation of smoke. However, the smoke ceases after this step. The third step kept the temperature at 783 K to achieve the complete combustion of carbon in the ash, hence resulting in a white ash [2].

3.2. The Influence of Leaching on the Chemical and Elemental Composition

Elemental analysis (CHN) of the untreated biomass and produced silica ash samples was performed to determine their chemical composition, where the focus was on carbon, hydrogen, and nitrogen (Table 2). Three measurements were carried out, an average of each set of analyses was considered and a standard deviation calculated. As expected, the biomass samples have a high carbon content, with the lowest found for untreated sugarcane leaves. Incidentally, sugarcane leaves have the highest ash content and the lowest loss on ignition when compared to the other biomass samples, as shown below. As expected, the thermo-chemical treatment reduces the content of the selected three elements to below 0.5 wt%, as indicated in Table 2.

Although citric acid is a weaker acid than sulfuric acid, it reduced the carbon content significantly by >99.7% for all biomasses, while the nitrogen contents were reduced by >68.0%, respectively. Citric acid leaching and subsequent combustion produced very white ashes, indicating almost quantitative combustion and removal of organic components. In comparison, pre-treatment of sugarcane leaves with a strong inorganic acid (7 wt% sulfuric acid; -SA) lead to a quantitative removal of carbon, and a reduction of >97.0% for nitrogen. In addition, the pre-treatment of sugarcane leaves with hot water (HW) resulted in 99.5% reduction of carbon and 92.2% for that of nitrogen.

The chemical analysis of the biomass samples was performed using XRF to study the influence of chemical treatment on the biomass samples. The results obtained are shown in Table 3. The leaching with hot citric acid resulted in white ashes with high silica contents of more than 94 wt% for all the sugarcane-based biomass samples and 93 wt% for the maize leaves sample.

The effect of hot water leaching (-HW) was investigated using sugarcane leaves, and it was found that even in the absence of an acid, the silica content in the formed ash increased to 87.6 wt% while the chlorine was reduced below the detection limit. The extraction of aluminum, iron, and sulfur was as effective as when using citric acid pre-treatment, but hot water leaching was not that effective in the reduction of phosphorus, alkali and alkali earth metals.

When comparing pre-treatments of sugarcane leaves with citric and sulfuric acid, a silica purity of 95.4 wt% was obtained from citric acid pre-treatment while sulfuric acid pre-treatment resulted in 99.3 wt% as illustrated in Table 3. This silica content obtained from sulfuric acid pre-treatment of the sugarcane leaves is incredible when compared to those from other agricultural residues. Cereal remnant, wheat straw, and miscanthus were all pre-treated with sulfuric acid and their resulting biogenic silica was reported to be 91.3 wt%, 92.8 wt% and 95.0 wt%, respectively [3]. Alyosef at al. [2] reported a high silica content of 97.7 wt% from the pre-treatment of rice husk but is still lower than the one reported in this work. Based on the results obtained from this study, it indicates that for applications where a high purity of biogenic silica is essential, only the silica obtained from the pre-treatment of sugarcane leaves with sulfuric acid can be utilized. The biogenic silica obtained through the pre-treatment of sugarcane leaves with citric acid pre-treatment can only be used if the employed thermochemical method is optimized through varying variables like time, temperature, acid concentration, etc. to obtain silica with an increased purity. Sulfuric acid leaching resulted in a reduction of potassium, manganese, iron, and chloride below the detection limits. Phosphorus and sulfur were both reduced by 93%, while aluminum and calcium were reduced by 97 and 99%, respectively. For citric acid, phosphorus, alkali earth metals, and sulfur were removed only between 71–83%, while potassium, iron, and aluminum contents were reduced by 95–97%.

It is known that acid treatment leads to the hydrolysis of hemicellulose and cellulose, the degree of which depends on the medium acidity as well as time and temperature. Sulfuric acid treatment is hence a more efficient hydrolysis catalyst, increasing the extractability of inorganics from biomass [26]. However, the extraction selectivity observed for the citric acid-treated sample indicates specific interactions between carboxylic acid groups of citric acid with some the metal ions. These interactions are known to occur by chelation [9,25–30].

The biogenic silica obtained from this study using citric acid as a leaching agent demonstrated a higher purity of silica when compared to the purity of other residues from sugarcane and maize reported in literature. For an example, Worathanakul et al. [31] pre-treated sugarcane bagasse (i.e., a mixture of fiber and pith) with 3M HCl to produce biogenic silica with a silica content of 89 wt% compared to 94.8 and 94.3 wt% for pith and fiber obtained in this work, respectively. On another study, [32] pre-treated sugarcane bagasse with 1M HCl and obtained an even lower silica content of 66 wt%. In yet another recent study a silica content of 85.6 wt% was obtained from the pre-treatment of sugarcane bagasse with 9% sulfuric acid [33]. Similarly, for maize leaves pre-treated with HCl, a silica content of 64 wt% was obtained by Lanning et al. [34], compared to 93 wt% obtained from this study. Considering the above argument regarding the effect of the acidity of the medium, this finding is somewhat surprising. However, it is noted that the biomass investigated here had been subjected to a sequence of drying, washing, and drying prior to acid treatment, which may affect the efficiency of acid treatment [2]. This demonstrates the incredible ability of hot citric acid pre-treatment in significantly removing/reducing organic compounds and accompanying metal oxides in the biomasses.

3.3. Phase Identification

XRD analysis was performed to identify different phases of the produced biogenic silica from the pre-treatment of sugarcane-based biomass (leaves, fiber and pith) and maize leaves as displayed in Figure 2. Judging from the broad peak with an equivalent Bragg angle of 2θ at $21.8°$ of the sugarcane-based samples, it can be deduced that the silica formed was amorphous with almost no traces of crystalline phases [12]. This finding indicates that the combustion protocol is indeed suitable for producing amorphous silica from sugarcane residues. Contrary to this, a presence of additional peaks for biogenic silica produced from maize leaves was observed. These peaks are in addition to the detected broad peak with an intensity of Bragg angle of $2\theta = 21.8°$ and the internal standard (CaF_2) peaks. This is an indication of the ash not being entirely amorphous. The peaks at $2\theta = 20.9°$ and $2\theta = 28.6°$ correspond to crystalline silica, quartz to be specific [35]. The presence of quartz in this biogenic silica is suspected to be due to the incomplete removal of the sand/impurities during the washing process.

3.4. Textural and Structural Properties

The textural properties of both sugarcane and maize-based biomass samples were obtained from nitrogen sorption measurements (Figure 3). The isotherms have a closed hysteresis loop and consist of a small knee at $p/p_0 = 0.05$ that is followed by a continuous increase in the nitrogen uptake. All the biomass ash samples followed a similar pattern categorized as class IV isotherms [16].

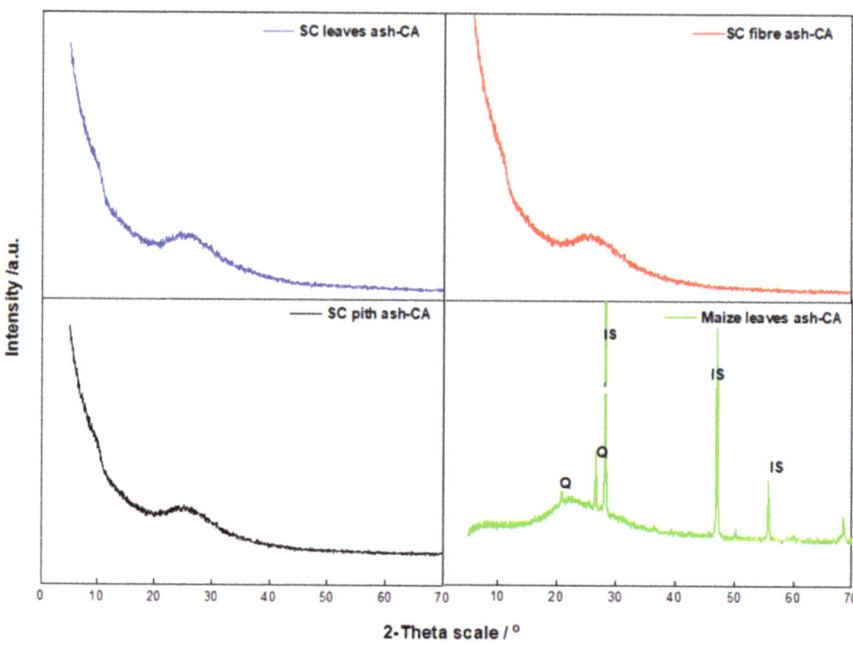

Figure 2. XRD patterns of sugarcane-based biomass (leaves, fiber, and pith) and maize leaves. IS is the internal standard (CaF_2) and Q is the quartz.

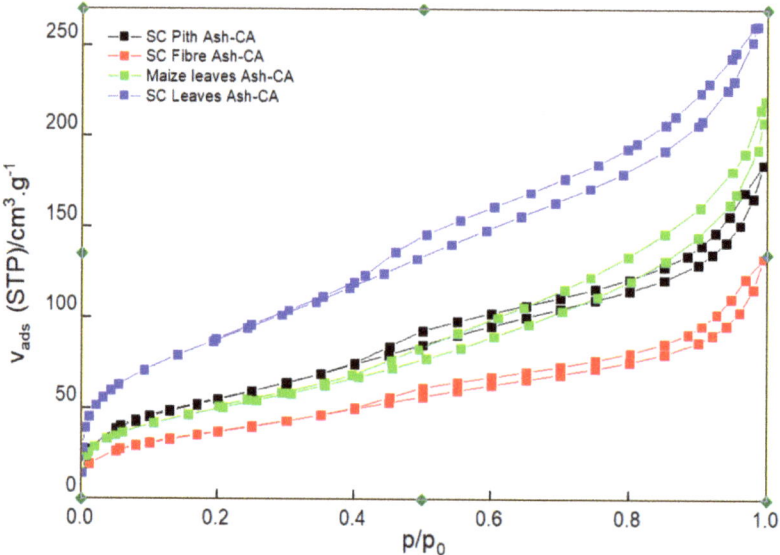

Figure 3. Nitrogen sorption isotherms of biogenic silica derived from sugarcane biomass (leaves, pith and fiber) and maize leaves.

The particle pores can either be micropores (≤2 nm), mesopores (2–50 nm) or macropores (≥50 nm). Each of these is characterized with a specific adsorption isotherm. The opening of micropores consists of a diameter of a few molecules and that causes the potential fields of the connection pore walls to overlap with one another [36]. The interaction between the adsorbent and the molecule is therefore automatically intensified, which results in a steep increase in adsorption quantity even at low relative pressures. The adsorption, however, levels off eventually due to the achieved saturation pressure [37], resulting in a long plateau in the isotherm over a wide range of relative pressures [35]. There is a presence of a hysteresis loop with a lower closure point at about $p/p_0 = 0.42$–0.45, which can be classified as H3 [37] as is the case for all four examined samples of biomass ash samples. This indicates that there is a presence of additional macropores and also stipulates that accessibility of the larger mesopores can only be through the smaller mesopores [4]. In the case of the sugarcane leaves ash sample (citric acid-treated), additional microporosity cannot be excluded. In addition, none of the isotherms reach a plateau at high $p/p_0 = 1$ values. This is a characteristic for incomplete pore filling by an adsorbate and suggests that larger meso- and macropores are also present in the materials [4,36].

The textural properties, including BET surface area (A_{BET}), average pore size (D_p), and pore volume (V_{meso}) of all porous silica products are given in Table 4. The citric acid pre-treated sugarcane leaves ash sample is characterized by the largest BET surface area of 323 $m^2 \cdot g^{-1}$. This is quite a competitive specific surface area when comparing it with that of other agricultural residues from the literature: rice husk ash and cereal remnants, after pre-treatment with citric acid, showed specific surface areas of 313 $m^2 \cdot g^{-1}$ [2] and 245 $m^2 \cdot g^{-1}$ [3], respectively. Citric acid leached sugarcane fiber, pith and maize leaves had BET surface areas of 136, 203 and 182 $m^2 \cdot g^{-1}$. Even though sugarcane fiber and the pith are from the same material (bagasse), their textural properties differ. The average pore diameter of the produced ash is 5 nm for sugarcane leaves, both sugarcane pith and fiber have 6 nm while the maize leaves have an average pore diameter of 7 nm. Sugarcane fiber has the lowest pore volume of 0.21 $cm^3 \cdot g^{-1}$, while sugarcane leaves show the highest value of 0.41 $cm^3 \cdot g^{-1}$.

Table 4. Textural properties of sugarcane and maize-based biomass ash samples (determined by nitrogen sorption).

Sample	A_{BET} ($m^2\,g^{-1}$)	D_p (nm)	V_{meso} ($cm^3\,g^{-1}$)
SC leaves Ash-CA	323	5.0	0.41
SC leaves Ash-SA	326	4.7	0.47
SC leaves Ash-HW	133	8.0	0.25
SC Pith Ash-CA	203	6.0	0.29
SC Fiber Ash-CA	136	6.0	0.21
Maize leaves Ash-CA	182	7.0	0.34

The effect of hot water and sulfuric acid leaching on the textural properties was also investigated using sugarcane leaves. Although the BET surface area of hot water leached sugarcane leaves is relatively low (133 $m^2\,g^{-1}$), with average pore diameter of 8.0 nm and a pore volume of 0.25 $cm^3\,g^{-1}$, it is noteworthy that it is similar to that of citric acid-treated sugarcane fiber. Compared to citric acid pre-treatment, sulfuric acid leaching resulted in a slight increase of BET surface area (0.9%) but decreased the pore diameter by 6% while the pore volume increased by 15%. It can be stated that the biogenic silica ashes produced from sugarcane leaves with either citric or sulfuric acid fall within the specifications of commercial products such as silica gels (A_{BET} = 250–1000 $m^2\,g^{-1}$, DP = 2–20 nm, narrow pore width distribution) [38].

The pore size distribution of the pre-treated biomass samples was also determined using the adsorption branch of the nitrogen sorption isotherms (Figure 4), indicating a broad pore size distribution of disordered meso and macropores.

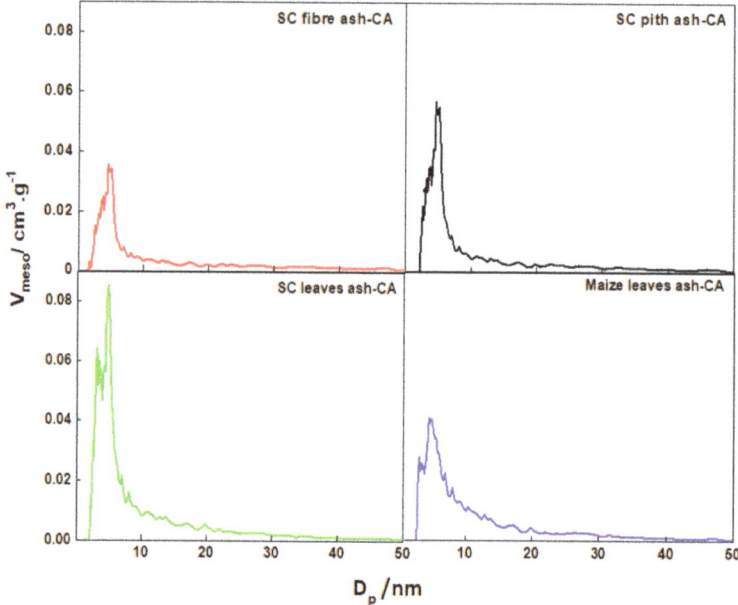

Figure 4. Pore size distribution of biogenic silica derived from sugarcane biomass (fiber, pith and leaves) and maize leaves (determined by an adsorption branch of nitrogen sorption isotherms).

Figure 5. SEM pictures of biogenic silica derived from sugarcane leaves after pre-treatment with citric acid and subsequent burning from different parts of inner epidermis with high (**A**,**B**), intermediate (**C**) and low magnification (**D**).

Figure 5A–D show the SEM images of the produced ash from sugarcane leaves leached with citric acid, demonstrating that the produced biogenic silica ash possesses numerous types of pore shapes and sizes. Figure 5A visualizes the evident porosity of the ash at higher magnification while Figure 5B indicates agglomeration of small particles. Intraparticle meso- and macropores can be observed from Figure 5A. Figure 5C (intermediate magnification) and Figure 5D (low magnification) show a broken inner epidermis from different parts with several shapes such as double rods and backbone structure. However, these fragments seem to have retained their original structure. In addition, a smooth surface is observed.

Figure 6A–D present the SEM images of biogenic silica ash resulting from the pre-treatment of maize leaves with citric acid, indicating the presence of different pore shapes and sizes. Figure 6A demonstrates intraparticle macropores in inner epidermis at a high magnification. Figure 6B shows agglomeration of small particles, while Figure 6C,D show the broken structure of the inner epidermis that has an intact silicified structure, and this is shown from different parts in different magnifications.

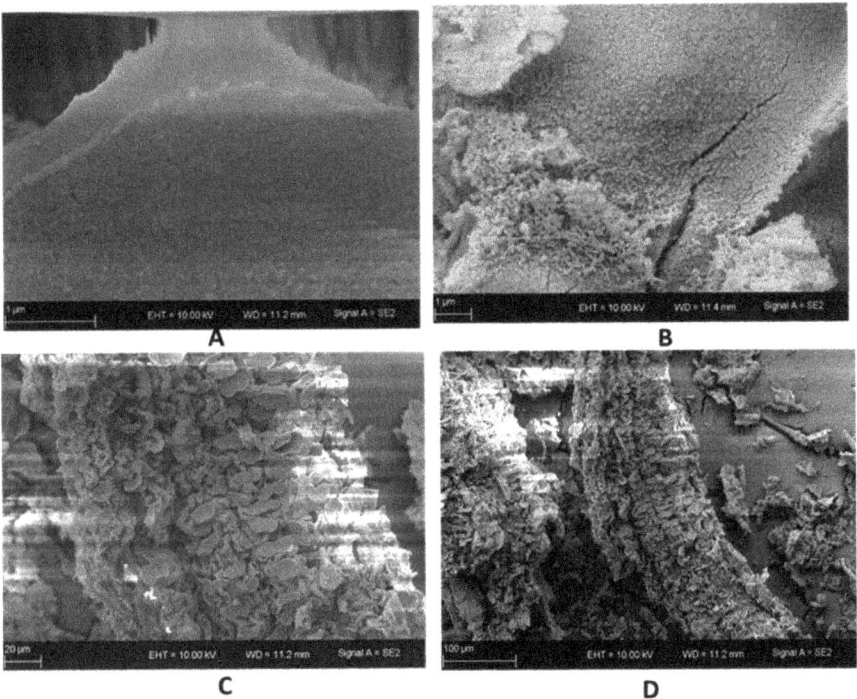

Figure 6. SEM pictures of biogenic silica derived from maize leaves after pre-treatment with citric acid and subsequent burning from different parts of inner epidermis with high (**A,B**), intermediate (**C**) and low magnification (**D**).

4. Conclusions

This study demonstrates the production of biogenic silica with high purity (99.3 wt% silica content) and exceptional textural properties. Different South African agricultural residues (sugarcane leaves, pith and fiber, as well as maize leaves) were utilized for the first time in the production of amorphous biogenic silica through a thermo-chemical procedure that employed hot citric acid leaching and subsequent combustion. Furthermore, the possibility of using sugarcane-based materials as an additional feedstock in the production of biogenic silica was demonstrated.

When compared to sulfuric acid, citric acid pre-treatment of sugarcane leaves resulted in fully amorphous silica with excellent quality, with 99.7% removal of carbon and 95.4 wt% silica content, which is higher than the purity obtained from other agricultural residues in the literature such as wheat straw, miscanthus and cereal remnant [3]. While the removal of inorganic impurities was somewhat less efficient with citric acid than with sulfuric acid pre-treatment (silica content of 99.3 wt% vs. 95.4 wt%), similar textural properties (323 m^2 g^{-1} surface area, 5.0 nm average pore diameter and 0.41 cm^3 g^{-1} pore volume) were still obtained from the produced amorphous biogenic silica. The advantage of citric acid lies in it being an organic and bio-based acid, which affects both costs of effluent treatment and material of construction when compared to strong mineral acids such as sulfuric or hydrochloric acid.

When comparing the citric acid pre-treatment of all the utilized biomasses, the sugarcane-based starting materials resulted in biogenic silica with higher purity (94.3–95.4 wt%) compared to that of maize leaves (93.0 wt%). Sugarcane leaves provided a material that consisted of best textural properties than sugarcane fiber and pith, as well as maize leaves.

Our study focused on biogenic silica production; hence, future research will investigate if an additional extraction step with citric acid could lead to further improvement regarding the removal of inorganic impurities.

Considering the high sugar production in South Africa with an export markets in Africa, the Middle East, Korea, Japan, and the North America, the continuous generation of sugarcane trash is predicted to increase. Therefore, sugarcane-based starting materials can be used to expand the feedstock for the production of biogenic silica. In addition, the production of biogenic silica from sugar cane trash will be a plus in solving disposal problems, and thus contribute to the mitigation of climate change and other environmental impacts that could result from the burning of its trash.

Finally, with growing potentials and opportunities of sugarcane agricultural residues in sugarcane-growing countries like South Africa, our future research will assess the effect of upscaling on the morphology of the products in terms of the potential of generating energy during combustion while producing biogenic silica, especially from sugarcane leaves.

Author Contributions: Conceptualization, S.W. and A.S.; Data curation, N.N.M., D.S. and S.W.; Investigation, N.N.M., Methodology, D.S. and D.E.; Supervision, D.E., S.A.I., J.P. and A.S.; Validation, D.E.; Visualization, D.E., S.A.I. and A.S.; Writing—original draft preparation, N.N.M.; Writing—review and editing, D.S., D.E., S.A.I., J.P., and A.S., funding acquisition, National Research Foundation of South Africa. All authors have read and agreed to the published version of the manuscript.

Funding: This research was funded in part by the National Research Foundation of South Africa, grant number 115465.

Acknowledgments: We appreciate EUROSA scholarship for enabling the collaboration that made this manuscript possible. We thank the Sugar Milling Research Institute (SMRI) for the provision of sugarcane pith and fiber, W. Bernhardt (UKZN) for sourcing sugarcane leaves and R. Denecke (University of Leipzig) for laboratory support.

Conflicts of Interest: The authors declare no conflict of interest.

References

1. Umeda, J.; Kondoh, K.; Michiura, Y. Process Parameters Optimization in Preparing High-Purity Amorphous Silica Originated from Rice Husks. *Mater. Trans.* **2007**, *48*, 3095–3100. [CrossRef]
2. Alyosef, H.A.; Eilert, A.; Welscher, J.; Ibrahim, S.S.; Denecke, R.; Schwieger, W.; Enke, D. Characterization of Biogenic Silica Generated by Thermo Chemical Treatment of Rice Husk. *Part. Sci. Technol.* **2013**, *31*, 524–532. [CrossRef]
3. Alyosef, H.A.; Schneider, D.; Wassersleben, S.; Roggendorf, H.; Weiß, M.; Eilert, A.; Denecke, R.; Hartmann, I.; Enke, D. Meso/Macroporous Silica from Miscanthus, Cereal Remnant Pellets, and Wheat Straw. *ACS Sustain. Chem. Eng.* **2015**, *3*, 2012–2021. [CrossRef]
4. Schneider, D.; Wassersleben, S.; Weiß, M.; Denecke, R.; Stark, A.; Enke, D. A Generalized Procedure for the Production of High-Grade, Porous Biogenic Silica. *Waste Biomass Valorization* **2020**, *11*, 1–15. [CrossRef]

5. Okoronkwo, E.A.; Imoisili, P.E.; Olubayode, S.A.; Olusunle, S.O.O. Development of Silica Nanoparticle from Corn Cob Ash. *Adv. Nanopart.* **2016**, *5*, 135–139. [CrossRef]
6. Alyosef, H.A.; Ibrahim, S.; Welscher, J.; Inayat, A.; Eilert, A.; Denecke, R.; Schwieger, W.; Münster, T.; Kloess, G.; Einicke, W.D.; et al. Effect of acid treatment on the chemical composition and the structure of Egyptian diatomite. *Int. J. Miner. Process.* **2014**, *132*, 17–25. [CrossRef]
7. Kootstra, A.M.J.; Beeftink, H.H.; Scott, E.L.; Sanders, J.P. Comparison of dilute mineral and organic acid pretreatment for enzymatic hydrolysis of wheat straw. *Biochem. Eng. J.* **2009**, *46*, 126–131. [CrossRef]
8. Yalçin, N.; Sevinç, V. Studies on silica obtained from rice husk. *Ceram. Int.* **2001**, *27*, 219–224. [CrossRef]
9. Mupa, M.; Hungwe, C.B.; Witzleben, S.; Mahamadi, C.; Muchanyereyi, N. Extraction of silica gel from Sorghum bicolour (L.) moench bagasse ash. *Afr. J. Pure Appl. Chem.* **2015**, *9*, 12–17. [CrossRef]
10. Umeda, J.; Kondoh, K. High-purity amorphous silica originated in rice husks via carboxylic acid leaching process. *J. Mater. Sci* **2008**, *43*, 7084–7090. [CrossRef]
11. Le Blond, J.S.; Horwell, J.C.; Williamson, B.J.; Oppenheimer, C. Generation of crystalline silica from sugarcane burning. *J Env. Monit.* **2010**, *12*, 1459–1470. [CrossRef]
12. Musić, S.; Filipović-Vinceković, N.; Sekovanić, L. Precipitation of amorphous SiO_2 particles and their properties. *Braz. J. Chem. Eng.* **2011**, *28*, 89–94. [CrossRef]
13. Sisuka, W. Annual South African Sugar Production Forecast to Grow Despite Revenue Pressures. Available online: https://apps.fas.usda.gov/newgainapi/api/report/downloadreportbyfilename?filename=Sugar%20Annual_Pretoria_South%20Africa%20-%20Republic%20of_4-15-2019.pdf (accessed on 11 January 2021).
14. Chandel, A.; Da Silva, S.; Carvalho, W.; Singh, O. Sugarcane bagasse and leaves: Foreseeable biomass of biofuel and bio-products. *J. Chem. Technol. Biotech.* **2011**, *87*, 11–20. [CrossRef]
15. Smithers, J. Review of sugarcane trash recovery systems for energy cogeneration in South Africa. *Renew. Sustain. Energy Rev.* **2014**, *32*, 915–925. [CrossRef]
16. Du Plessis, J. *Maize Production*; Department of Agriculture South Africa: Pretoria, South Africa, 2003.
17. Wilkonson, D. Industrial guidelines for burning sugarcane. In *SASRI Information Sheets*; Sugar Research Institute: Durban, South Africa, 2013; pp. 1–2.
18. Street, R.A. *Technology and Applications of Amorphous Silicon*; Springer: Berlin, Germany, 2000.
19. Joshi, H.H.; Gertz, R.E.; Carvalho, M.D.G.; Beall, B.W. Use of silica desiccant packets for specimen storage and transport to evaluate pneumococcal nasopharyngeal carriage among nepalese children. *J. Clin. Microbiol.* **2008**, *46*, 3175–3176. [CrossRef]
20. Joiner, A. A silica toothpaste containing blue covarine: A new technological breakthrough in whitening. *Int. Dent. J.* **2009**, *59*, 284–288. [CrossRef]
21. Yu, D.P.; Hang, Q.L.; Ding, Y.; Zhang, H.Z.; Bai, Z.G.; Wang, J.J.; Zou, Y.H.; Qian, W.; Xiong, G.C.; Feng, S.Q. Amorphous silica nanowires: Intensive blue light emitters. *Appl. Phys. Let.* **1998**, *73*, 3076–3078. [CrossRef]
22. Carmona, V.; Oliveira, R.; Silva, W.; Mattoso, L.; Marconcini, J. Nanosilica from rice husk: Extraction and characterization. *Ind. Crop. Prod.* **2013**, *43*, 291–296. [CrossRef]
23. Barrett, E.; Joyner, L.; Halenda, P. The determination of pore volume and area distributions in Porous Substances. I. computations from nitrogen isotherms. *J. Am. Chem. Soc.* **1951**, *73*, 373–380. [CrossRef]
24. Kumar, A.; Wang, L.; Dzenis, Y.A.; Jones, D.D.; Hanna, M.A. Thermogravimetric characterization of corn stover as gasification and pyrolysis feedstock. *Biomass Energy* **2008**, *32*, 460–467. [CrossRef]
25. Yang, H.; Yan, R.; Chen, H.; Lee, D.; Zheng, C. Characteristics of hemicellulose, cellulose and lignin pyrolysis. *Fuel* **2007**, *86*, 1781–1788. [CrossRef]
26. Liou, T. Preparation and characterization of nano-structured silica from rice husk. *Mater. Sci. Eng.* **2004**, *364*, 313–323. [CrossRef]
27. Fang, F. *Pretreatment Techniques for Biofuels and Biorefineries*, 1st ed.; Springer: Berlin/Heidelberg, Germany, 2015; pp. 158–159.
28. Umeda, J.; Imai, H.; Kondoh, K. Polysaccharide hydrolysis and metallic impurities removal behavior of rice husks in citric acid leaching treatment. *Trans. JWRI* **2009**, *38*, 13–18.
29. Faizul, C.; Chik, A.; Bari, M.; Noorina, H. Extraction of silica from palm ash using organic acid leaching treatment. *Key Eng. Mat.* **2013**, *7*, 3690–3695. [CrossRef]
30. Boochapun, S.; Lamamorphanth, W.; Kamwilaisak, K. The acid hydrolysis of sugarcane leaves as a biofeedstook for bioethanol production. *Adv. Mater. Res.* **2014**, *931–932*, 194–199. [CrossRef]
31. Worathanakul, P.; Payubnop, W.; Muangpet, A. Characterization for post-treatment effect of bagasse ash for silica extraction. *World Acad. Sci. Eng. Technol.* **2009**, *3*, 339–341. [CrossRef]
32. Chindaprasirt, P.; Rattanasak, U. Eco-production of silica from sugarcane bagasse ash for use as a photochromic pigment filler. *Sci. Rep.* **2020**, *10*, 1–8. [CrossRef]
33. Sholeh, M.; Rochmadi, R.; Sulistyo, H.; Budhijanto, B. Synthesis of precipitated silica from bagasse ash as reinforcing filler in rubber. In *IOP Conference Series: Materials Science and Engineering*; IOP Publishing: Kuala Lumpur, Malaysia, 2020; Volume 778, p. 012012. [CrossRef]
34. Lanning, F.; Ponnaiya, B.; Crumpton, C. The chemical nature of silica in plants. *Plant Physiol.* **1958**, *33*, 339–343. [CrossRef]
35. Chisholm, J. Comparison of Quartz Standards for X-ray Diffraction Analysis: HSE A9950 (Sikron F600) and NIST SRM 1878. *Ann. Occup. Hyg.* **2005**, *49*, 351–358. [CrossRef]

36. Kumar, R.; Bhattacharjee, B. Porosity, pore size distribution and in situ strength of concrete. *Cem. Concr. Res.* **2003**, *33*, 155–164. [CrossRef]
37. Thommes, M.; Kaneko, K.; Neimark, A.V.; Olivier, J.P.; Rodriguez-Reinoso, F.; Rouquerol, J.; Sing, K.S. Physisorption of gases, with special reference to the evaluation of surface area and pore size distribution (IUPAC Technical Report). *Pure Appl. Chem.* **2015**, *87*, 1051–1069. [CrossRef]
38. Ferch, H. Pulverförmige amorphe synthetische Kieselsäure-Produkte Herstellung und Charakterisierung. *Chem. Ing. Tech.* **1976**, *48*, 922–933. [CrossRef]

Review

Recent Progress on the Development of Engineered Silica Particles Derived from Rice Husk

Jinyoung Chun [1] and Jin Hyung Lee [2,*]

1. Energy & Environment Division, Korea Institute of Ceramic Engineering & Technology (KICET), Jinju 52851, Gyeongnam, Korea; jchun@kicet.re.kr
2. Convergence R&D Division, Korea Institute of Ceramic Engineering & Technology (KICET), Cheongju 28160, Chungbuk, Korea
* Correspondence: leejinh1@kicet.re.kr; Tel.: +82-43-913-1502

Received: 26 November 2020; Accepted: 17 December 2020; Published: 21 December 2020

Abstract: The development of engineered silica particles by using low-cost renewable or waste resources is a key example of sustainability. Rice husks have emerged as a renewable resource for the production of engineered silica particles as well as bioenergy. This review presents a state-of-the-art process for the development of engineered silica particles from rice husks via a bottom-up process. The first part of this review focuses on the extraction of Si from rice husks through combustion and chemical reactions. The second part details the technologies for synthesizing engineered silica particles using silicate obtained from rice husks. These include technologies for the precipitation of silica particles, the control of morphological properties, and the synthesis of ordered porous silica particles. Finally, several issues that need to be resolved before this process can be commercialized are addressed for future research.

Keywords: rice husk; rice husk ash; silica; engineered particle; bottom-up process; silica extraction; valorization; agricultural byproduct; sustainable material; biomass; renewable material

1. Introduction

Global rice and paddy production in 2018 was approximately 996 million tons as reported by the Food and Agricultural Organization [1]. The countries with the largest volume of rice and paddy production are located in Asia (China, India, Indonesia, Bangladesh, Vietnam, and Thailand). Rice husk is a residue produced during the rice mill process and on average, accounts for 20% of the paddy produced. The rice husk output in 2018 was approximately 199 million tons. Many countries use rice husks as a renewable energy resource for power generation [2]. The heating value of the rice husk is 15 MJ kg^{-1}, and there is an energy potential of 2985 PJ available per year [3]. Currently, rice husks are burnt in simple incinerators for resident energy, industrial streams, and thermal power plants in most Asian countries. Several Asian countries such as India, China, and Thailand are operating gasification power plants using rice husks [4,5]. Gasification power plants using rice husks have power capacities ranging between 20 and 400 kWe; however, they are still in the demonstration stage [6]. One ton of rice husk generates 800 kWh of electric power [7] and after generating electric power, approximately 0.195 tons of rice husk ash is produced as a byproduct. High ash content in rice husks causes operational problems and consequently renders their thermal conversion difficult and expensive. Therefore, the valorization of rice husk ash for value-added material applications is important for improving the economic return of the entire process. In addition, the extraction of inorganic compounds from rice husks before energy generation could be worthwhile to reduce the burden of the energy production process.

Rice husk ash mainly contains amorphous silica (SiO_2) and other metallic impurities. Rice husk-derived silica has gained increasing interest as a renewable source. Engineered silica particles have recently been intensively studied for bio-applications [8–10], energy storage [11,12], bioremediation [13], and as construction materials [14–17]. By increasing the utilization of engineered silica particles, the synthesis of engineered silica from renewable resources is considered to enhance sustainability. The use of rice husk silica for synthesizing engineered silica particles has advantages not only in the economy but also in mitigating environmental issues [18].

However, there is lack of comprehensive articles on the recovery and synthesis of silica particles derived from rice husks. Therefore, this review presents the recent progress in the development of engineered silica particles derived from rice husks.

2. Purified Silica Extraction from Rice Husk

2.1. Combustion to Remove Organic Contents

The most widely used method for obtaining silica from rice husks is direct combustion, resulting in the production of rice husk ash, which contains 85–95% silica [19–21]. The direct combustion of rice husk can produce thermal energy and can be used to generate steam, which subsequently drives the blades of a turbine to produce electricity. However, the direct combustion of rice husks generates greenhouse gases and emits significant quantities of particulate matter [22,23]. The emissions from rice husk burning contain CO_2, CH_4, CO, NO_x, SO_2, $PM_{2.5}$, and PM_{10} of black carbon. Hence, when using direct combustion, it is important to always use both a dust collector and a gas absorber. The emission of CO_2 from rice husk combustion is "net zero" because the hull reduces CO_2 in nature. It also replaces the use of a fossil fuel. Therefore, the use of rice husk to produce energy is encouraged only when it is installed with emission control devices.

The phase change of silica depends on the combustion temperature. The silica in rice husks is a non-crystalline phase. As the combustion temperature increases over 600 °C, the phase transformation to tridymite and cristobalite starts [24]. However, the crystallization temperature varied depending on the chemical composition of rice husk ash [25]. It was also reported that pretreated rice husk remained amorphous up to 1000 °C. At higher combusting temperatures, the physical properties of silica also change. Zareihassangheshlaghi et al. compared the physical properties of rice husk ashes synthesized at 700 and 900 °C [26]. At a higher combustion temperature (900 °C), fewer metal impurities were observed. This might be due to the formation of more volatile and less stable phases that can be easily released into the gas phase. At this temperature, alkaline earth metal oxides can be refractory. At higher temperatures, the size of ash particles increases. Furthermore, the mesopores and micropores in the ash diminish after combustion at 900 °C. In rice husk ash, alkali metal impurities, such as K, P, Ca, and Mg, incorporate and form sticky alkali silicate at high combustion temperatures. This causes ash melting and agglomeration, which increases the size of particles and diminishes the meso- and micro-pores [27].

Previous studies report that combustion conditions could define the characteristics of silica; however, changes in characteristics could be within a narrow range: the specific surface area (11–39 $m^2 g^{-1}$), purity (29.7–96.7 wt %), and crystallinity (completely amorphous or partially crystalline) [25]. The characteristics of the shape, particle size, pore, and uniformity cannot be controlled by changing the combustion conditions. Therefore, chemical treatment should be followed to synthesize the engineered silica particles.

2.2. Principle of Chemically Extracting Silica from Rice Husk

The main components of rice husk—cellulose, hemicellulose, lignin and inorganics—can be separated by their thermo-chemical properties. Figure 1 shows the phase changes of rice husk components depending on pH. Under acidic conditions, cellulose and hemicellulose are dissolved in the aqueous phase and can be separated from solid residues, mainly lignin and inorganics [28].

During the hydrolysis of polysaccharides, metallic impurities, such as K_2O, P_2O_5, CaO, and MgO, were discarded by the washing step. Alkaline hydrothermal conditions induce the cleavage of ester linkages in lignin, which dissolves the lignin. After alkaline hydrothermal treatment, lignin can be removed from the solid residue [29,30]. The alkaline thermal treatment also dissolves silica and partially dissolves xylan to the liquid phase. The solubility of amorphous silica rapidly increases above pH 9.14 [31]. It can be again solidified by decreasing the pH. During solidification, silica particles can be engineered to have specific properties fit for their intended purpose. Residual organic compounds in the silica can be thermally removed at temperatures over 575 °C.

Figure 1. Phase change of rice husk components depending on pH.

2.3. Acid Leaching to Obtain High Purity Silica

Acid leaching to remove organic matter and metallic impurities from rice husks was conducted before or after the combustion process. Acid leaching can produce a higher purity and specific surface area when compared to without acid leaching [19,32–34]. Lee et al. compared the performances of three acid solutions—sulfuric acid, hydrogen chloride, and oxalic acid—to remove the organic matter and metallic impurities [19]. Sulfuric acid effectively dissolved and removed both cellulose and hemicellulose but was not effective in removing lignin. Both hydrogen chloride and oxalic acid mainly reduced hemicellulose. The removal of metallic impurities was investigated depending on the acid solutions. However, the capability of acid solutions to remove organic and metallic impurities in rice husk varied depending on the solution's concentration, and the reaction time and temperature [35]. Traditionally, the acid leaching of rice husk was performed with sulfuric acid, hydrogen chloride [36], and nitric acid [33]. The use of a strong acid solution is significantly hazardous to the environment and human life. In addition, strong acid leaching produces not only soluble sugars but also smaller compounds such as aliphatic carboxylic acids and furans, which are inhibitors to microbes and enzymes. Therefore, attempts to replace these strong acid agents with environmentally harmless agents have been reported [19,37–40].

The environmentally harmless agents used to remove organic matter in rice husks were citric acid, ionic liquid, and deionized water. The use of citric acid could remove metallic impurities such as Na, K, Ca, Mg, Fe, Cu, etc., through a chelate reaction between carboxyl groups and metal elements [37,38]. Ionic liquids are green solvents and are effective in dissolving polysaccharides. The ionic liquids used to remove polysaccharides and metallic impurities from rice husks were 1-butyl-3-methylimidazolium chloride and 1-butyl-3-methylimidazolium hydrogen sulfate. The dissolved cellulose can be recovered by mixing with distilled water and further used to produce fermentable sugar [41]. Trinh et al. reported that ionic liquid-treated cellulose significantly changed its crystallinity, surface morphology, and composition, resulting in improved enzymatic digestibility. Conventionally, lignocellulose in rice husks is simply burned to generate energy, which leads to air pollution. However, the use of ionic liquid-treated cellulose to produce biofuel does not produce any pollutants. Therefore, using ionic liquids improves both the purity of silica and the comprehensive utilization of lignocellulose. Shen et al. used deionized water to leach metallic impurities, but water leaching was effective only on the external surface [40]. Mochidzuki et al. performed water leaching under pressurized conditions using a batch autoclave and steam explosion [42]. Both autoclave hot-water and steam-explosion-treated silica showed improved purities, even comparable to those obtained with the hydrochloric acid treatment. The pressurized water treatment dissolved some portion of the silica in hot water and changed its structure, which is more applicable to the synthesis of water-glass-like materials.

Most studies reported that acid leaching could improve the amorphousness of silica [19,39,43]. Alkali metals in rice husks facilitate the initiation of the formation of cristobalite, which causes the phase transformation of the silica [44]. Acid leaching could remove alkaline metals, which prevent the phase transformation of the silica. In addition, Real et al. reported that the leaching of rice husks with an acid solution before their combustion would yield silica powder with a high specific surface area [32]. However, if the acid leaching was performed after combustion, the specific surface area of silica would be poor.

2.4. Alkali Extraction to Obtain Silicate

Silica can be extracted from rice husk by solubilizing it in an alkali solution and precipitating in an acidic medium. In general, sodium hydroxide was used to extract silica from the rice husk as sodium silicate. The following formula represents the chemical reaction:

$$SiO_2\ (s) + 2NaOH\ (l) \rightarrow Na_2SiO_3\ (l) + H_2O\ (l) \tag{1}$$

Various hydrothermal conditions were used for the alkali extraction of silica. Most studies have prepared rice husk ash or acid-treated rice husk before the alkali extraction. Typically, thermal treatment is performed before alkali extraction to remove lignin. Otherwise, both lignin and silica were extracted from rice husks by alkali solution, which affected the quality of sodium silicate. In addition to the thermal treatment, organosolv fractionation can be applied to separate lignin by organic solvents such as ethanol [45] and 1,4-butanediol [46]. Rice husk ash was dissolved in sodium hydroxide solution at 80 °C [47], 90 °C [21,48], 100 °C [49], and even 150 °C [50]. In addition, a considerable amount of research has reported the performance of alkali extraction at room temperature [51–55]. In this case, rice husk ash was ground down to micron-sized particles and reacted for 24 h. The concentrations of sodium hydroxide were varied in the range of 1–10 M, which determined the reaction time. Bazargan et al. showed the removal of lignin and silica from the rice husk, not rice husk ash, by the assistance of sodium hydroxide and hydrogen peroxide [56]. However, recoveries of silica and lignin in alkaline peroxide solution were only 75 and 60%, respectively. Alkali extraction degrades lignin into phenolic compounds, such as benzoic acid, ferulic acid, and coniferyl aldehyde, and those cannot be utilized in further process.

3. Synthesis of Engineered Silica Particles Using Silicate Extracted from Rice Husks

As mentioned before, there is a limitation in changing the structure and shape of silica by changing the combustion conditions. Therefore, we focused on various methods for the synthesis of engineered silica particles using a sodium silicate solution extracted from rice husks. Such methods are based on a bottom-up process in which nano- or micron-sized particles are formed through chemical reactions of precursors at the atomic or molecular level. Although the overall synthetic procedure is relatively complex compared to methods based on a top-down process, the bottom-up process enables the precise and uniform control of the morphological properties of products. Accordingly, in modern industries that require biomass-derived silica, it is expected that research on the development of advanced synthetic techniques based on bottom-up processes will be more actively conducted in the future.

3.1. Simple Precipitation of Silica Particles by the Addition of Acidic Reagents

To obtain silica particles from a silicate solution, the pH of the solution should be adjusted to certain ranges to induce the precipitation of silica. Because the solubility of silica increases as the pH of the solution increases, the use of acidic reagents is the basic requirement for lowering the pH of an alkaline silicate solution [57]. The precipitation reaction of silica is simply described by the following equation, where H_aX is an acidic molecule:

$$Na_2SiO_3 + nH_aX \rightarrow SiO_2 + nNa_aX + H_2O \tag{2}$$

Neutralization of sodium silicate solution using sulfuric acid is a common method for the precipitation of silica particles [58–63]. For example, in the paper reported by Ghorbani et al. [60], sodium silicate solution prepared from rice husk was titrated with diluted sulfuric acid to pH 7 under vigorous stirring. The solution was further stirred for 24 h and then aged for 48 h at room temperature. After filtration and washing, the obtained silica gel was freeze-dried overnight. Amorphous silica particles prepared by this process exhibited an aggregated form of primary particles with an average size of ~200 nm, and the products showed a relatively high surface area of 409 $m^2\ g^{-1}$. The agglomerated forms of silica particles were identically observed in other studies, where the silicate solution was titrated using sulfuric acid, although the sizes of the primary particles were different in each case [61–63]. The optimized process for the synthesis of pure silica nanoparticles reported by Nassar et al. was as follows [62]: (i) Dried rice husk was acid-leached by 2 M nitric acid solution; (ii) After washing and drying, the dried residue was calcined at 600 °C; (iii) The obtained rice husk ash was refluxed with 2 M NaOH solution; (iv) The sodium silicate solution was separated by filtration; (v) The silicate solution was titrated with 1 M sulfuric acid until the pH reached 7; (vi) The obtained silica gel was aged for 24 h, and then calcined at 800 °C. The final product prepared by this process exhibited peanut-like or irregularly shaped particles (Figure 2a,b). The TEM image shows irregular shaped particles composed of tiny nanoparticles that are 10–50 nm in size. Adam et al. reported the sol–gel synthesis of silica nanoparticles from rice husk using a template-free approach [64]. They simply titrated sodium silicate, which was obtained from rice husk, with nitric acid until the pH reached 9.0. After aging for 2 days, the yellowish gel was recovered by centrifugation and washed with distilled water. Because no template molecules were used, a further calcination step was not required. Through this process, amorphous silica nanoparticles assembled from primary particles of tens of nanometers were synthesized. Their surface area was 245 $m^2\ g^{-1}$. Davarpanah et al. also used nitric acid for the titration of sodium silicate extracted from rice husk [65]. The pH of the solution was adjusted to 5.0, and the precipitated gel was aged for 24 h. Figure 2c shows that the obtained silica was irregularly shaped nanoparticles composed of 10 to 20 nm-sized primary particles. These morphological properties of precipitated silica particles have also been reported in previous studies using hydrochloric acid for pH adjustment [66–69].

The rise of environmental and safety issues has led to attempts to use organic acids instead of hazardous strong acids. In a study by Kalapathy et al., a silicate solution obtained from rice husk was titrated using citric acid or oxalic acid [70]. When the pH of the solution was adjusted to 4.0 or 7.0, amorphous silica was successfully obtained. In contrast to the case of using hydrochloric acid, no impurity peaks were observed in the XRD measurements. Moreover, no significant difference in the silica yields were observed regardless of the type of acid. The experimental results reported by Liou et al. showed that the silica particles produced by oxalic acid or citric acid exhibited higher yields than the silica particles produced by hydrochloric acid or sulfuric acid [33]. However, the residual sodium content in the silica particles was relatively high when citric acid or oxalic acid was used for the precipitation reaction. The surface areas and pore volumes of silica produced by citric acid or oxalic acid were also lower than those of silica produced by hydrochloric acid (Figure 3).

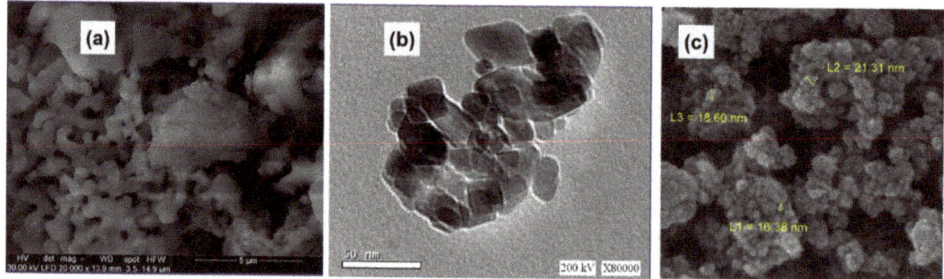

Figure 2. (a) SEM and (b) TEM images of silica nanoparticles prepared under the optimum condition reported by Nassar et al. (reprinted with permission from [62], Copyright 2019, Elsevier); and (c) SEM image of silica particles prepared by the titration of sodium silicate solution with nitric acid (reprinted with permission from [65], Copyright 2019, Elsevier).

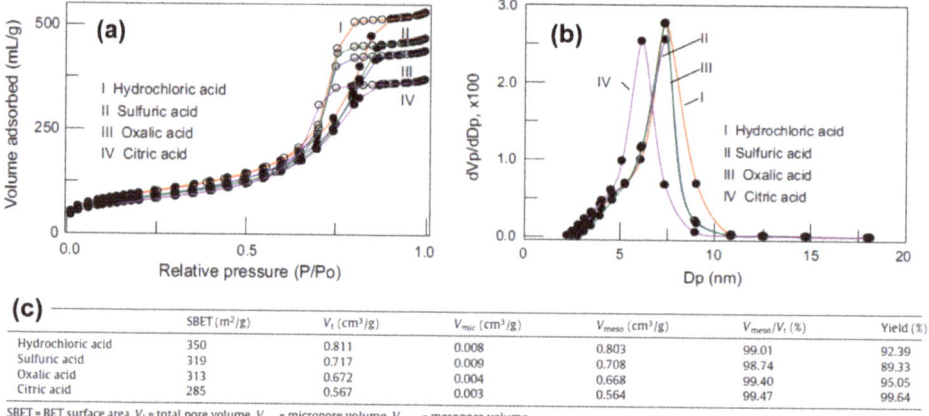

Figure 3. (a) Adsorption–desorption isotherm; (b) differential pore size distribution; and (c) BET (Brunauer–Emmett–Teller) surface area, pore volume, and extraction yield of silica samples prepared through various acid treatments (reprinted with permission from [33], Copyright 2011, Elsevier).

3.2. Control of Morphological Properties of Silica Particles

Although the precipitation of silica by adding acidic reagents is a simple and convenient method, it cannot precisely control the morphological properties of silica. Most of the silica prepared by this approach exhibited irregularly shaped large particles assembled by small-sized primary particles. Therefore, advanced synthetic methods that can control the shape and surface properties of silica are required to apply rice husk-derived silica particles to modern industries. In this section, we review previous studies that used (i) organic co-solvents, (ii) polymer additives, and (iii) a water-in-oil emulsion system to control the morphology of silica particles.

Zulkifli et al. used ethanol as a co-solvent at the precipitation step of silicate [71]. In this study, sodium silicate solution extracted from rice husks was mixed with different amounts of ethanol. Then, the mixed solution was titrated with phosphoric acid. As a result, while irregular and highly aggregated particles were obtained without ethanol, dispersed spherical particles were synthesized with increasing amounts of ethanol. At the optimized condition (ratio of sodium silicate: H_2O:ethanol = 1:1:0.25 at pH 7), uniform silica nanoparticles with low aggregation were obtained, and their sizes ranged from 75 to 252 nm. The silica powders exhibited a BET surface area of ~364 m² g⁻¹. It was surmised that the formation of spherical particles originated from the formation of self-emulsion droplets consisting of silicate anions upon the addition of ethanol to an aqueous silicate solution [72]. A similar phenomenon was also observed in a study by Zulfiqar et al. [21]. The neutralization of the silicate solution with phosphoric acid in the presence of ethanol led to the formation of spherical silica particles. Furthermore, as the concentration of sodium silicate increased, the average size of silica particles precipitated at room temperature also increased continuously (Figure 4). Silica particles made from sodium silicate solution containing 0.7 wt % white rice husk (WRH) had a unimodal size distribution of 181 ± 17 nm. When the concentration of WRH in the sodium silicate solution increased to 5.6 wt %, the silica particles exhibited a tri-modal size distribution (352 ± 77 nm, 1.8 ± 0.3 µm, and 7.1 ± 1.3 µm). On the other hand, when the precipitation reaction progressed at 65 °C, the particle sizes decreased as the concentration of silicate solution increased.

Figure 4. SEM images of silica particles produced at 25 °C by using sodium silicate solution containing (**a,b**) 1.4 wt %, (**c,d**) 2.8 wt %, and (**e–h**) 5.6 wt % white rice husk (WRH) (reprinted with permission from [21], Copyright 2015, Elsevier)

The use of acetone, an organic solvent, has also been shown to be effective in producing spherical silica particles. Rajan et al. [73] added 40 mL of acetone to 100 mL of sodium silicate solution extracted from rice husk. Afterwards, the pH of the solution was adjusted to 7 using 5 M acetic acid. Through this optimized condition, they synthesized well defined spherical silica particles in the range of 200 to 400 nm.

Several studies have reported the synthesis of spherical silica particles using a polymer additive. In particular, the spherical shape control was achieved when polymers with sufficient ethylene oxide (EO) chains were used as additives [74–77]. It is considered that the interaction of the EO chains of polymers with the silicate species stabilizes the silicate during the solidification process, which leads to the formation of spherical particles [78]. In a study by Li et al., [74,75], they first dissolved polyethylene glycol (PEG, molecular weight = 20,000) in a sodium silicate solution. Then, by the titration of the silicate solution using phosphoric acid, spherical silica nanoparticles were obtained. Similar results were also reported by Le et al. [76]. When PEG (molecular weight = 10,000) was used as an additive, spherical silica nanoparticles could be obtained from the rice husk-derived silicate solution. On the other hand, by using a Pluronic P-123 polymer instead of PEG, Shahnani et al. prepared silica microspheres from a rice husk-derived sodium silicate [77]. A certain amount of Pluronic P-123 was dissolved in 2 M hydrochloric acid solution and then sodium silicate was added to the mixture. The overall reactions progressed under acidic conditions. After filtration and washing, the porous silica microspheres were obtained by calcination at 550 °C. The non-aggregated silica microspheres obtained by this process are shown in Figure 5. The silica microspheres had a BET surface area of 445 m^2 g^{-1} and a pore volume of 0.298 cm^3 g^{-1}, respectively.

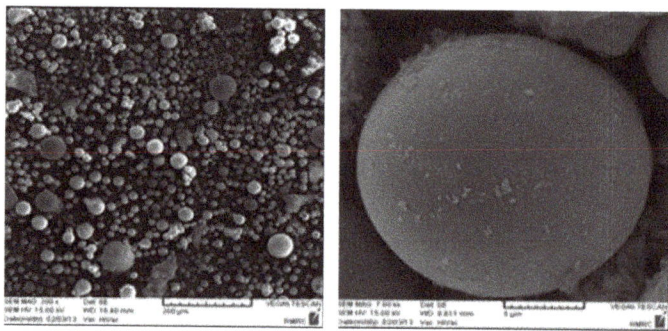

Figure 5. SEM images of the silica microspheres (reprinted with permission from [77], Copyright 2018, Elsevier).

The use of an emulsion system is a well known method for producing spherical particles. Hasan et al. reported a study that adapted this system for the preparation of spherical silica from a rice husk-derived silicate solution [79]. In the water-in-oil emulsion composed of water and toluene, cetyltrimethylammonium bromide (CTAB) and n-butanol were used as the surfactant and co-surfactant, respectively (Figure 6). After the formation of the reverse micelle, silicate was hydrolyzed by the urea at the micelle interface. The obtained silica particles exhibited a distinct spherical shape with a surface area and pore volume of 227 m^2 g^{-1} and 1.24 cm^3 g^{-1}, respectively.

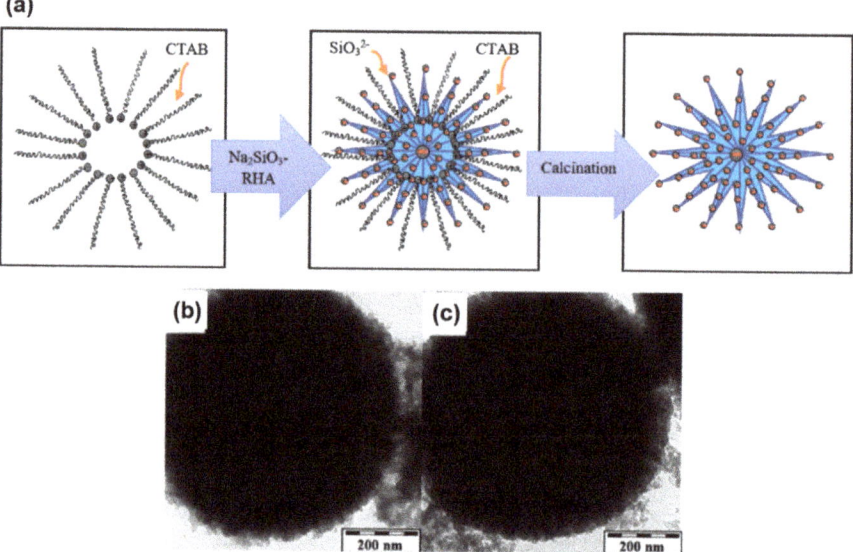

Figure 6. (a) Formation mechanism of silica sphere using Na$_2$SiO$_3$-RHA (rice husk ash) and CTAB (cetyltrimethylammonium bromide) in the water-in-oil emulsion system. TEM images of silica particles made from (b) commercial sodium silicate and (c) rice husk-derived sodium silicate (reprinted with permission from [79], Copyright 2018, Elsevier).

3.3. Synthesis of Ordered Porous Silica Particles from Rice Husk

Since a mesoporous aluminosilicate using a soft-template synthetic method was developed by the Mobil Oil Corporation in 1992 [80], numerous studies have been conducted to synthesize various types of ordered porous materials. In the case of silica-based materials, starting with the MCM (Mobil Composition of Matter) series, research on the development of various types of mesoporous silica, such as SBA (Santa Barbara Amorphous), KIT (Korea Advanced Institute of Science and Technology), FDU (Fudan University), and MSU (Michigan State University) series, have been conducted. At the same time, the requirement for cost-effective processes has led to the search for low-cost silica precursors that can replace silicon alkoxide reagents. This is one of the main reasons why rice husk-derived silicate has attracted attention as a raw material for the synthesis of mesoporous silica.

In the last decade, MCM-type silica has been successfully produced from rice husks in many studies. Conventional MCM-41 silica has a uniform hexagonal pore structure with a pore size of 2–4 nm. CTAB, a cationic surfactant, has been typically used as a structure-directing agent for MCM-41. CTAB forms rod-like micelles in the aqueous solution and aligns into a hexagonal array. The negatively charged silicate species above pH 10 preferentially interact with the positively charged surfactant heads. Accordingly, after the removal of the surfactant by calcination, mesoporous silica with a hexagonal pore structure was obtained. An example of the synthetic method in detail with reference to the reported paper is as follows [81]. First, considering the specific molar composition of SiO_2, NaOH, CTAB, and H_2O, a certain amount of CTAB was dissolved in distilled water. Then, this solution was added to the sodium silicate solution extracted from rice husk and stirred at 80 °C for 24 h. After the titration of the mixed solution using nitric acid to a pH of 10.0, it was aged for 48 h at the same temperature. The resulting gel was washed with distilled water and acetone, and subsequently calcined at 600 °C. Through this process, Ramalingam et al. synthesized MCM-41 silica particles using rice husk as the silica precursor (Figure 7); it showed a hexagonal pore structure with a monomodal pore size of 2.3 nm and a high surface area of 1115 $m^2\ g^{-1}$ [81].

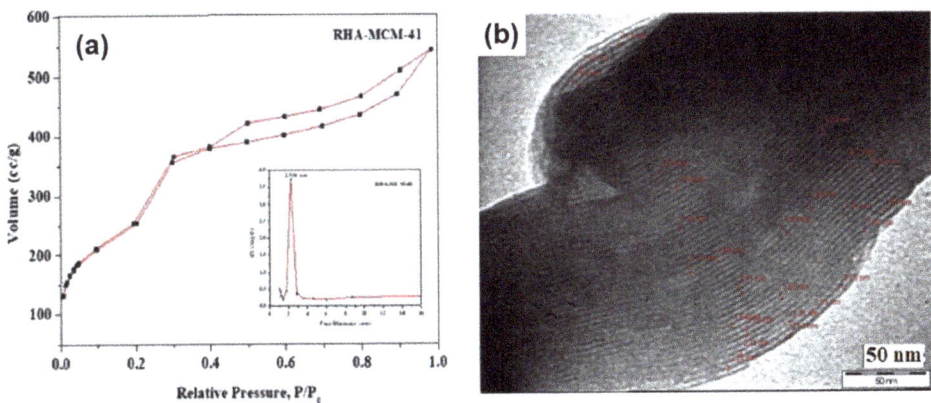

Figure 7. (a) N_2 adsorption–desorption isotherm of rice husk ash-derived MCM (Mobile Composition of Matter)-41. The inset shows the corresponding pore size distribution; and (b) the TEM image of rice husk ash-derived MCM-41 (reprinted with permission from [81], Copyright 2020, Elsevier).

Unlike MCM-41 silica, MCM-48 silica has a three-dimensional pore structure with cubic I_{a3d} symmetry. To obtain MCM-48 silica, the structure of the micelle formed by the self-assembly of surfactants must be changed. Several studies satisfied this condition by utilizing a cationic-neutral surfactant mixture system [82–84] or changing the surfactant concentration [85]. Subsequently, MCM-48 silica was successfully synthesized using sodium silicate solution extracted from rice husk. Looking at a specific example, Bhagiyalakshmi et al. used a mixture of CTAB and polyoxyethylene lauryl ether

(LE-4) as the structure directing agents [83]. Sodium silicate solution was added to the surfactant mixture (CTAB and LE-4) dissolved in an aqueous solution at 80 °C. Then, the mixed solution was titrated using acetic acid until the pH reached 10, and heated at 100 °C for 48 h. After filtration and washing, the final product was obtained by calcination at 550 °C. The obtained MCM-48 silica showed a bicontinuous I_{a3d} cubic phase with a surface area of 1124 $m^2 g^{-1}$ and a main pore size of 3.9 nm. Morphological properties, such as the surface area, pore volume, and main pore size of the rice husk-derived MCM-41 and MCM-48 silica reported to date are summarized in Table 1 [81–95].

Table 1. Pore diameter, surface area, and pore volume of MCM-type silica prepared from rice husk-derived silicate solution.

Silica Type	Pore Diameter (nm)	Surface Area ($m^2 g^{-1}$)	Pore Volume ($cm^3 g^{-1}$)	Ref.
MCM-41	2.9	800	0.93	[86]
	2.86	943	–	[87]
	3.54	1101	0.96	[83]
	3.51	1099	0.96	[84]
	3.28	903	–	[88]
	2.3	1115	0.92	[89]
	3.6	602	0.49	[90]
	2.8–3.1	545–1210	0.36–1.00	[85]
	2.92	797	0.57	[91]
	3.16	1347	0.906	[92]
	3.8	500.5	0.45	[93]
	3.0–3.4	552–769	1.025–1.167	[94]
	2.71	972.5	0.87	[95]
MCM-48	2.3	1115	0.92	[81]
	4.02	1024	2.58	[82]
	3.89	1124	0.98	[83]
	2.6	1059	0.68	[84]
	2.5	815	0.75	[85]

SBA-15 is one of the most common types of mesoporous silica and was first developed at the University of California [96]. SBA-15 has a two-dimensional hexagonal pore structure. Its pore size can be controlled in the range of 4–12 nm. It can be further increased up to 30 nm by using additional organic additives. One important feature of SBA-15 is that it has thicker pore walls compared to MCM-41. Thus, SBA-15 is more stable under high temperature and hydrothermal conditions [97]. In order to obtain SBA-15 having these advantages from low-cost raw materials, a number of studies using silicate extracted from rice husk have been reported. For the synthesis of SBA-15, Pluronic P-123, a nonionic triblock copolymer, is used as a structure directing agent. SBA-15 is assembled by the $N^0H^+X^-I^+$ mechanism, where N, H, X, and I indicate nonionic surfactant, hydrogen, halide, and silica source, respectively. Therefore, in difference with a case of MCM-41, the synthetic reaction proceeds under acidic conditions. Under these synthetic conditions, Henao et al. obtained SBA-15, which had a monomodal pore size of 7.6 nm with a high surface area of 604 $m^2 g^{-1}$, from rice husk-derived silicate solution (Figure 8) [98]. Chareonpanich et al. synthesized SBA-15 from rice husk via an ultrasonic technique [99]; their SBA-15 exhibited a highly ordered hexagonal pore arrangement with a pore size of 9.5 nm. In the meantime, by using the Pluronic F-127 polymer instead of P-123, SBA-16-type silica was also successfully obtained from rice husk [100]. The obtained SBA-16 showed a three-dimensional cubic pore structure with a pore size of ~8.0 nm. The morphological properties of rice husk-derived SBA-type silica reported in previous studies are summarized in Table 2 [83,84,98–102].

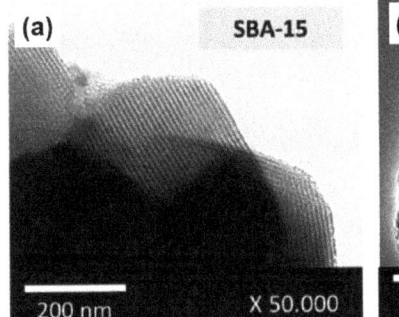

Figure 8. (a,b) TEM images of SBA (Santa Barbara Amorphous)-15-type mesoporous silica made from rice husk-derived silicate solution (reprinted with permission from [98], Copyright 2020, Elsevier).

Table 2. Pore diameter, surface area, and pore volume of SBA-type silica prepared from rice husk-derived silicate solution.

Silica Type	Pore Diameter (nm)	Surface Area (m^2 g^{-1})	Pore Volume (cm^3 g^{-1})	Ref.
SBA-11	3.8/7.7/12.9	356	0.792	[98]
SBA-15	9.5	530–860	0.96–1.27	[99]
	4.8	790	0.82	[101]
	5.8	712	0.68	[83]
	7.4	780	0.95	[84]
	7.7	1095	1.705	[102]
	7.6	604	1.192	[98]
SBA-16	5.8–8.2	775–840	–	[100]

The synthesis of large-pore-sized ordered mesoporous silica from rice husk was reported by Chun et al. [103]. In this study, Pluronic P-123 and trimethylbenzene (TMB) were used as templates and pore expanders, respectively. Pluronic P-123, TMB, and rice husk-derived silicate solution were assembled in a neutral environment with the addition of acetic acid. Through the N^{0}I^0 mechanism, amorphous silica with a well defined mesocellular foam structure was successfully obtained [104]. It has uniform mesopores with a size of approximately 30 nm and a large pore volume of 1.77 cm^3 g^{-1} (Figure 9a–c). Interestingly, when the concentration of sodium silicate extracted from rice husk increased, the usage of acetic acid increased to adjust the neutral pH, the pore size of silica was further expanded to approximately 60 nm (Figure 9d–f). The authors suggested that the large amounts of acetate promote the oxolation reaction and increase the hydrophobicity of silica, which led to the further pore expansion of silica [103,105].

Although they are not pure silica materials, several types of zeolites have also been synthesized using silicate extracted from rice husks. Zeolites are microporous aluminosilicate materials. Therefore, additional alumina resources had to be used for zeolite synthesis. Using rice husk-derived silicate and sodium aluminate reagent as a raw silica and alumina material, respectively, the synthesis of zeolite A [106], zeolite Y in sodium form (NaY) [107,108], ZSM(Zeolite Socony Mobil)-5 [109], ZSM-12 [110], ZSM-48 [111], and Linde Type J zeolite [112] have been reported in previous studies. The synthesis of various kinds of zeolite from the rice husk-derived silicate solution is covered in detail in a previous review [113].

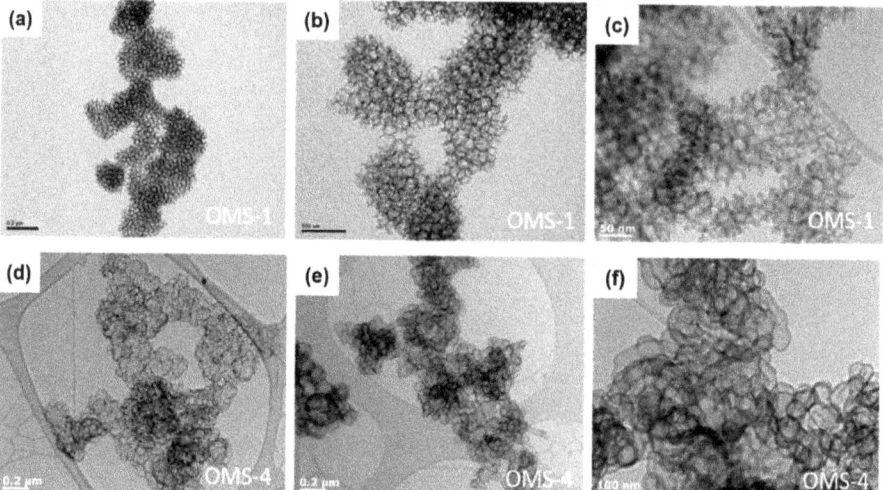

Figure 9. TEM images of large pore-sized ordered mesoporous silica made from rice husk-derived silicate: (**a–c**) mesocellular foam type with a pore size of ~30 nm and (**d–f**) an expanded pore size of ~60 nm (reprinted with permission from [103], Copyright 2020, Elsevier).

4. Conclusions and Perspectives

To date, we covered the purification of the silica component from rice husk and its use for the synthesis of engineered silica particles in previous studies. In particular, silica particles obtained by the bottom-up process using liquid silicate extracted from rice husk have been dealt with in detail. Morphologically controlled and ordered porous silica particles can be manufactured through various methods based on the bottom-up process. These engineered silica particles have unique properties including uniform shapes and sizes, large surface areas and pore volumes, and precisely controlled mesopores. Within the scope of this review, the following conclusions can be drawn:

- The widely used method for obtaining silica from rice husks is direct combustion. The characteristics of shape, particle size, pore, and uniformity can be controlled by changing the combustion conditions but its changes could be within a narrow range. Therefore, chemical treatment should be followed to synthesize the engineered silica particles.
- The main components of rice husk—cellulose, hemicellulose, lignin and inorganics—can be separated by their thermo-chemical properties.
- Acid leaching using a strong acid solution (e.g., H_2SO_4, HCl, and HNO_3) is effective to produce silica with high purity and surface area. However, these reagents are significantly hazardous to the environment and human life. Therefore, attempts to use environmentally harmless agents, such as citric acid and ionic liquid, have been reported.
- Silica can be extracted from rice husk by solubilizing it in an alkali solution and precipitating in an acidic medium. In general, sodium hydroxide was used to extract silica from rice husk as sodium silicate.
- The development of synthetic methods based on bottom-up processes enables the precise and uniform control of the morphological properties of silica products. These methods generally use a sodium silicate solution extracted from rice husks as a raw material.
- The neutralization of sodium silicate solution using acidic solution (e.g., H_2SO_4, HCl, HNO_3, organic acid) is a common method for the precipitation of silica particles. However, it is still not possible to accurately control the morphological properties of silica.

- The use of (i) organic co-solvents (e.g., ethanol and acetone), (ii) polymer additives (e.g., PEG), and (iii) a water-in-oil emulsion system enables the control of the morphological properties, such as the shape and size, of silica particles.
- Various ordered porous silica particles including MCM-type, SBA-type, and mesocellular foam structure have been successfully obtained from a rice husk-derived silicate solution with additional structure directing agents (e.g., CTAB, Pluronic P-123, etc.).

Based on these properties, they are expected to be utilized in potential value-added applications, such as heterogeneous catalysts, CO_2 capture, adsorbents for aqueous pollutants, biomolecular delivery, and cosmetic ingredients. The rice husk-derived engineered silica will further increase their value in modern society because they are manufactured from sustainable biomass resources.

For the practical use of engineered silica particles made from rice husks, several important issues must be addressed. As described in the previous chapter, the combustion of rice husks to remove organic components generates greenhouse gases and emits significant quantities of particulate matter. The use of strong acids to obtain high-purity silica is significantly hazardous to the environment and human life. Moreover, the overall synthesis procedure is complex, as shown in Figure 10a. As a result, the unit price of the final product is higher than those of silica particles made from mineral resources. We believe that overcoming these obstacles in the synthetic process is a prerequisite for the commercialization of engineered silica particles made from rice husks. One strategy for the preparation of rice husk-derived engineered silica through an environmentally friendly and cost-competitive process is proposed in Figure 10b. The direct extraction of silicate from rice husks using alkaline solution and thereafter titration using weak acids with polymer additives or co-solvents simplifies the overall process, while the use and emission of harmful substances can be minimized. However, the direct extraction of silicate using alkaline solution also leads to the extraction of organic components such as xylan. These organic components make it difficult to obtain engineered silica particles with high purity. Therefore, further research is needed to overcome these unresolved problems, and through this, it is expected that the commercialization of rice husk-derived silica will be one step closer.

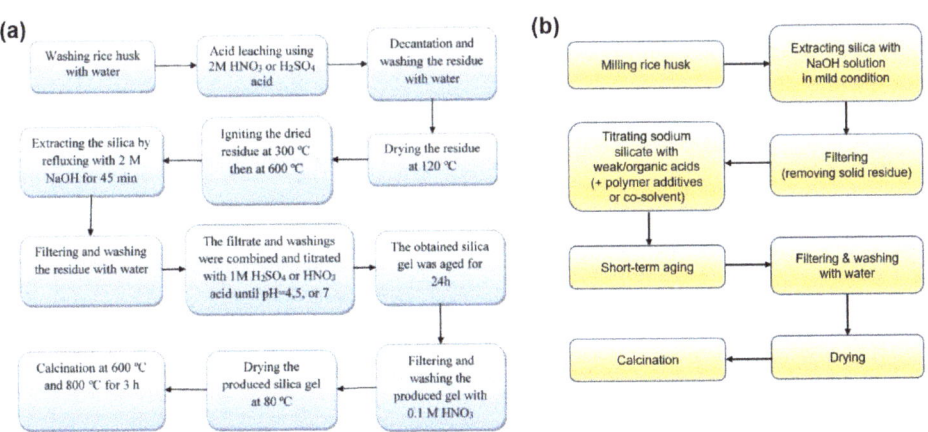

Figure 10. (a) General process (reprinted with permission from [62], Copyright 2019, Elsevier); and (b) the simplified and environmentally friendly process for the preparation of engineered silica particles from rice husk.

Author Contributions: Conceptualization, investigation and writing—original draft, project administration J.H.L.; investigation and writing—original draft, writing—review and editing, J.C. All authors have read and agreed to the published version of the manuscript.

Funding: This research was supported by the R&D program of Korea Institute of Energy Technology Evaluation and Planning (KETEP) grant funded by the Ministry of Trade, Industry and Energy (MOTIE), the Republic of Korea (No. 20183030091950).

Conflicts of Interest: The authors declare no conflict of interest.

References

1. FAOSTAT. Available online: http://www.fao.org/faostat/en/#data/QC (accessed on 12 November 2020).
2. Bhattacharyya, S.C. Viability of off-grid electricity supply using rice husk: A case study from South Asia. *Biomass Bioenergy* **2014**, *68*, 44–54. [CrossRef]
3. Quispe, I.; Navia, R.; Kahhat, R. Energy potential from rice husk through direct combustion and fast pyrolysis: A review. *Waste Manag.* **2017**, *59*, 200–210. [CrossRef] [PubMed]
4. Asadullah, M. Barriers of commercial power generation using biomass gasification gas: A review. *Renew. Sustain. Energy Rev.* **2014**, *29*, 201–215. [CrossRef]
5. Hiloidhari, M.; Baruah, D. Crop residue biomass for decentralized electrical power generation in rural areas (part 1): Investigation of spatial availability. *Renew. Sustain. Energy Rev.* **2011**, *15*, 1885–1892. [CrossRef]
6. Assanee, N.; Boonwan, C. State of the art of biomass gasification power plants in Thailand. *Energy Procedia* **2011**, *9*, 299–305. [CrossRef]
7. Pode, R. Potential applications of rice husk ash waste from rice husk biomass power plant. *Renew. Sustain. Energy Rev.* **2016**, *53*, 1468–1485. [CrossRef]
8. Lin, Y.-S.; Hurley, K.R.; Haynes, C.L. Critical considerations in the biomedical use of mesoporous silica nanoparticles. *J. Phys. Chem. Lett.* **2012**, *3*, 364–374. [CrossRef]
9. Liu, B.; Li, C.; Cheng, Z.; Hou, Z.; Huang, S.; Lin, J. Functional nanomaterials for near-infrared-triggered cancer therapy. *Biomater. Sci.* **2016**, *4*, 890–909. [CrossRef]
10. Wang, Y.; Zhao, Q.; Han, N.; Bai, L.; Li, J.; Liu, J.; Che, E.; Hu, L.; Zhang, Q.; Jiang, T. Mesoporous silica nanoparticles in drug delivery and biomedical applications. *Nanomedicine* **2015**, *11*, 313–327. [CrossRef]
11. Jung, D.S.; Ryou, M.-H.; Sung, Y.J.; Park, S.B.; Choi, J.W. Recycling rice husks for high-capacity lithium battery anodes. *Proc. Natl. Acad. Sci. USA* **2013**, *110*, 12229–12234. [CrossRef]
12. Shen, Y. Rice husk silica-derived nanomaterials for battery applications: A literature review. *J. Agric. Food Chem.* **2017**, *65*, 995–1004. [CrossRef] [PubMed]
13. Kumari, B.; Singh, D. A review on multifaceted application of nanoparticles in the field of bioremediation of petroleum hydrocarbons. *Ecol. Eng.* **2016**, *97*, 98–105. [CrossRef]
14. Van Tuan, N.; Ye, G.; Van Breugel, K.; Copuroglu, O. Hydration and microstructure of ultra high performance concrete incorporating rice husk ash. *Cem. Concr. Res.* **2011**, *41*, 1104–1111. [CrossRef]
15. Martirena, F.; Monzó, J. Vegetable ashes as supplementary cementitious materials. *Cem. Concr. Res.* **2018**, *114*, 57–64. [CrossRef]
16. Kang, S.-H.; Hong, S.-G.; Moon, J. The use of rice husk ash as reactive filler in ultra-high performance concrete. *Cem. Concr. Res.* **2019**, *115*, 389–400. [CrossRef]
17. Kang, S.-H.; Kwon, Y.-H.; Hong, S.-G.; Chun, S.; Moon, J. Hydrated lime activation on byproducts for eco-friendly production of structural mortars. *J. Clean Prod.* **2019**, *231*, 1389–1398. [CrossRef]
18. Shen, Y. Rice husk silica derived nanomaterials for sustainable applications. *Renew. Sustain. Energy Rev.* **2017**, *80*, 453–466. [CrossRef]
19. Lee, J.H.; Kwon, J.H.; Lee, J.-W.; Lee, H.-s.; Chang, J.H.; Sang, B.-I. Preparation of high purity silica originated from rice husks by chemically removing metallic impurities. *J. Ind. Eng. Chem.* **2017**, *50*, 79–85. [CrossRef]
20. Fernandes, I.J.; Calheiro, D.; Kieling, A.G.; Moraes, C.A.; Rocha, T.L.; Brehm, F.A.; Modolo, R.C. Characterization of rice husk ash produced using different biomass combustion techniques for energy. *Fuel* **2016**, *165*, 351–359. [CrossRef]
21. Zulfiqar, U.; Subhani, T.; Husain, S.W. Towards tunable size of silica particles from rice husk. *J. Non-Cryst. Solids* **2015**, *429*, 61–69. [CrossRef]
22. Junpen, A.; Pansuk, J.; Kamnoet, O.; Cheewaphongphan, P.; Garivait, S. Emission of air pollutants from rice residue open burning in Thailand, 2018. *Atmosphere* **2018**, *9*, 449. [CrossRef]

23. Lasko, K.; Vadrevu, K. Improved rice residue burning emissions estimates: Accounting for practice-specific emission factors in air pollution assessments of Vietnam. *Environ. Pollut.* **2018**, *236*, 795–806. [CrossRef] [PubMed]
24. Nakata, Y.; Suzuki, M.; Okutani, T.; Kikuchi, M.; Akiyama, T. Preparation and properties of SiO_2 from rice hulls. *J. Ceram. Soc. Jpn.* **1989**, *97*, 842–849. [CrossRef]
25. Beidaghy Dizaji, H.; Zeng, T.; Hartmann, I.; Enke, D.; Schliermann, T.; Lenz, V.; Bidabadi, M. Generation of high quality biogenic silica by combustion of rice husk and rice straw combined with pre-and post-treatment strategies—A review. *Appl. Sci.* **2019**, *9*, 1083. [CrossRef]
26. Zareihassangheshlaghi, A.; Beidaghy Dizaji, H.; Zeng, T.; Huth, P.; Ruf, T.; Denecke, R.; Enke, D. Behavior of metal impurities on surface and bulk of biogenic silica from rice husk combustion and the impact on ash-melting tendency. *ACS Sustain. Chem. Eng.* **2020**, *8*, 10369–10379. [CrossRef]
27. Alyosef, H.A.; Eilert, A.; Welscher, J.; Ibrahim, S.S.; Denecke, R.; Schwieger, W.; Enke, D. Characterization of biogenic silica generated by thermo chemical treatment of rice husk. *Part. Sci. Technol.* **2013**, *31*, 524–532. [CrossRef]
28. Pedersen, M.; Meyer, A.S. Lignocellulose pretreatment severity–relating pH to biomatrix opening. *New Biotechnol.* **2010**, *27*, 739–750. [CrossRef]
29. Pedersen, M.; Viksø-Nielsen, A.; Meyer, A.S. Monosaccharide yields and lignin removal from wheat straw in response to catalyst type and pH during mild thermal pretreatment. *Process Biochem.* **2010**, *45*, 1181–1186. [CrossRef]
30. Kristensen, J.B.; Thygesen, L.G.; Felby, C.; Jørgensen, H.; Elder, T. Cell-wall structural changes in wheat straw pretreated for bioethanol production. *Biotechnol. Biofuels* **2008**, *1*, 5. [CrossRef]
31. Alexander, G.B.; Heston, W.; Iler, R.K. The solubility of amorphous silica in water. *J. Phys. Chem.* **1954**, *58*, 453–455. [CrossRef]
32. Real, C.; Alcala, M.D.; Criado, J.M. Preparation of silica from rice husks. *J. Am. Ceram. Soc.* **1996**, *79*, 2012–2016. [CrossRef]
33. Liou, T.-H.; Yang, C.-C. Synthesis and surface characteristics of nanosilica produced from alkali-extracted rice husk ash. *Mater. Sci. Eng. B* **2011**, *176*, 521–529. [CrossRef]
34. Chen, P.; Bie, H.; Bie, R. Leaching characteristics and kinetics of the metal impurities present in rice husk during pretreatment for the production of nanosilica particles. *Korean J. Chem. Eng.* **2018**, *35*, 1911–1918. [CrossRef]
35. Chakraverty, A.; Mishra, P.; Banerjee, H. Investigation of combustion of raw and acid-leached rice husk for production of pure amorphous white silica. *J. Mater. Sci.* **1988**, *23*, 21–24. [CrossRef]
36. Vayghan, A.G.; Khaloo, A.; Rajabipour, F. The effects of a hydrochloric acid pre-treatment on the physicochemical properties and pozzolanic performance of rice husk ash. *Cem. Concr. Compos.* **2013**, *39*, 131–140. [CrossRef]
37. Umeda, J.; Kondoh, K. High-purity amorphous silica originated in rice husks via carboxylic acid leaching process. *J. Mater. Sci.* **2008**, *43*, 7084–7090. [CrossRef]
38. Umeda, J.; Kondoh, K. Process optimization to prepare high-purity amorphous silica from rice husks via citric acid leaching treatment. *Trans. JWRI* **2008**, *37*, 13–17.
39. Chen, H.; Wang, W.; Martin, J.C.; Oliphant, A.J.; Doerr, P.A.; Xu, J.F.; DeBorn, K.M.; Chen, C.; Sun, L. Extraction of lignocellulose and synthesis of porous silica nanoparticles from rice husks: A comprehensive utilization of rice husk biomass. *ACS Sustain. Chem. Eng.* **2013**, *1*, 254–259. [CrossRef]
40. Shen, J.; Liu, X.; Zhu, S.; Zhang, H.; Tan, J. Effects of calcination parameters on the silica phase of original and leached rice husk ash. *Mater. Lett.* **2011**, *65*, 1179–1183. [CrossRef]
41. Trinh, L.T.P.; Lee, Y.J.; Lee, J.-W.; Lee, H.-J. Characterization of ionic liquid pretreatment and the bioconversion of pretreated mixed softwood biomass. *Biomass Bioenergy* **2015**, *81*, 1–8. [CrossRef]
42. Mochidzuki, K.; Sakoda, A.; Suzuki, M.; Izumi, J.; Tomonaga, N. Structural behavior of rice husk silica in pressurized hot-water treatment processes. *Ind. Eng. Chem. Res.* **2001**, *40*, 5705–5709. [CrossRef]
43. Ahmad Alyosef, H.; Schneider, D.; Wassersleben, S.; Roggendorf, H.; Weiß, M.; Eilert, A.; Denecke, R.; Hartmann, I.; Enke, D. Meso/macroporous silica from miscanthus, cereal remnant pellets, and wheat straw. *ACS Sustain. Chem. Eng.* **2015**, *3*, 2012–2021. [CrossRef]
44. Moroz, I.K.; Maslennikova, G. Thermal transformations of silica. *Glass Ceram.* **1985**, *42*, 559–564. [CrossRef]

45. Kim, T.H.; Ryu, H.J.; Oh, K.K. Improvement of organosolv fractionation performance for rice husk through a low acid-catalyzation. *Energies* **2019**, *12*, 1800. [CrossRef]
46. Zhang, H.; Ding, X.; Chen, X.; Ma, Y.; Wang, Z.; Zhao, X. A new method of utilizing rice husk: Consecutively preparing d-xylose, organosolv lignin, ethanol and amorphous superfine silica. *J. Hazard. Mater.* **2015**, *291*, 65–73. [CrossRef]
47. Tchakouté, H.K.; Rüscher, C.H.; Kong, S.; Kamseu, E.; Leonelli, C. Geopolymer binders from metakaolin using sodium waterglass from waste glass and rice husk ash as alternative activators: A comparative study. *Constr. Build. Mater.* **2016**, *114*, 276–289. [CrossRef]
48. Geraldo, R.H.; Fernandes, L.F.; Camarini, G. Water treatment sludge and rice husk ash to sustainable geopolymer production. *J. Clean. Prod.* **2017**, *149*, 146–155. [CrossRef]
49. Tchakouté, H.K.; Rüscher, C.H.; Kong, S.; Ranjbar, N. Synthesis of sodium waterglass from white rice husk ash as an activator to produce metakaolin-based geopolymer cements. *J. Build. Eng.* **2016**, *6*, 252–261. [CrossRef]
50. Jahan, M.S.; Haris, F.; Rahman, M.M.; Samaddar, P.R.; Sutradhar, S. Potassium hydroxide pulping of rice straw in biorefinery initiatives. *Bioresour. Technol.* **2016**, *219*, 445–450. [CrossRef]
51. Bernal, S.A.; Rodríguez, E.D.; de Gutiérrez, R.M.; Provis, J.L.; Delvasto, S. Activation of metakaolin/slag blends using alkaline solutions based on chemically modified silica fume and rice husk ash. *Waste Biomass Valoriz.* **2012**, *3*, 99–108. [CrossRef]
52. Bernal, S.; Rodríguez, E.; Mejía de Gutiérrez, R.; Provis, J.L. Performance at high temperature of alkali-activated slag pastes produced with silica fume and rice husk ash based activators. *Mater. Constr.* **2015**, *65*, e049. [CrossRef]
53. Kamseu, E.; à Moungam, L.B.; Cannio, M.; Billong, N.; Chaysuwan, D.; Melo, U.C.; Leonelli, C. Substitution of sodium silicate with rice husk ash-NaOH solution in metakaolin based geopolymer cement concerning reduction in global warming. *J. Clean. Prod.* **2017**, *142*, 3050–3060. [CrossRef]
54. Mejía, J.; de Gutiérrez, R.M.; Puertas, F. Rice husk ash as a source of silica in alkali-activated fly ash and granulated blast furnace slag systems. *Mater. Constr.* **2013**, *63*, 361–375.
55. Mejía, J.M.; de Gutiérrez, R.M.; Montes, C. Rice husk ash and spent diatomaceous earth as a source of silica to fabricate a geopolymeric binary binder. *J. Clean. Prod.* **2016**, *118*, 133–139. [CrossRef]
56. Bazargan, A.; Wang, Z.; Barford, J.P.; Saleem, J.; McKay, G. Optimization of the removal of lignin and silica from rice husks with alkaline peroxide. *J. Clean. Prod.* **2020**, *260*, 120848. [CrossRef]
57. Soltani, N.; Bahrami, A.; Pech-Canul, M.; González, L. Review on the physicochemical treatments of rice husk for production of advanced materials. *Chem. Eng. J.* **2015**, *264*, 899–935. [CrossRef]
58. Muñoz-Aguado, M.-J.; Gregorkiewitz, M. Sol–gel synthesis of microporous amorphous silica from purely inorganic precursors. *J. Colloid Interface Sci.* **1997**, *185*, 459–465. [CrossRef]
59. Schlomach, J.; Kind, M. Investigations on the semi-batch precipitation of silica. *J. Colloid Interface Sci.* **2004**, *277*, 316–326. [CrossRef] [PubMed]
60. Ghorbani, F.; Sanati, A.M.; Maleki, M. Production of silica nanoparticles from rice husk as agricultural waste by environmental friendly technique. *Environ. Stud. Persian Gulf* **2015**, *2*, 56–65.
61. Kamath, S.R.; Proctor, A. Silica gel from rice hull ash: Preparation and characterization. *Cereal Chem.* **1998**, *75*, 484–487. [CrossRef]
62. Nassar, M.Y.; Ahmed, I.S.; Raya, M.A. A facile and tunable approach for synthesis of pure silica nanostructures from rice husk for the removal of ciprofloxacin drug from polluted aqueous solutions. *J. Mol. Liq.* **2019**, *282*, 251–263. [CrossRef]
63. Vaibhav, V.; Vijayalakshmi, U.; Roopan, S.M. Agricultural waste as a source for the production of silica nanoparticles. *Spectrochim. Acta A Mol. Biomol. Spectrosc.* **2015**, *139*, 515–520. [CrossRef]
64. Adam, F.; Chew, T.-S.; Andas, J. A simple template-free sol–gel synthesis of spherical nanosilica from agricultural biomass. *J. Sol-Gel Sci. Technol.* **2011**, *59*, 580–583. [CrossRef]
65. Davarpanah, J.; Sayahi, M.H.; Ghahremani, M.; Karkhoei, S. Synthesis and characterization of nano acid catalyst derived from rice husk silica and its application for the synthesis of 3, 4-dihydropyrimidinones/thiones compounds. *J. Mol. Struct.* **2019**, *1181*, 546–555. [CrossRef]
66. Song, S.; Cho, H.-B.; Kim, H.T. Surfactant-free synthesis of high surface area silica nanoparticles derived from rice husks by employing the Taguchi approach. *J. Ind. Eng. Chem.* **2018**, *61*, 281–287. [CrossRef]

67. Azat, S.; Korobeinyk, A.; Moustakas, K.; Inglezakis, V. Sustainable production of pure silica from rice husk waste in Kazakhstan. *J. Clean. Prod.* **2019**, *217*, 352–359. [CrossRef]
68. Nayak, P.; Datta, A. Synthesis of SiO_2-nanoparticles from rice husk ash and its comparison with commercial amorphous silica through material characterization. *Silicon* **2020**. [CrossRef]
69. Ngoc, T.M.; Man, T.M.; Phong, M.T.; Nam, H.M.; Hieu, N.H. Fabrication of tubular ceramic-supported malic acid cross-linked poly (vinyl alcohol)/rice husk ash-silica nanocomposite membranes for ethanol dehydration by pervaporation. *Korean J. Chem. Eng.* **2019**, *36*, 584–590. [CrossRef]
70. Kalapathy, U.; Proctor, A.; Shultz, J. An improved method for production of silica from rice hull ash. *Bioresour. Technol.* **2002**, *85*, 285–289. [CrossRef]
71. Zulkifli, N.S.C.; Ab Rahman, I.; Mohamad, D.; Husein, A. A green sol–gel route for the synthesis of structurally controlled silica particles from rice husk for dental composite filler. *Ceram. Int.* **2013**, *39*, 4559–4567. [CrossRef]
72. Jung, C.Y.; Kim, J.S.; Chang, T.S.; Kim, S.T.; Lim, H.J.; Koo, S.M. One-step synthesis of structurally controlled silicate particles from sodium silicates using a simple precipitation process. *Langmuir* **2010**, *26*, 5456–5461. [CrossRef] [PubMed]
73. Rajan, R.; Zakaria, Y.; Shamsuddin, S.; Hassan, N.F.N. Robust synthesis of mono-dispersed spherical silica nanoparticle from rice husk for high definition latent fingermark development. *Arab. J. Chem.* **2020**, *13*, 8119–8132. [CrossRef]
74. Li, D.; Chen, D.; Zhu, X. Reduction in time required for synthesis of high specific surface area silica from pyrolyzed rice husk by precipitation at low pH. *Bioresour. Technol.* **2011**, *102*, 7001–7003. [CrossRef] [PubMed]
75. Li, D.; Zhu, X. Short-period synthesis of high specific surface area silica from rice husk char. *Mater. Lett.* **2011**, *65*, 1528–1530. [CrossRef]
76. Thuc, C.N.H.; Thuc, H.H. Synthesis of silica nanoparticles from Vietnamese rice husk by sol–gel method. *Nanoscale Res. Lett.* **2013**, *8*, 1–10.
77. Shahnani, M.; Mohebbi, M.; Mehdi, A.; Ghassempour, A.; Aboul-Enein, H.Y. Silica microspheres from rice husk: A good opportunity for chromatography stationary phase. *Ind. Crop. Prod.* **2018**, *121*, 236–240. [CrossRef]
78. Hwang, J.; Lee, J.H.; Chun, J. Facile approach for the synthesis of spherical mesoporous silica nanoparticles from sodium silicate. *Mater. Lett.* **2021**, *283*, 128765. [CrossRef]
79. Hasan, R.; Chong, C.; Bukhari, S.; Jusoh, R.; Setiabudi, H. Effective removal of Pb (II) by low-cost fibrous silica KCC-1 synthesized from silica-rich rice husk ash. *J. Ind. Eng. Chem.* **2019**, *75*, 262–270. [CrossRef]
80. Kresge, C.; Leonowicz, M.; Roth, W.J.; Vartuli, J.; Beck, J. Ordered mesoporous molecular sieves synthesized by a liquid-crystal template mechanism. *Nature* **1992**, *359*, 710–712. [CrossRef]
81. Ramalingam, R.J.; Appaturi, J.N.; Pulingam, T.; Al-Lohedan, H.A.; Al-dhayan, D.M. In-situ incorporation of ruthenium/copper nanoparticles in mesoporous silica derived from rice husk ash for catalytic acetylation of glycerol. *Renew. Energy* **2020**, *160*, 564–574. [CrossRef]
82. Jang, H.T.; Park, Y.; Ko, Y.S.; Lee, J.Y.; Margandan, B. Highly siliceous MCM-48 from rice husk ash for CO_2 adsorption. *Int. J. Greenh. Gas Control* **2009**, *3*, 545–549. [CrossRef]
83. Bhagiyalakshmi, M.; Yun, L.J.; Anuradha, R.; Jang, H.T. Utilization of rice husk ash as silica source for the synthesis of mesoporous silicas and their application to CO2 adsorption through TREN/TEPA grafting. *J. Hazard. Mater.* **2010**, *175*, 928–938. [CrossRef] [PubMed]
84. Bhagiyalakshmi, M.; Yun, L.J.; Anuradha, R.; Jang, H.T. Synthesis of chloropropylamine grafted mesoporous MCM-41, MCM-48 and SBA-15 from rice husk ash: Their application to CO_2 chemisorption. *J. Porous Mater.* **2010**, *17*, 475–484. [CrossRef]
85. Ahmad-Alyosef, H.; Uhlig, H.; Münster, T.; Kloess, G.; Einicke, W.; Gläser, R.; Enke, D. Biogenic silica from rice husk ash–Sustainable sources for the synthesis of value added silica. *Chem. Eng. Trans.* **2014**, *37*, 667–672.
86. Chiarakorn, S.; Areerob, T.; Grisdanurak, N. Influence of functional silanes on hydrophobicity of MCM-41 synthesized from rice husk. *Sci. Technol. Adv. Mater.* **2007**, *8*, 110–115. [CrossRef]
87. Artkla, S.; Kim, W.; Choi, W.; Wittayakun, J. Highly enhanced photocatalytic degradation of tetramethylammonium on the hybrid catalyst of titania and MCM-41 obtained from rice husk silica. *Appl. Catal. B* **2009**, *91*, 157–164. [CrossRef]
88. Suyanta, S.; Kuncaka, A. Utilization of rice husk as raw material in synthesis of mesoporous silicates MCM-41. *Indones. J. Chem.* **2011**, *11*, 279–284. [CrossRef]

89. Appaturi, J.N.; Adam, F. A facile and efficient synthesis of styrene carbonate via cycloaddition of CO_2 to styrene oxide over ordered mesoporous MCM-41-Imi/Br catalyst. *Appl. Catal. B* **2013**, *136*, 150–159. [CrossRef]
90. Renuka, N.; Praveen, A.; Anas, K. Influence of CTAB molar ratio in tuning the texture of rice husk silica into MCM 41 and SBA-16. *Mater. Lett.* **2013**, *109*, 70–73. [CrossRef]
91. Areerob, T.; Grisdanurak, N.; Chiarakorn, S. Utilization of rice husk silica as adsorbent for BTEX passive air sampler under high humidity condition. *Environ. Sci. Pollut. Res.* **2016**, *23*, 5538–5548. [CrossRef]
92. Nguyen, N.T.; Chen, S.-S.; Nguyen, N.C.; Nguyen, H.T.; Tsai, H.H.; Chang, C.T. Adsorption of methyl blue on mesoporous materials using rice husk ash as silica source. *J. Nanosci. Nanotechnol.* **2016**, *16*, 4108–4114. [CrossRef] [PubMed]
93. Costa, J.A.S.; Sarmento, V.H.; Romão, L.P.; Paranhos, C.M. Adsorption of organic compounds on mesoporous material from rice husk ash (RHA). *Biomass Convers. Biorefin.* **2020**, *10*, 1105–1120. [CrossRef]
94. Purnawira, B.; Purwaningsih, H.; Ervianto, Y.; Pratiwi, V.; Susanti, D.; Rochiem, R.; Purniawan, A. Synthesis and characterization of mesoporous silica nanoparticles (MSNp) MCM 41 from natural waste rice husk. *IOP Conf. Ser. Mater. Sci. Eng.* **2019**, *541*, 012018. [CrossRef]
95. Kamari, S.; Ghorbani, F. Extraction of highly pure silica from rice husk as an agricultural by-product and its application in the production of magnetic mesoporous silica MCM–41. *Biomass Convers. Biorefin.* **2020**. [CrossRef]
96. Zhao, D.; Feng, J.; Huo, Q.; Melosh, N.; Fredrickson, G.H.; Chmelka, B.F.; Stucky, G.D. Triblock copolymer syntheses of mesoporous silica with periodic 50 to 300 angstrom pores. *Science* **1998**, *279*, 548–552. [CrossRef] [PubMed]
97. Chaudhary, V.; Sharma, S. An overview of ordered mesoporous material SBA-15: Synthesis, functionalization and application in oxidation reactions. *J. Porous Mater.* **2017**, *24*, 741–749. [CrossRef]
98. Henao, W.; Jaramillo, L.; López, D.; Romero-Sáez, M.; Buitrago-Sierra, R. Insights into the CO_2 capture over amine-functionalized mesoporous silica adsorbents derived from rice husk ash. *J. Environ. Chem. Eng.* **2020**, *8*, 104362. [CrossRef]
99. Chareonpanich, M.; Nanta-Ngern, A.; Limtrakul, J. Short-period synthesis of ordered mesoporous silica SBA-15 using ultrasonic technique. *Mater. Lett.* **2007**, *61*, 5153–5156. [CrossRef]
100. Ho, S.T.; Dinh, Q.K.; Tran, T.H.; Nguyen, H.P.; Nguyen, T.D. One-step synthesis of ordered Sn-substituted SBA-16 mesoporous materials using prepared silica source of rice husk and their selectively catalytic activity. *Can. J. Chem. Eng.* **2013**, *91*, 34–46. [CrossRef]
101. Bhagiyalakshmi, M.; Do Park, S.; Cha, W.S.; Jang, H.T. Development of TREN dendrimers over mesoporous SBA-15 for CO_2 adsorption. *Appl. Surf. Sci.* **2010**, *256*, 6660–6666. [CrossRef]
102. Pimprom, S.; Sriboonkham, K.; Dittanet, P.; Föttinger, K.; Rupprechter, G.; Kongkachuichay, P. Synthesis of copper–nickel/SBA-15 from rice husk ash catalyst for dimethyl carbonate production from methanol and carbon dioxide. *J. Ind. Eng. Chem.* **2015**, *31*, 156–166. [CrossRef]
103. Chun, J.; Gu, Y.M.; Hwang, J.; Oh, K.K.; Lee, J.H. Synthesis of ordered mesoporous silica with various pore structures using high-purity silica extracted from rice husk. *J. Ind. Eng. Chem.* **2020**, *81*, 135–143. [CrossRef]
104. Kim, S.-S.; Pauly, T.R.; Pinnavaia, T.J. Non-ionic surfactant assembly of ordered, very large pore molecular sieve silicas from water soluble silicates. *Chem. Commun.* **2000**, 1661–1662. [CrossRef]
105. Boissière, C.; Martines, M.A.; Tokumoto, M.; Larbot, A.; Prouzet, E. Mechanisms of pore size control in MSU-X mesoporous silica. *Chem. Mater.* **2003**, *15*, 509–515. [CrossRef]
106. Wajima, T.; Kiguchi, O.; Sugawara, K.; Sugawara, T. Synthesis of zeolite-A using silica from rice husk ash. *J. Chem. Eng. Jpn.* **2009**, *42*, s61–s66. [CrossRef]
107. Wittayakun, J.; Khemthong, P.; Prayoonpokarach, S. Synthesis and characterization of zeolite NaY from rice husk silica. *Korean J. Chem. Eng.* **2008**, *25*, 861–864. [CrossRef]
108. Mohamed, R.; Mkhalid, I.; Barakat, M. Rice husk ash as a renewable source for the production of zeolite NaY and its characterization. *Arab. J. Chem.* **2015**, *8*, 48–53. [CrossRef]
109. Vempati, R.K.; Borade, R.; Hegde, R.S.; Komarneni, S. Template free ZSM-5 from siliceous rice hull ash with varying C contents. *Microporous Mesoporous Mater.* **2006**, *93*, 134–140. [CrossRef]
110. Loiha, S.; Prayoonpokarach, S.; Songsiriritthigun, P.; Wittayakun, J. Synthesis of zeolite beta with pretreated rice husk silica and its transformation to ZSM-12. *Mater. Chem. Phys.* **2009**, *115*, 637–640. [CrossRef]

111. Wang, H.P.; Lin, K.S.; Huang, Y.; Li, M.; Tsaur, L. Synthesis of zeolite ZSM-48 from rice husk ash. *J. Hazard. Mater.* **1998**, *58*, 147–152. [CrossRef]
112. Ng, E.-P.; Lim, G.K.; Khoo, G.-L.; Tan, K.-H.; Ooi, B.S.; Adam, F.; Ling, T.C.; Wong, K.-L. Synthesis of colloidal stable Linde Type J (LTJ) zeolite nanocrystals from rice husk silica and their catalytic performance in Knoevenagel reaction. *Mater. Chem. Phys.* **2015**, *155*, 30–35. [CrossRef]
113. Shen, Y.; Zhao, P.; Shao, Q. Porous silica and carbon derived materials from rice husk pyrolysis char. *Microporous Mesoporous Mater.* **2014**, *188*, 46–76. [CrossRef]

Publisher's Note: MDPI stays neutral with regard to jurisdictional claims in published maps and institutional affiliations.

© 2020 by the authors. Licensee MDPI, Basel, Switzerland. This article is an open access article distributed under the terms and conditions of the Creative Commons Attribution (CC BY) license (http://creativecommons.org/licenses/by/4.0/).

MDPI
St. Alban-Anlage 66
4052 Basel
Switzerland
Tel. +41 61 683 77 34
Fax +41 61 302 89 18
www.mdpi.com

Sustainability Editorial Office
E-mail: sustainability@mdpi.com
www.mdpi.com/journal/sustainability

www.ingramcontent.com/pod-product-compliance
Lightning Source LLC
LaVergne TN
LVHW070743100526
838202LV00013B/1295

www.ingramcontent.com/pod-product-compliance
Lightning Source LLC
LaVergne TN
LVHW070156120526
838202LV00013BA/1266